Fourth
Edition

Design of
Reinforced
Concrete
Structures
IS:456-2000

Fourth Edition

Design of
Reinforced Concrete Structures
IS:456-2000

N Krishna Raju
BE, MSc (Engg), PhD, MIE, MI (Struct E)

Emeritus Professor of Civil Engineering
MS Ramaiah Institute of Technology
Bengaluru
Karnataka

CBS

CBS Publishers & Distributors Pvt Ltd

New Delhi • Bengaluru • Chennai • Kochi • Kolkata • Mumbai

Bhopal • Bhubaneswar • Hyderabad • Jharkhand • Nagpur • Patna • Pune • Uttarakhand • Dhaka (Bangladesh)

Design of
Reinforced
Concrete
Structures

ISBN: 978-93-85915-36-9

Copyright © Author and Publisher

Fourth Edition: 2016
Reprint: 2019

Third Edition: 2003
Reprint: 2004, 2005, 2006, 2007, 2008, 2009, 2010, 2011, 2012, 2013, 2014, 2015

Second Edition: 1990
Reprint: 1994, 1996, 1998, 1999, 2000, 2001

First Edition: 1984
Reprint: 1985, 1987, 1988

Published by Satish Kumar Jain and produced by Varun Jain for

CBS Publishers & Distributors Pvt Ltd

4819/XI Prahlad Street, 24 Ansari Road, Daryaganj, New Delhi 110 002, India.
Ph: 23289259, 23266861, 23266867   Website: www.cbspd.com
Fax: 011-23243014   e-mail: delhi@cbspd.com; cbspubs@airtelmail.in.

Corporate Office: 204 FIE, Industrial Area, Patparganj, Delhi-110092
Ph: 4934 4934   Fax: 4934 4935   e-mail: publishing@cbspd.com; publicity@cbspd.com

Branches

- **Bengaluru:** Seema House 2975, 17th Cross, K.R. Road, Banasankari 2nd Stage, Bengaluru 560 070, Karnataka
  Ph: +91-80-26771678/79   Fax: +91-80-26771680   e-mail: bangalore@cbspd.com
- **Chennai:** 7, Subbaraya Street, Shenoy Nagar, Chennai 600 030, Tamil Nadu
  Ph: +91-44-26680620, 26681266   Fax: +91-44-42032115   e-mail: chennai@cbspd.com
- **Kochi:** 42/1325, 1326 Power House Road, Opp. KSEB Power House, Ernakulam 682 018, Kochi, Kerala
  Ph: +91-484-4059061-65   Fax: +91-484-4059065   e-mail: kochi@cbspd.com
- **Kolkata:** 6/B, Ground Floor, Rameswar Shaw Road, Kolkata-700 014, West Bengal
  Ph: +91-33-22891126, 22891127, 22891128   e-mail: kolkata@cbspd.com
- **Mumbai:** 83-C, Dr E Moses Road, Worli, Mumbai-400018, Maharashtra
  Ph: +91-22-24902340/41   Fax: +91-22-24902342   e-mail: mumbai@cbspd.com

Representatives

- **Bhopal**   0-8319310552   • **Bhubaneswar** 0-9911037372   • **Hyderabad** 0-9885175004   • **Jharkhand** 0-9811541605
- **Nagpur**   0-9021734563   • **Patna**   0-9334159340   • **Pune**   0-9623451994   • **Uttarakhand** 0-9716462459
- **Dhaka (Bangladesh)** 0-1912003485

Printed at Manipal Technologies Limited, Manipal, India

Preface to the Fourth Edition

Rapid developments in the field of high performance concrete technology during the last few years have opened up new vistas in the reinforced concrete industry. Innovations in the technology of cements, admixtures and concretes have resulted in the production of high strength/high performance concretes of grades in the range of M40 to M100. Revolutionary developments in the field of nanotechnology has resulted in the production of nanoconcretes with superior properties, with a direct impact on the reinforced concrete construction industry.

Phenomenal developments have taken place in the codification process for the design of reinforced concrete structural elements in the USA, England, Europe and Australia. One of the major developments has been the implementation of Euro codes with a number of innovative provisions from 2004 for European nations. The Euro codes have influenced the revision of the British and American codes on design of structural concrete resulting in the release of comprehensive national codes for design of concrete structures incorporating limit state concepts.

The significant developments during the last decade in the domain of reinforced concrete structures witnessed in the developed countries should serve as a stimulus for rapid revision of the Indian standard codes on concrete and reinforced concrete design. The present Indian standard code, IS:456-2000 dealing with plain and reinforced concrete will also be revised shortly to keep pace with the significant developments in the field of concrete production and its use in structural concrete.

At present, the Indian standard code, IS:456-2000 (to be revised) is not comprehensive and does not cover the entire gamut of design of the various types of structural concrete elements. This thoroughly revised edition attempts to consolidate the design features of reinforced concrete structural elements in a concise and analytical format with emphasis on the Indian standard code specifications. Various chapters are arranged with topics of practical importance in a logical manner so that the new monograph should serve as a useful text for graduate and postgraduate students of civil and structural engineering stream and teachers, and also a valuable reference for practising engineers.

Chapter 1 presents the general features like the basic principles of reinforced concrete followed by a brief historical evolution, structural design philosophy, structural applications, basic elements and systems, latest national codes and loading standards.

Chapter 2 discusses the traditional elastic or working stress theory of reinforced concrete sections in flexure with derivations of mathematical formulae to facilitate the analysis of reinforced concrete flexural members along with a variety of numerical exercises.

Chapter 3 is devoted to the study of the evolution of universally adopted limit state method of design involving various limit states, safety factors, characteristic strength/load and reliability. At present limit state design is adumbrated in the national codes of most of the countries since it incorporates the important limit states of strength and serviceability.

The design of reinforced concrete structural elements at the limit state of collapse under flexure, is examined in detail in Chapter 4. The computation of flexural strength of different types of sections like rectangular, flanged, singly reinforced and doubly reinforced are presented with several numerical examples.

Chapter 5 is dedicated to the study of shear strength of reinforced concrete members. Various types of shear failure mechanisms of reinforced concrete structural elements are examined with relevant examples. The design of shear reinforcements conforming to the Indian standard code specifications is presented with several numerical examples.

The limit state design of reinforced concrete structural members subjected to torsion, shear and flexure is treated in detail in Chapter 6. The design equations are presented and their application is illustrated with several numerical examples with emphasis on the provision of detailing of reinforcements for torsion according to the specifications of the Indian standard code.

Chapter 7 throws light into the study of bond and anchorage in reinforced concrete members. The mechanism of bond, flexural bond stress computations and factors like anchorage development, code requirements for bond in structural concrete members are discussed in detail. Splicing of reinforcements and the use of design aids like SP:16 for checking development length is emphasised in this chapter.

The integrated concepts of serviceability limit states of reinforced concrete structural elements like cracking and deflections at working loads are examined in detail in Chapter 8. Computation of crack widths and control of cracking in reinforced concrete structural members are examined in detail in the light of code specifications. Numerous examples are presented for the computation of crack widths with emphasis on the empirical method of control of cracking in structural concrete members.

Chapter 9 deals with the limit state design of various types of reinforced concrete beams like, singly reinforced, doubly reinforced, cantilever, Tee, Ell and continuous types, fortified with numerical design examples. The use of bending moment and shear force coefficients specified in the Indian standard code is illustrated with examples. The design aspect of deep beams is also presented with numerical examples.

Chapter 10 is devoted to the limit state design of reinforced concrete slabs covering the various types like one-way, two-way, cantilever, continuous and flat slabs with different types of edge conditions. The use of design coefficients for the design of slabs is illustrated with examples. The application of yield line theory for the analysis and design of slabs is treated in detail for various shapes of slabs with different types of edge conditions.

Chapter 11 deals with the design aspects of reinforced concrete columns and footings. The various failure modes of reinforced concrete columns are examined along with the classification of columns based on types of loading, reinforcements and slenderness ratio. The design of columns under axial and biaxial loading using the SP:16 design charts is illustrated with examples. The designs of slender and helically reinforced columns are included in this chapter. The designs of independent and combined footings are explained with examples.

The designs of various types of staircases used in buildings are treated in detail in Chapter 12. The theory and design of different types of staircases like dog-legged, open well, spiral and helicoidal types are covered with numerical examples. The designs of helicoidal and tread–riser type staircases are examined in detail with examples.

Chapter 13 covers the analysis and design of various types of retaining walls like cantilever, counterfort, buttressed and anchored types. The various structural components of retaining walls are discussed and their structural designs and detailing are illustrated with number of numerical examples.

Deep foundations like piles and raft foundations generally used for multistoreyed buildings are treated in detail in Chapter 14. Different types like *cast-in-situ* and precast concrete piles and their designs according to the specifications of the relevant Indian standard codes are presented with numerical design examples. The design of pile caps for connecting groups of piles is also covered with examples.

Chapter 15 deals with the design of structural concrete members subjected to direct tension like ties of trusses, Virendeel girders, walls of circular water tanks, hopper bottoms of bunkers and silos. The specifications of the Indian and British standard code for the design of tension members are presented along with the detailed designs of several types of tension members with special emphasis on detailing of reinforcements.

The traditional working stress method of design has been retained in the Indian standard code and hence Chapter 16 deals with the design aspects of various common types of reinforced concrete structural elements like slabs, beams, retaining walls, staircases, domes, water tanks of different types are covered in the light of code specifications. Detailed designs of each of these structural elements are presented with several numerical examples.

Each chapter is fortified with a variety of solved and unsolved numerical examples to help the students preparing for university and competitive examinations. The list of references will immensely help students regarding the source material.

Both review and objective type questions included at the end of each chapter will help the examiners to set model question papers to test the basic understanding of the principles of design covered in each of the chapters by the students opting for this course.

In keeping with the spirit of *drawing is the language of the engineer*, numerous illustrations have been included, which, it is hoped will help in a clearer understanding of the subject matter. SI units have been adopted throughout the text for all design examples. Finally, the author welcomes constructive criticism and useful suggestions, which will immensely help in updating and improving the contents of the book.

N Krishna Raju

Preface to the First Edition

In recent years, the philosophy of design of reinforced concrete structures has been modified with the introduction of the limit state concepts. Following the revision of the British code: CP:110-1972 and the American code ACI:318-1977, the Indian standard code IS:456 has been thoroughly revised in 1978. The code permits the use of the traditional working stress method of design and in keeping with recent developments, it also provides the limit state method of design for reinforced concrete structures.

This book presents the design of reinforced concrete structures conforming to the latest revised Indian standard code IS:456-1978 and SI units have been adopted for all design examples. The topics covered are intended to meet the requirements of the curricula of most of the engineering institutions in India. The book is basically design oriented with more emphasis on types of design with minimum extend of theory, presented wherever required for application in design. The various design steps are identified and provided in a logical sequence.

Working drawings showing the reinforcement details are provided for most of the design examples. The working stress method is presented, first to help the beginner to understand the basic principles of design followed by the limit state method and yield line analysis of slabs. The examples for practice provided at the end of each chapter is intended to help the students preparing for university examinations.

The references provided at the end of the book have been freely used in the preparation of this book and are gratefully acknowledged. The author is grateful to his wife, Pramila, for extending the fullest cooperation in the preparation of the typescript. Finally, the author welcomes constructive criticism and suggestions which will immensely help in updating the contents of the book.

N Krishna Raju

Acknowledgements

I have received immense help from several sources during the preparation of the fourth revised edition of this book and gratefully acknowledge the various societies, journals, associations, building standards of various countries and several authors for the reproduction of salient design data, charts, tables, figures and reference material mentioned throughout the text.

The excellent works of authors like PC Verghese, RE Rowe, N Subramanian, R Park and T Paulay, S Unnikrishna Pillai and Devadas Menon have served as valuable reference material in the preparation of this book. Very useful inputs have been received from the fellow teachers and my ex-students in the revision of the various chapters of the book. The complete details of the source material used in the preparation of the text matter are listed as references at the end of each chapter.

I wish to record my gratitude to Mrs Pramila Raju for her kind cooperation in shouldering the domestic responsibilities and helping me to concentrate on the compilation of the source material for the preparation of the revised text.

Finally, I also express gratitude to my colleagues, students, friends, family members and the publishers for their kind encouragement, cooperation and timely help extended during the preparation of this monograph.

N Krishna Raju

Contents

List of Symbols

A	Cross-sectional area
A_c	Concrete cross-sectional area
a	Lever arm
b	Breadth of beam, or shorter dimension of a rectangular column
b_{ef}	Effective width of slab
b_f	Effective width of flange
b_w	Breadth of web or rib
D	Overall depth of beam or slab or diameter of column; dimension of a rectangular column in the direction under consideration
D_f	Thickness of flange
DL	Dead load
d	Effective depth
d'	Depth of compressive reinforcement from the highly compressed face
$E_c,\ E_{cm}$	Modulus of elasticity of concrete
E_{ce}	Effective modulus of elasticity of concrete
EL	Earthquake load
E_s	Modulus of elasticity of steel
e	Eccentricity
F	Resisting force
f_{ck}	Characteristic cube compressive strength of concrete
f_c'	Cylinder compressive strength
f_{ctm}	Tensile strength of concretre
f_{ct}	Split tensile strength of concrete
f_{cr}	Modulus of rupture of concrete (flexural strength of concrete)
f_{ct}	Split tensile strength of concrete
f_d	Design strength

f_y	Characteristic tensile strength of steel
f_{ck}	Characteristic tensile strength of concrete
g	Gravity load or dead load
h	Overall height of retaining wall
h_s	Height of stem
I	Second moment of area or moment of inertia
I_{ef}	Effective moment of inertia
I_{gr}	Moment of inertia of gross section excluding reinforcement
I_r	Moment of inertia of cracked section
j	Lever arm factor
K	Stiffness of member
k	Constant or coefficient or factor
L_d	Development length
LL	Live load
L	Length of a beam or column between adequate lateral restraints or the unsupported length of a column
L_{ef}	Effective span of beam or slab
L_x	Length of shorter side of slab
L_y	Length of longer side of slab
L_{ex}	Effective span length along xx axis
L_{ey}	Effective span length along yy axis
L_n	Clear span face to face of supports
L_1	Span in the direction in which moments are determined, c/c of supports
L_2	Span transverse to L_1, centre to centre of supports
L_0	Distance between points of zero moments in a beam
M	Bending moment
M_r	Moment of resistance
m	Modular ratio
n	Neutral axis depth
n_a	Actual neutral axis depth
n_c	Critical neutral axis depth
P	Axial load on a compression member
p	Safe bearing capacity of soil or intensity of pressure
p_t	Percentage reinforcement in tension
p_c	Percentage reinforcement in compression
q	Live load
Q	Design coefficient
r	Radius
s_v	Spacings of stirrups
T	Torsional moment
V	Shear force

w	Distributed load per unit area
W	Total load or concentrated load
WL	Wind load
LL	Live Load
x_u	Neutral axis depth
Z	Modulus of section
δ	Displacement
γ_f	Partial safety factor for load
γ_m	Partial safety factor for material
μ	Coefficient of friction or coefficient of orthotropy
σ_{cbc}	Permissible stress in concrete in bending compression
σ_{cc}	Permissible stress in concrete in direct compression
σ_{sc}	Permissible stress in steel in compression
σ_{st}	Permissible stress in steel in tension
σ_{sv}	Permissible tensile stress in shear reinforcement
τ_{bd}	Design bond stress
τ_c	Shear stress in concrete
$\tau_{c,max}$	Maximum shear stress in concrete with shear reinforcement
τ_v	Nominal shear stress
ϕ	Diameter of bar
ψ_{cs}	Shrinkage curvature
ε	Support to span ratio
ε_{cc}	Strain in concrete
ε_{cd}	Drying shrinkage strain
ε_{ca}	Autogenous shrinkage strain
ε_{cs}	Total shrinkage strain
ε_{sc}	Strain in steel
θ	Creep coefficient
ν	Poisson's ratio
α, β	Angles or ratio
λ	Multiplying factor

Introduction to Reinforced Concrete

1.1 EVOLUTION OF REINFORCED CONCRETE

Reinforced concrete is the most widely used material widely used for multifarious structural applications in the world today and it has the unique distinction of being the universally suitable material to be used in different types of environmental conditions. Reinforced concrete is a composite material comprising concrete and steel and the versatility of the material is such that it can be used at site in cast *in situ* form, precast to the required shapes in a factory and it can be designed to have suitable strength by varying the contents of the concrete ingredients. It has also the distinction being the most durable material for use in structures located in dry, windy, freezing and underwater situations. The worldwide consumption of this unique material exceeds several billion cubic metres per year and with rapid innovations in the quality concrete and steel, reinforced concrete has achieved the distinction of the most economical material for varied types of structural applications.

The development in the field of reinforced concrete is attributed to the continuous research work done during the last 150 years. Soon after the invention of Portland cement by Joseph Aspdin in 1845, Joseph Monier[1], a French gardener embedded iron and steel rods in concrete and made garden posts and tubs of reinforced concrete heralding the birth of reinforced concrete in 1849. The early history of reinforced concrete is replete with innovative research by investigators like Francois Coignet[2], Joseph Lambot[3], Arthur Talbot[4], Frederick Turneaure[5], Morsch[6], Oscar Faber[7], Charles Whitney[8] and others who toiled incessantly for the development and widespread use of reinforced concrete as an economical structural material.

Early 20th century witnessed the development of reliable design and construction procedures and paved the way for extensive use of reinforced concrete in construction industry throughout the world. The amenability of concrete to be cast in various shapes and with attractive surface characteristics is a salient feature for preference of this material by architects and engineers in comparison with other materials. Reinforced concrete is a structural material with desirable properties like mould-ability, strength, elasticity, durability, fire, impermeability, good resistance to static, fatigue and dynamic loads.

The success of reinforced concrete as a composite material is mainly attributed to the excellent bond between concrete and reinforcements which ensure strain compatibility so that the external loads on the structural elements is shared by steel and concrete without disruption of the composite material.

1.2 APPLICATIONS FOR REINFORCED CONCRETE

Reinforced concrete has more or less replaced steel as the primary construction material mainly due to its universal adaptability, versatility, coupled with resistance to fire, wind and corrosion resulting in negligible maintenance costs. The primary applications of reinforced concrete as an economical structural material in various types of structures are described below:

1. Horizontal structural elements like slabs and beams of different types such as simply supported, continuous, flat, grid or coffered slabs, Tee and continuous beams.
2. Vertical structural elements like columns, piers, abutments, towers and pylons.
3. Curved structural elements like domes and shells of different shapes.
4. Soil retaining structures like cantilever- and counterfort-type retaining walls.
5. Water storage structures like ground and overhead water tanks of different shapes.
6. Material storage structures like bins and silos.
7. Transmission structures like electric poles.
8. Energy structures like chimneys, cooling towers of thermal stations.
9. Railway track supporting elements like sleepers.
10. Foundation structures like footings, rafts, piles, wells and caissons.
11. Reinforced concrete rocker bearings for supporting bridge girders.
12. Box culverts for bridges.
13. Water transmission structures like pipes and aqueducts.

1.3 REINFORCED CONCRETE STRUCTURAL SYSTEMS

The various types of reinforced structures are basically assemblages of different type of structural elements developed to perform the function of resisting various types of forces. The primary structural systems can be broadly classified as:

1. One-dimensional elements comprising of beams, columns, arches etc.
2. Two-dimensional elements such as slabs, grids, shells etc.
3. Three-dimensional elements like pipes, domes, dams, retaining walls, silos and chimneys.

The most common types of structural systems are described below:
1. One-way slabs
2. Cantilever slabs
3. Two-way slabs
4. Continuous slabs
5. Beam and slab floors
6. Coffered or grid floors
7. Flat slab floors
8. Multistorey building frames

9. Arched structures
10. Shell structures
11. Cylindrical structures
12. Folded plate roofs.

1.4 NATIONAL CODES AND HANDBOOKS

Most of the developed countries have prescribed their specific national codes based on the extensive research and practical knowledge. These codes serve as guidelines for the design of reinforced concrete structures. The principal objectives of the codes can be summarized as:

1. Provision of adequate safety by ensuring strength, serviceability and durability codifying design procedure, design tables to facilitate easy computations.
2. Protection of structural engineers from any liability due to failure of structures caused due to inadequate design and improper materials and lack of proper supervision during construction.
3. To provide a uniform set of design guidelines to be followed by various structural designers and engineers in the country.
4. To provide simple design procedures, design tables and formulae for easy computations.

The national building codes are periodically revised to reflect the improvements in the quality of materials and design practices evolved as a result of comprehensive research investigations conducted in various institutions in the country and abroad.

In India, the design of reinforced concrete structures should conform to the Indian national code IS:456-2000[9]. The corresponding national codes of the leading countries generally referred to are listed as:

1. British code: BS EN:1992-1-1, Euro code-2, Design of concrete structure, general rules and rules for buildings, British Standards Institution, 2004[10].
2. American code: ACI:318M-11(metric), Building code requirements for structural concrete, American Concrete Institute, 2005[11].
3. AS:3800-1988, Concrete structures, Standards Association of Austraila, 1988[12].
4. CSA standard: A23.1-00/A23.2-00, Concrete materials and methods of concrete construction/methods of test for concrete, Canadian Standards Associations, Toronto, 2000[13].

In addition to the abovementioned codes, the following special publications and handbooks are very useful in the design offices of structural engineers:

1. SP:16-1980[14]: Design aids for reinforced concrete.
2. SP:34-1987[15]: Handbook of concrete reinforcement and detailing.
3. SP:10262-1982[16]: Recommended guidelines for concrete mix design.
4. Handbook of concrete engineering by Mark Fintel[17].
5. Reynolds' reinforced concrete designers' handbook[18].

1.5 PHILOSOPHY OF STRUCTURAL DESIGN

The design philosophy of structural concrete elements has seen significant changes during the last century due to the research investigations by several engineering scientists like Rowe *et al*[19] and Bate[20]. Elastic design was first introduced in 1900 followed by the ultimate strength design by 1950 and now most of the codes of developed and developing countries follow the limit state design philosophy comprising the following essential requirements:

A satisfactory structural design should ensure the three basic criteria of strength, serviceability and stability. In addition, the structural designer should also consider aesthetics and economy. The structural designer and the architect should coordinate so that the structures designed are not only aesthetically superior, but are strong enough to safely sustain the designed loads without any distress during life time of the structure.

According to Moseley *et al*[21], the structural design process may be considered as a series of interrelated and overlapping stages. In their simplest forms consisting of:

- Conceptual design in which a range of potential structural forms and materials are considered.
- Preliminary design which will typically involve simple and approximate hand computations to assess the viability of a range of alternative conceptual solutions.
- Detailed design to conclude comprehensive analysis and calculations for the selected scheme of design using suitable computer software.

1.6 LOADING STANDARDS

The design of reinforced concrete structural elements requires a knowledge of various types of loads acting on the members. The various types of loads to be considered as prescribed in the Indian standard code are detailed below:

1.6.1 Dead Loads

Dead loads comprise the self-weight of the material used in the structure such as brick, stones, concrete, wood and various other materials. The unit weight of most of the commonly used building materials specified in the Indian standard code IS:875-1987(part 1)[22] are given in Table 1.1.

1.6.2 Live or Imposed Loads

Live or imposed loads are transient loads changing with time due to people using the structure. These are prescribed in IS:875-1987(part 2)[23] and are given in Table 1.2.

1.6.3 Wind Loads

The wind loads acting on the structure depending upon the location of the structure in India is specified in IS:875-1987(part 3)[24]. For multistorey structures, water tank towers, chimneys and other types of tall structures, wind loads should be considered in design. Wind loads acting on the structure are influenced by the plan dimensions, height of the structure, location and the design wind pressure as specified in the Indian standard code.

The code prescribes basic wind speeds in various zones by dividing the country into 6 zones. The design wind pressure is computed as

$$p_x = V_x^2$$

Table 1.1: Unit weight of building materials (IS:875-1987)	
Type of material	*Unit weight*
Brick masonry	18.85 to 22.0 kN/m³
Stone masonry	20.40 to 26.50 kN/m³
Cement concrete	22.00 to 23.50 kN/m³
Reinforced concrete	22.75 to 26.50 kN/m³
Cement mortar	20.40 kN/m³
Terrazo (10 mm thick)	0.24 kN/m³
Mastic asphalt (10 mm thick)	0.215 kN/m²
Brick wall (100 mm thick)	1.91 kN/m²
Brick wall (100 mm thick)	3.84 kN/m²
Laterite	20.40 to 23.55 kN/m³
Marble	26.70 kN/m³
Granite	25.90 to 27.45 kN/m³
Asbestos cement sheeting	0.118 to 0.130 kN/m²
Common burnt clay bricks	15.70 to 18.85 kN/m³
Hollow concrete blocks	1.41 kN/m³
Solid concrete blocks	17.65 kN/m³
Plain concrete (IS:456-2000)	24 kN/m³
Reinforced concrete	25 kN/m³
Rubble masonry	20.8 kN/m³
Concrete tile flooring (25 mm thick)	0.5 kN/m²
Dry soil	13.85 to 18.05 kN/m³
Moist soil	15.70 to 19.60 kN/m³
Fine sand (dry)	15.10 to 15.70 kN/m³
Aggregate (stone-dry)	15.70 to 18.35 kN/m³
Teakwood	6.28 kN/m³

where,

p_x = design wind pressure in N/mm² at a height Z, and

V_x = design wind velocity in m/s at a height Z.

Wind load F, acting in a direction normal to the individual structural element or cladding unit and is computed using the relation

$$F = (C_{pe} - C_{pi})Ap_d$$

where,

C_{pe} = external pressure coefficient

C_{pi} = internal poressure coefficient

A = surface area of structural element or cladding unit, and

p_d = design wind pressure.

The values of internal and external pressure coefficients depend upon the type of structure and are compiled in a tabular form in the Indian standard code.

1.6.4 Snow Loads and Local Combinations

In locations where snow is encountered, structures located in such zones have to be designed to resist the snow loads prevailing in the region and also the various load combinations. The snow loads are prescribed in IS:875-1987(part 4)[25] and the special loads and their combinations are specified in IS:875-1987(part 5)[26].

Table 1.2: Live loads on structures [(IS:876-1987 (part 2)]		
Loading class	Types of floors	Minimum live load kN/m²
2	Floors in dwelling houses, tenements, hospital wards, bedrooms and private sitting rooms in hostels and dormitories	2
2.5	Office floors other than entrance hall floors of light workrooms	2.5–4.0
3.0	Floors of banking halls, office entrance halls and reading rooms	3.0
4.0	Shop floors used for display and sale of merchandise, floors of workrooms, floors of classrooms, restaurants, machinery halls, power station etc. where not occupied by plant or equipment	4.0
5.0	Floors of warehouses, workshops, factories and other buildings or similar category for light weight loads, office floors for storage and filling purposes. Assembly floor space without fixed seating, public rooms in hotels, dance halls and waiting halls	5.0
7.5	Floors of warehouses, workshops, factories and other buildings or parts of buildings of similar category for medium weight loads	7.5
10.0	Floors of warehouses, workshops, factories and other buildings or parts of buildings of similar category for heavy weight loads, floors of book stores and libraries	10.0
	Garages (light)	
	Floors used for garages for vehicles not exceeding 25 kN gross weight	
	Slabs	4.0
	Beams	2.5
	Garages (heavy)	
	Floors used for garages for vehicles not exceeding 40 kN gross weight	7.5
	Stair cases	
	Stairs, landings and corridors for class 2, but not liable to overcrowding	3.0
	Stairs, landings and corridors for class 2 loading but liable to overcrowding and for all other classes	5.0
	Balcony	
	Balconies not liable to overcrowding for class 2 loading	3
	Loading for other classes	5
	Balconies liable to overcrowding	5
	Roofs	
	Types of roofs	Live load in plan kN/m²
	Flat, sloping or curved roofs with slopes upto and including 10 degrees	
	a. Access provided	1.5
	b. Access not provided, except for maintenance	0.75
	c. Sloping roof with slope greater than 10° – 0.75 kN/m² less 0.001 kN/m² for every increase in slope over 10° upto and including 20° and 0.002 kN/m² for every degree increase in slope over 20°	

1.6.5 Seismic or Earthquake Loads

Earthquake or seismic loads should be considered in the design of structures located in specific zones in the Indian subcontinent which experience earthquake resulting in lateral loads on the structures. Based on seismic studies, India has been divided into five zones depending upon the severity of the intensity of earthquake prevalent in the zone. Horizontal seismic forces induced due to the earthquake as specified in

IS:1893-2002[27] and these forces should be considered in the design of reinforced concrete structures located in the respective zones. The horizontal seismic force is computed as

$$F_{eq} = [\alpha\beta\lambda G]$$

where,

α = horizontal seismic coefficient depending on location with values of 0.08, 0.05, 0.04, 0.02 and 0.01 for zones V, IV, III, II and I respectively

β = a coefficient depending on soil formation system ranging from 1.0 to 1.5

λ = a coefficient depending upon the importance of the structure varying from 1.5 to 1.0

G = dead load above the section considered.

Structures located in zones III to V are grouped under severe earthquake zones and it is mandatory to design them as specified in IS:4326-1993[28].

In the limit state method, the Indian Code specifies that for earthquake resistant structures, the following load combinations should be considered in design:

a. at limit state of collapse   (1.5 DL + 1.2 LL + 1.2 EL)

b. at limit state of serviceability (1.0 DL + 0.8 LL + 0.8 EL)

where, DL = dead load, LL = live load and EL = earthquake load.

1.7 MATERIALS FOR REINFORCED CONCRETE

1.7.1 Concrete

(a) Grades of concrete

The Indian standard code IS:456-2000 specifies that the concrete mix proportion should be selected so as to ensure the workability of fresh concrete and durability of hardened concrete. The determination of the proportions of cement, aggregates and water to achieve the desired strength is made by:

1. designing the concrete by any well established method, or
2. adopting nominal mix proportions.

The code specifies that the design mix should be preferred to nominal mix since the design mix concrete exhibits better performance compared to nominal mix concrete.

For reinforced concrete work, the most commonly used grades of concrete are M20 to M30. The revised Indian standard code IS:456-2000 prescribes M20 as the minimum grade of concrete for reinforced concrete work while M10 and M15 grades may be used for plain concrete constructions.

(b) Tensile strength of concrete

The tensile strength of concrete is expressed in terms of flexural strength (f_{cr}) and its magnitude is expressed in the form of an empirical relation expressed as

$$f_{cr} = 0.7\sqrt{f_{ck}} \ \text{N/mm}^2$$

where, f_{ck} = characteristic cube compressive strength of concrete in N/mm^2.

(c) Modulus of elasticity of concrete

The short term static modulus of elasticity of concrete required for computations of deflections of structural concrete members is a function of the characteristic strength and is expressed as

$$E_c = 5000\sqrt{f_{ck}}\ \text{N/mm}^2$$

(d) Shrinkage of concrete

The total shrinkage of concrete depends upon the constituents of concrete, size of member and environmental conditions. The approximate value of the total shrinkage strain in concrete for design purposes may be taken as 0.0003.

(e) Creep of concrete

The creep of concrete depends upon several factors such as strength of concrete at loading, stress in concrete and duration of loading. In the absence of test data, the ultimate creep strain may be estimated from values of creep coefficient (ratio of ultimate creep strain to elastic strain).

Table 1.3: Creep coefficient of concrete (IS:456-2000)

Age of loading	Creep coefficient
7 days	2.2
28 days	1.6
1 year	1.1

These values are helpful in computations of long term deflections of structural concrete members compiled in Table 1.3.

(f) Durability of concrete

Durability of concrete is influenced by various factors such as type of environment, cement content, water/cement ratio, workmanship, cover to the reinforcement, shape and size of structural concrete member. IS:456-2000 code categorises the exposure conditions into five types and minimum cement content and maximum water/cement ratios to be used in concrete for different types of exposure conditions to ensure durability of concrete are given in Tables 1.4 and 1.5.

Table 1.4: Environmental exposure condition (Table 3, IS:456-2000)

Sl. No.	Environment	Exposure conditions
1.	Mild	Concrete surfaces protected against weather or aggressive conditions, except those situated in costal area
2.	Moderate	Concrete surfaces sheltered from severe rain of freezing whilst wet Concrete exposed to condensation and rain Concrete continuously under water Concrete in contact or buried under non-aggressive soil/ground water Concrete surfaces sheltered from saturated salt air in coastal area
3.	Severe	Concrete surfaces exposed to severe rain, alternate wetting and drying or occasional freezing whilst wet or severe condensation Concrete completely immersed in sea water Concrete exposed to coastal environment
4.	Very severe	Concrete surfaces exposed to sea water spray, corrosive fumes or severe freezing conditions whilst wet Concrete in contact with or buried under aggressive sub-soil/ground water
5.	Extreme	Surface of members in tidal zone Members in direct contact with liquid/solid aggressive chemicals

Table 1.5: Minimum cement content, maximum water–cement ratio and minimum grade of concrete for different exposure with normal weight aggregate of 20 mm maximum size

Sl. No.	Exposure	Plain concrete			Reinforced concrete		
		Minimum cement content kg/m³	Maximum free water-cement ratio	Minimum grade of concrete	Minimum cement content kg/m³	Maximum free water-cement ratio	Minimum grade of concrete
1.	Mild	220	0.60	–	300	0.55	M20
2.	Moderate	240	0.60	M15	300	0.50	M25
3.	Severe	250	0.50	M20	320	0.45	M30
4.	Very severe	260	0.45	M20	340	0.45	M35
5	Extreme	280	0.40	M25	360	0.40	M40

Notes:
1. Cement content prescribed in Table 1.5 is irrespective of the grades of cement and it is inclusive of additions mentioned in Table 5.2. The additions such as fly ash or ground granulated blast furnace slag may be taken into account in the concrete composition with respect to the cement content and water–cement ratio if the suitability is established and as long as the maximum amounts taken into account do not exceed the limit of Pozzolona and slag specified in IS:1489–1989(part I)[29] and IS:455–1989[30].
2. Minimum grade for plain concrete under mild exposure condition is not specified.

(g) Freezing and thawing of concrete

For concrete subjected to freezing and thawing, it is preferable to use air entrained concrete to resist the destructive effects of severe exposure conditions.

Table 1.6: Air entrained concrete

Nominal maximum size of aggregate (mm)	Entrained air (per cent by volume)
20	5 ± 1%
40	4 ± 1%

The Indian standard code IS:456-2000 recommends the percentages of entrained air for nominal maximum size of aggregates of 20 and 40 mm as outlined in Table 1.6.

(h) Concrete exposed to sulphate attack

In the case of concrete used for marine structures where the sea water has excessive sulphates, the Indian standard code recommends the use of different types of cements from ordinary Portland to sulphate resisting Portland depending upon the sulphate content. The minimum cement content and the corresponding free water/cement ratios are compiled in Table 1.7.

1.7.2 Steel Reinforcements

Steel bars of different types and strength are used to reinforce concrete in the tension zone of flexural members to resist tension since concrete is weak in tension and in members subjected to compression to enhance the load carrying capacity.

The four main types of steel reinforcements generally used in structural concrete members are designated as follows:

1. Mild and medium tensile steel bars conforming to the specifications of IS:432 (part 1)[31].
2. High strength deformed steel bars (HYSD bars) conforming to the specifications of IS:1786-1985[32].
3. Hard drawn steel wire fabric conforming to the specifications of IS:1566-1982[33].

Table 1.7: Requirements for concrete exposed to sulphate attack (Table 4, IS:456-2000)

Sl. No.	Class	Concentration of sulphates expressed as SO_3			Types of cement	Dense, fully compacted concrete, made with 20 mm nominal maximum size aggregates complying with IS:383[35]	
		In soil		In ground water		Minimum cement content kg/m^3	Maximum free water/ cement ratio
		Total SO_3 per cent	SO_3 in 2:1 water content soil extract g/l	g/l			
1.	I	Traces (< 0.2)	Less than 1.0	Less than 0.3	Ordinary Portland cement or Portland slag cement or Portland Pozzolana cement	280	0.55
2.	II	0.2 to 0.5	1.0 to 1.9	0.3 to 1.2	Ordinary Portland cement or Portland slag cement or Portland Pozzolana cement	330	0.50
					Supersulphated cement or sulphate resisting Portland cement	310	0.50
3.	III	0.5 to 1.0	1.9 to 3.1	1.2 to 2.5	Supersulphated cement or sulphated resisting Portland cement	330	0.50
					Portland Pozzolana cement or Portland slag cement	350	0.45
4.	IV	1.0 to 2.0	3.1 to 5.0	2.5 to 5.0	Supersulphated or sulphate resisting Portland cement	370	0.45
5.	V	More than 2.0	More than 5.0	More than 5.0	Sulphate resisting Portland cement or supersulphated cement with protective coatings	400	0.40

Notes:
1. Cement content given in Table 1.7 is irrespective of grades of cement.
2. Use of supersulphated cement is generally restricted where the prevailing temperature is above 40°C.
3. Supersulphated cement gives an acceptable life provided that the concrete is dense and prepared with a water–cement ratio of 0.4 or less, in mineral acids, down to pH 3.5.
4. The cement contents given in col 6 of Table 1.7 are the minimum recommended. For SO_3 contents near the upper limit of any class, cement contents above these minimum are advised.
5. For severe conditions, such as thin sections under hydrostatic pressure on one side only and sections partly immersed, considerations should be given to a further reduction of water–cement ratio.
6. Portland slag cement conforming to IS:455 with slag content more than 50 per cent exhibits better sulphate resisting properties.
7. Where chloride is encountered along with sulphates in soil or ground water, ordinary Portland cement with C_3A content from 5 to 8 per cent shall be desirable to be used in concrete, instead of sulphate resisting cement. Alternatively, Portland slag cement conforming to IS:455 having more than 50 per cent slag or a blend of ordinary Portland cement and slag may be used provided sufficient information is available on performance of such blended cements in these conditions.

4. Structural steel conforming to grade A10, IS:2062[34], which covers the various types of rolled steel sections.

Nearly 90 per cent of the steel reinforcement used in structural concrete members comprises high strength deformed bars which has superior bond characteristics with concrete due to protruding ribs on the surface.

In general, reinforcement used in reinforced concrete work should be free from loose mill scale, rust oil, mud and other substances since they reduce the bond between steel and concrete, reducing the composite action.

The characteristic yield strength of most of the steel reinforcements may be assumed as the minimum yield or 0.2 per cent proof stress in the case of steel bars not exhibiting a definite yield point.

The nominal diameters of steel bars generally available in Indian market are 5, 6, 8, 10, 12, 16, 18, 20, 22, 25, 28, 32, 36, 40, 45 and 50 mm.

The most widely used type of reinforcement is the high strength deformed bars (HYSD bars) with a specified yield strength of 415 N/mm^2, which have protruding ribs on the surface which helps to enhance the bond strength between concrete and steel.

1.8 REINFORCEMENT SPECIFICATIONS IN STRUCTURAL CONCRETE MEMBERS

1.8.1 Reinforcement in Slabs

In reinforced concrete slabs, the minimum reinforcement required in either direction, spacing and cover requirements as specified in Indian standard code IS:456-2000 is illustrated in Fig. 1.1.

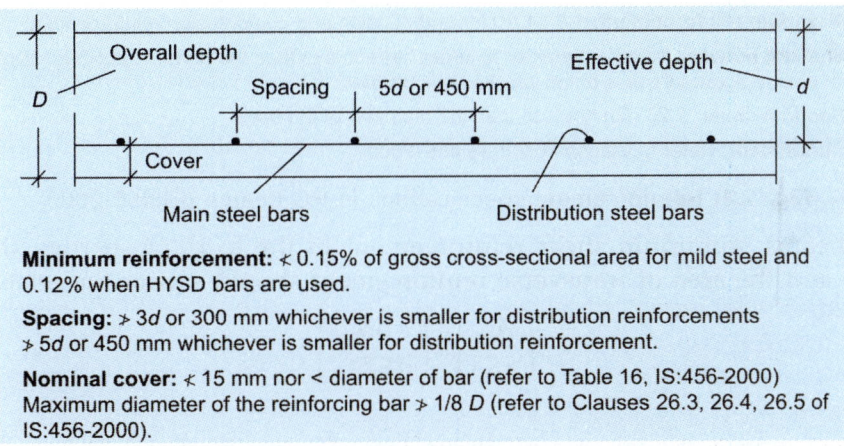

Minimum reinforcement: ≮ 0.15% of gross cross-sectional area for mild steel and 0.12% when HYSD bars are used.

Spacing: ≯ 3d or 300 mm whichever is smaller for distribution reinforcements ≯ 5d or 450 mm whichever is smaller for distribution reinforcement.

Nominal cover: ≮ 15 mm nor < diameter of bar (refer to Table 16, IS:456-2000) Maximum diameter of the reinforcing bar ≯ 1/8 D (refer to Clauses 26.3, 26.4, 26.5 of IS:456-2000).

Fig. 1.1: Reinforcement specifications in RC slabs (IS:456-2000)

1.8.2 Reinforcement in Beams

In the case of reinforced concrete beams, the minimum area of tension reinforcement according to the IS code specification is given by the relation

$$\left(\frac{A_s}{bd}\right) = \left(\frac{0.85}{f_y}\right)$$

where,

A_s = minimum area of tension reinforcement

b = breadth of beam or breadth of the web flanged beams

d = effective depth

f_y = characteristic strength of reinforcement (N/mm^2)

Figure 1.2 shows the minimum and maximum quantity of reinforcement to be provided for mild steel and HYSD bars together with the cover requirements. IS:456-2000, Clause 26.2.3 provides the guidelines for curtailment of tension reinforcement in flexural members. When the overall depth of web in a beam exceeds 750 mm, side face consisting of 0.1 per cent of web area should be distributed equally or the two faces at a spacing not exceeding 300 mm or web thickness whichever is less.

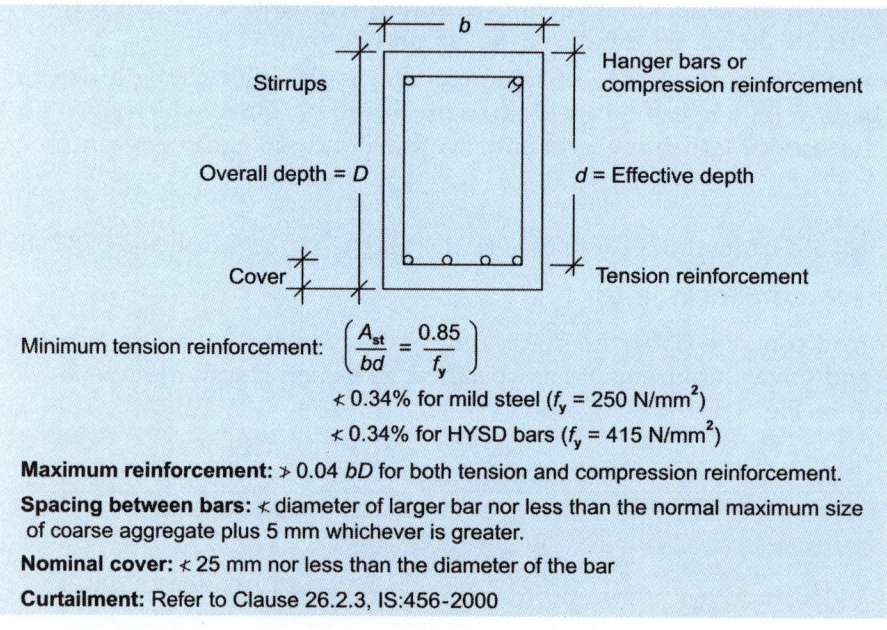

Minimum tension reinforcement: $\left(\dfrac{A_{st}}{bd} = \dfrac{0.85}{f_y} \right)$

≮ 0.34% for mild steel (f_y = 250 N/mm^2)

≮ 0.34% for HYSD bars (f_y = 415 N/mm^2)

Maximum reinforcement: ≯ 0.04 bD for both tension and compression reinforcement.

Spacing between bars: ≮ diameter of larger bar nor less than the normal maximum size of coarse aggregate plus 5 mm whichever is greater.

Nominal cover: ≮ 25 mm nor less than the diameter of the bar

Curtailment: Refer to Clause 26.2.3, IS:456-2000

Fig. 1.2: Reinforcement specifications in RC beams (IS:456-2000)

In all beams, minimum shear reinforcement in the form of stirrups should be provided and the area of transverse reinforcement should satisfy the relation

$$\left(\frac{A_{sv}}{bS_v} \right) \geq \left(\frac{0.4}{0.87 f_y} \right)$$

where,

A_{sv} = total cross-sectional area of stirrup legs effective in shear

S_v = stirrups spacing along the length of the member

b = breadth of beam or breadth of web of flanged beam

f_y = characteristic strength of stirrup reinforcement in N/mm^2 which shall not be taken greater than 415 N/mm^2

1.8.3 Reinforcement in Columns

In practice, columns of square, rectangular and circular cross-section are generally used. Main longitudinal reinforcements are provided in columns along with lateral ties to prevent buckling of the main bars. Details of minimum and maximum quantity of reinforcement, minimum number of bars and their diameter, cover requirements etc. are shown in Fig. 1.3[*].

---

[*] The reader may refer to Clauses 26.5.3.1 and 26.5.3.2, IS:456-2000 for further details.

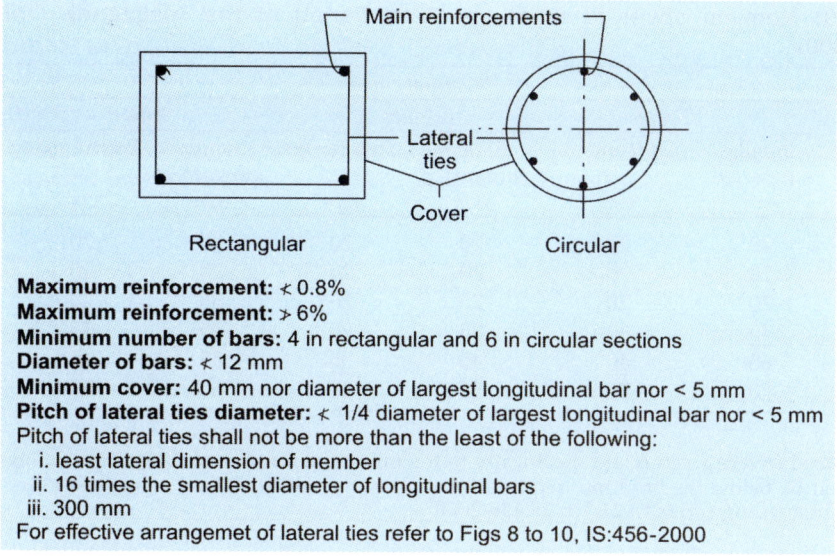

Rectangular   Circular

Maximum reinforcement: ≮ 0.8%
Maximum reinforcement: ≯ 6%
Minimum number of bars: 4 in rectangular and 6 in circular sections
Diameter of bars: ≮ 12 mm
Minimum cover: 40 mm nor diameter of largest longitudinal bar nor < 5 mm
Pitch of lateral ties diameter: ≮ 1/4 diameter of largest longitudinal bar nor < 5 mm
Pitch of lateral ties shall not be more than the least of the following:
   i. least lateral dimension of member
   ii. 16 times the smallest diameter of longitudinal bars
   iii. 300 mm
For effective arrangemet of lateral ties refer to Figs 8 to 10, IS:456-2000

Fig. 1.3: Reinforcement specifications in RC columns (IS:456-2000).

1.9 SPECIFICATIONS OF COVER REQUIREMENTS FOR STEEL REINFORCEMENTS

In reinforced concrete structures, sufficient cover of concrete is required to protect the steel reinforcements from exposure to aggressive environmental conditions and consequent rusting and deterioration of the cross-sectional area. Nominal cover is the design depth of concrete cover to all steel reinforcements including links.

Minimum values of the nominal cover for reinforcement depends upon the exposure conditions and the IS Code requirements for cover are given in Table 1.8. In case of columns, the nominal cover for longitudinal bars should be not less than the diameter of the bar. However, in the case of columns of minimum dimension of 200 mm or less and whose bars do not exceed 12 mm, a nominal cover of 25 mm may be used. In the case of RCC footings under the column, the minimum cover is prescribed as 50 mm.

The Indian standard code also provides for nominal cover requirements to meet specified period of fire resistance as shown in Table 1.9. The values indicate the minimum cover requirements for beams, slabs, ribs and columns for fire resistance varying from 0.5 to 4 hours.

Table 1.8: Nominal cover to meet durability requirements (IS:456-2000, Table 16)

Exposure	Nominal concrete cover in mm not less than
Mild	20
Moderate	30
Severe	45
Very severe	50
Extreme	75

Notes:
1. For main reinforcement up to 12 mm diameter bar for mild exposure, the nominal cover may be reduced by 5 mm.
2. Unless specified otherwise, actual concrete cover should not deviate from the required nominal cover by +10 mm.
3. For exposure condition 'severe' and 'very severe', reduction of 5 mm may be made, where concrete grade is M35 and above.

Table 1.9: Nominal cover to meet specified period of fire resistance (Table 16A, IS:456-2000)

Fire resistance	Nominal cover						
	Beams		Slabs		Ribs		Columns
	Simply supported	Continuous	Simply supported	Continuous	Simply supported	Continuous	
h	mm	mm	mm	mm	mm	mm	mm
0.5	20	20	20	20	20	20	40
1.0	20	20	20	20	20	20	40
1.5	20	20	25	20	<u>35</u>	20	40
2.0	<u>40</u>	30	<u>35</u>	25	45	<u>35</u>	<u>40</u>
3.0	60	<u>40</u>	45	<u>35</u>	55	45	40
4.0	70	50	55	45	65	55	40

Notes:
1. The nominal covers given relate specifically to the minimum number of dimensions in Fig. 1.1.
2. Cases that lie below the bold line require attention to the additional measures necessary to reduce the risk of spalling (refer to 21.3.1, IS:456-2000).

REFERENCES

1. Monier J, History of Concrete and Cement, Inventors library records, Paris, 1849.
2. Coignet F, First Reinforced Concrete Three-Storeyed Building, Library archives, Paris, 1853 (Acquired a French patent for reinforced concrete in 1855).
3. Lambot J, Concrete boat with embedded iron bars exhibited in Paris exhibition, French library archives, 1854.
4. Talbot NA, Testing of Reinforced Concrete Beams, Bulletin No. 1, 1904.
5. Turneaure F, Cyclopedia of Civil Engineering, American Technical School, 1908.
6. Morsch E and Goodrich EP, *Concrete and Steel Construction*, 3 edn, 1909–1910.
7. Faber O and Bowie PG, *Reinforced Concrete—Theory and Practice*, Vol 1 & 2, London, 1920.
8. Whitney AC and Hool AG, Concrete Design Manual, London, 1921.
9. IS:456-2000, Indian Standard Code of Practice for Plain and Reinforced Concrete (4th Revision), BIS, New Delhi, July 2000.
10. BS EN:1992-1-1, Euro Code-2, Design of Concrete Structures, General Rules & Rules for Buildings, British Standards Institution, London, 2004.
11. ACI:318M-11, Building Code Requirements for Structural Concrete, American Concrete Institute, Farmington Hills, Michigan, 2005, p 443.
12. AS:3800-1988, Concrete Structures, Standards Association of Australia, 1988.
13. CSA Standard A23.1-00/A23.2-00, Concrete Materials and Methods of Concrete Construction/Methods of Test for Concrete, Canadian Standard Associations, Toronto, 2000.
14. SP:16-1980, Design Aids for Reinforced Concrete to IS:456, BIS, New Delhi, 1980.
15. SP:34-1987, Handbook of Concrete Reinforcement and Detailing, BIS, New Delhi, 1987.
16. IS:10262-1982 (Reaffirmed in 1999), Recommended Guidelines for Concrete Mix Design, BIS, New Delhi, 1982.

17. Fintel M, *Handbook of Concrete Engineering*, 2 edn, CBS Publishers, New Delhi, 1986, p 892.

18. Reynolds CE, Steedman JC and Thvelfall AJ, *Reinforced Concrete Design Handbook*, 11 edn, Psychology Press, London, 2008, p 401.

19. Rowe RE, Cranston WB and Best BC, New concepts in the design of structural concrete, *Structural Engineer*, Vol. 43, 1965, pp 339–403.

20. Bate SCC, Why limit state design? *Concrete*, March 1968, pp 103–108.

21. Moseley B, Bungey J and Hulse R, *Reinforced Concrete to Euro Code*-2, 7 edn, Palgrave Macmillan, London, 2012.

22. IS:875-1987 (Part 1), Code of Practice for Design Loads (other than Earthquake) for Buildings and Structures (Part 1), Dead Loads (2nd Revision), BIS, New Delhi, 1989.

23. IS:875-1987 (Part 2), Code of Practice for Design Loads (other than Earthquake) for Buildings and Structures (Part 2), Imposed Loads (2nd Revision), BIS, New Delhi, 1989.

24. IS:875-1987 (Part 3), Code of Practice for Design Loads (other than Earthquake) for Buildings and Structures (Part 3), Wind Loads (2nd Revision), BIS, New Delhi, 1989.

25. IS:875-1987 (Part 4), Code of Practice for Design Loads (other than Earthquake) for Buildings and Structures (Part 4), Snow Loads (2nd Revision), BIS, New Delhi, 1989.

26. IS:875-1987 (Part 5), Code of Practice for Design Loads (other than Earthquake) for Buildings and Structures (Part 5), Special Loads and Combinations (2nd Revision), BIS, New Delhi, 1989.

27. IS:1893-2002, Criteria for Earthquake Design of Structures (Part 1), General Provisions and Buildings (5th Revision), BIS, New Delhi, 2005.

28. IS:4326-1993, Code of Practice for Earthquake Resistant Design and Construction of Buildings, BIS, New Delhi, 1993 (Reaffirmed 1998).

29. IS:1489, 1991 (Part 1), Specification for Portland Pozzolana Cement, BIS, New Delhi, 1991.

30. IS:455-1989, Specifications for Portland Slag Cement (4th Revision), BIS, New Delhi, 1989.

31. IS:432-1982 (Part 1), Specifications for Mild Steel and Medium Tensile Bars for Concrete Reinforcement (3rd Revision), BIS, New Delhi, 1982.

32. IS:1786-1985, Specifications for High Strength Deformed Steel Bars for Concrete Reinforcement (3rd Revision), BIS, New Delhi, 1985.

33. IS:1566-1982, Specifications for Hard Drawn Steel Wire Fabric for Concrete Reinforcement, BIS, New Delhi, 1982.

34. IS:2062-1992, Specifications for Steel for General Structural Purposes (4th Revision), BIS, New Delhi, 1992.

35. IS:383-1970, Specifications for Coarse and Fine Aggregates from Natural Sources for Concrete (2nd Revision), BIS, New Delhi, 1970.

REVIEW QUESTIONS

1. Briefly review the evolution of reinforced concrete as a successful construction material, specifying the main reasons for its composite action.
2. List the pioneers responsible for developing design rules for reinforced concrete.
3. Discuss briefly the various applications of reinforced concrete in structural members.
4. What are the various classifications of reinforced concrete structural elements?
5. List the common types of structural systems used in reinforced concrete structures.
6. Mention the principle objective of the various national codes on reinforced concrete.
7. What are the different types of loads to be considered in the design of RC structures?
8. Mention briefly the main ingredients of reinforced concrete.
9. Explain the terms: grade, tensile strength, modulus of elasticity, shrinkage, creep and durability with respect to concrete.
10. Write a brief note on the Indian standard code reinforcement specifications in structural concrete members.

OBJECTIVE QUESTIONS

1. Reinforced concrete is a successful composite material because
   a. concrete is strong in compression
   b. steel is strong in tension
   c. bond strength is strong between concrete and steel
2. Reinforced concrete is preferred to steel in the construction of bridges due to
   a. high strength of reinforced concrete
   b. low maintenance costs
   c. faster construction using reinforced concrete
3. Reinforced concrete is preferred for compression members of buildings because
   a. steel is strong in tension
   b. concrete is strong in compression
   c. Economy in construction costs
4. Reinforced concrete is preferred to steel in the construction of bridges in coastal zones mainly due to
   a. high strength of reinforced concrete
   b. low maintenance costs
   c. faster construction using reinforced concrete
5. The strain compatibility in reinforced concrete is attributed to
   a. high strength of steel
   b. high durability of concrete
   c. high bond strength between steel and concrete

6. In multistoreyed structures, overall costs can be reduced by using
   a. steel
   b. reinforced concrete
   c. plain concrete
7. Structural concrete members should meet the criterion of
   a. strength
   b. strength and serviceability
   c. serviceability
8. In designing tall reinforced concrete structures like chimneys and towers, the prominent load to be considered in design is
   a. dead load
   b. live load
   c. wind load
9. Minimum amount of tensile reinforcements in beams should be provided mainly to
   a. ensure adequate strength
   b. prevent sudden failure
   c. improve durability
10. According to the Indian standard code, the minimum number of reinforcing bars in a circular column should be not less than
   a. 8
   b. 4
   c. 6

Elastic or Working Stress Theory of Reinforced Concrete Sections

2.1 ELASTIC THEORY OF RC SECTIONS

The earliest codified design philosophy by various countries is based on the elastic or working stress theory proposed by Francois Coignet[1] of France, who acquired a patent for reinforced concrete in 1855. The concept of bond between steel rods and the surrounding concrete was propounded by the American lawyer, Thaddeus Hyatt as mentioned by Turneaure[2] in 1877. Later Koenan of Germany developed the design rules in 1886 as mentioned by Taylor and Thompson[3]. These fundamental concepts were incorporated by the French Commission as reported by Faber and Bowie[4] in their design rules for reinforced concrete in 1907 followed by the American Concrete Institute and the American Society of Civil Engineers, jointly developing the first design code for reinforced concrete in 1909 as reported by Adams and Mathews[5]. In the early 20th century, these design principles were recognized as the elastic or working stress theory of reinforced concrete sections[6,7].

The success of the elastic theory is attributed to the good bond[8,9] between concrete and steel resulting in the composite behaviour of the material in an elastic manner under service loads. Working stress method is used not only for reinforced concrete, but also for steel, timber and other metallic structures.

2.2 BASIC ASSUMPTIONS IN ELASTIC THEORY

In the elastic theory, the materials are assumed to behave in a linear elastic manner and the required safety of the structure is ensured by restricting the stresses in concrete and steel to permissible stresses obtained by applying suitable factor of safety to the characteristic strength of the materials. The resulting permissible or working stresses under service loads will be well within the linear elastic range of the materials. The basic assumptions incorporated in the elastic theory of flexure according to the Indian standard code IS:456-2000[10] are as follows:

1. At any cross-section, plane sections before bending remain plane after bending indicating that strain varies linearly over the depth of the section.

2. All tensile stresses are resisted by the reinforced and none by concrete except in the uncracked phase where concrete can resist a small magnitude of tension as permitted depending upon the grade of concrete as specified in the code.
3. The stress–strain relationship of steel and concrete under working loads is a straight line implying that stresses are linearly proportional to strains for both concrete and steel.
4. The modular ratio $m = (E_s/E_c)$ has the value $(280/3\sigma_{cbc})$.

2.3 NEUTRAL AXIS DEPTH AND MOMENT OF RESISTANCE OF SECTIONS

Referring to a rectangular section, shown in Fig. 2.1, subjected to a moment 'M' under service loads,

let σ_{cbc} = compressive stress developed in concrete
      σ_{st} = tensile stress developed in steel
      A_{st} = area of tensile reinforcement
        d = effective depth
        b = width of the member
        n = neutral axis depth factor
     nd = depth of neutral axis
      m = modular ratio $(280/3\sigma_{cbc})$
       j = lever arm coefficient $(1 - n/3)$
      C = compressive force in concrete
      T = tensile force in steel
     M = moment of resistance of the section

Since concrete is weak in tension, concrete below the neutral axis is neglected in computations. Below neutral axis the steel is converted into an equivalent area of concrete by multiplying the steel area by modular ratio and this area contributes to the tensile force for equilibrium of the section.

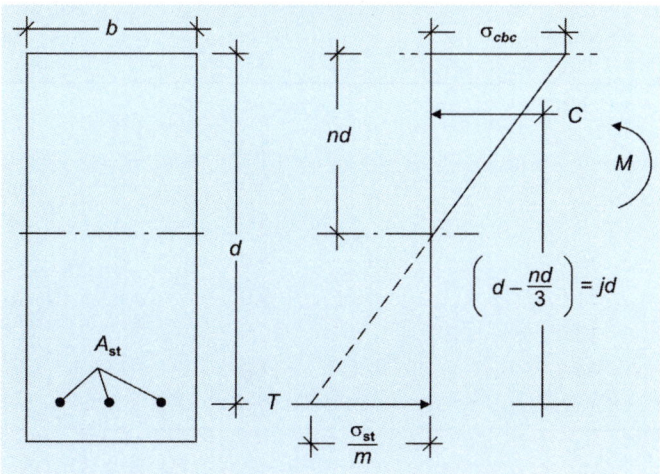

Fig. 2.1: Stress distribution in rectangular section

Referring to the stress distribution shown in Fig. 2.1, we have the relation

$$\left[\frac{\sigma_{cbc}}{(\sigma_{st}/m)}\right] = \left[\frac{nd}{d - nd}\right] = \left(\frac{n}{1 - n}\right)$$

Neutral axis depth factor $n = \left[\dfrac{1}{1 + (\sigma_{st} / m\sigma_{cbs})}\right]$

From moment equilibrium of the section, we have

$$M = C[d - (nd/3)]$$
$$= Cd[1 - (n/3)$$
$$= 0.5\sigma_{cbc}b{\cdot}nd{\cdot}d[1 - (n/3)]$$

substituting $j = (1 - n/3) =$ lever arm coefficient

$$M = 0.5\sigma_{cbc}bnjd^2$$
$$M = Qbd^2$$

where $Q = $ (a constant) $= (0.5\sigma_{cbs}nj)$

Hence $d = \sqrt{\dfrac{M}{Qb}}$ (2.1)

Also $M = \sigma_{st}A_{st}(d - nd/3) = \sigma_{st}A_{st}jd$

$\therefore$ $A_{st} = \left(\dfrac{M}{\sigma_{st}jd}\right)$ (2.2)

From Eqs (2.1) and (2.2), we can estimate the depth required for the concrete section and corresponding area of steel required to resist a given moment of the section.

The permissible stresses in concrete and steel are given in Tables 2.1 and 2.2.

Table 2.1: Permissible stresses in concrete (all values in N/mm²) (IS:456-2000, Table 21)

Grade of concrete	Permissible stress in compression		Permissible stress in bond (average) for plain bars in tension
	Bending σ_{cbc}	Direct σ_{cc}	τ_{hd}
M10	3.0	2.5	–
M15	5.0	4.0	0.6
M20	7.0	5.0	0.8
M25	8.5	6.0	0.9
M30	10.0	8.0	1.0
M35	11.5	9.0	1.1
M40	13.0	10.0	1.2
M45	14.5	11.0	1.3
M50	16.0	12.0	1.4

Notes:
1. The values of permissible shear stress in concrete are given in IS:456-2000, Table 23.
2. The bond stress (τ_{hd}) given in column 4 shall be increased by 25% for bars in compression.

Table 2.2: Permissible stresses in steel reinforcement (IS:456-2000, Table 22)

Sl. No.	Type of stress in steel reinforcement	Permissible stress in N/mm^2		
		Mild steel bars conforming to grade I of IS:432 (part I)	Medium tensile steel conforming to IS:432 (part I)	High yield strength deformed bars conforming to IS:1786 (grade Fe 415)
1.	Tension (σ_{st} or σ_{sv})			
	a. Up to and including 20 mm	140	Half the guaranteed yield stress subject to a maximum of 190	230
	b. Over 20 mm	130		230
2.	Compression in column bars (σ_{sc})	130	130	190
3.	Compression in bars in a beam or slab when the compressive resistance of concrete is taken into account:	The calculated compressive stress in the surrounding concrete multiplied by 1.5 times the modular ratio or σ_{sc} whichever is lower		
	a. Up to and including 20 mm	140	Half the guaranteed yield stress subject to a maximum of 190	190
	b. Over 20 mm	130		190

Notes:
1. For high yield strength deformed bars of grade Fe 500, the permissible stress in direct tension and flexural tension shall be $0.55f_y$. The permissible stress of shear and compression reinforcement shall be as for grade Fe 415.
2. For welded wire fabric conforming to IS:1566, the permissible value in tension σ_{st} is 230 N/mm^2.
3. For the purpose of this standard (IS:456-2000), the yield stress of steels for which there is no clearly defined yield point should be taken to be 0.2% proof stress.
4. When mild steel conforming to grade II, IS:432 (part I) is used, the permissible stresses shall be 90% of the permissible stress in col. 3, or if the design details have already been worked out on the basis of mild steel conforming to grade I, IS:432 (part I); the area of reinforcement shall be increased by 10% of that required for grade I steel.

The values of design coefficients for different grades of concrete M15, M20, M25 and grades of steel Fe 250, Fe 415 and Fe 500 are given in Table 2.3. These coefficients are very useful in the analysis of stresses in reinforced concrete sections and also in the design office for rapid working stress design of reinforced concrete sections subjected to flexure.

Table 2.3: Design coefficients

σ_{cbc} (N/mm^2)	m	σ_{st} (N/mm^2)	n	j	Q
5	19	140	0.400	0.866	0.874
		230	0.292	0.903	0.659
		280	0.256	0.914	0.584
7	13	140	0.400	0.870	1.220
		230	0.284	0.900	0.910
		280	0.250	0.920	0.800
8.5	11	140	0.400	0.870	1.480
		230	0.288	0.900	1.100
		280	0.250	0.920	0.980
10	9	140	0.400	0.870	1.740
		230	0.288	0.900	1.300
		280	0.250	0.920	1.150

2.4 STRESSES IN SINGLY REINFORCED RECTANGULAR SECTIONS

Consider the singly reinforced rectangular beam section shown in Fig. 2.2, subjected to a bending moment M. The corresponding *cracked transformed section* is shown in Fig. 2.2(c). The concrete on the tension side below the neutral axis is neglected. The neutral axis is located by the line passing through the centroid of transformed section.

Equating the first moment of the areas on the compression and tension side, we have the relation

$$\frac{b(kd)^2}{2} = mA_{st}(d - kd)$$

Solving the quadratic equation to determine the value of kd, the acceptable root for the value of $k(0 < k < 1)$ is expressed as

$$k = \sqrt{2\rho m + (\rho m)^2} - \rho m,$$

where ρm = reinforcement ratio $\left(\dfrac{A_{st}}{bd}\right)$

The second moment of area of the cracked transformed section is given by the relation

$$I_{cr} = \left[\frac{b(kd)^3}{3} = mA_{st}(d - kd)^2\right]$$

Knowing the neutral axis depth and second moment of area, the stress in concrete and steel in the composite section subjected to a moment M can be evaluated using the flexure relation, $f = [My/I]$. The stresses can also be directly determined by considering the equilibrium of forces acting on the compression side from Fig. 2.2(d). The tensile and compressive forces are computed as

$$C = \left(\frac{bkd}{2}\right)f_c$$

and $T = A_{st}f_{st}$ and the lever arm factor $j = (1 - k/3)$

$M = (Cjd) = (Tjd)$

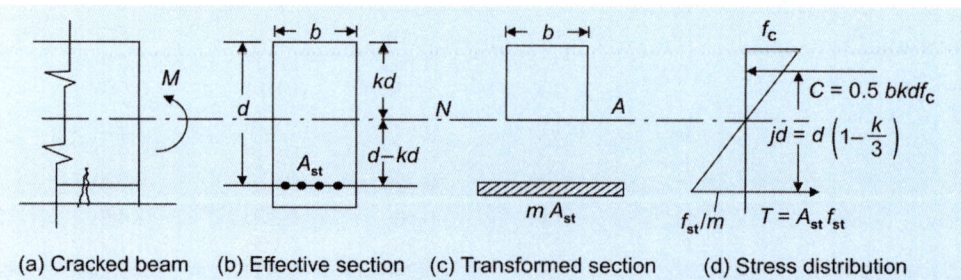

(a) Cracked beam   (b) Effective section   (c) Transformed section   (d) Stress distribution

Fig. 2.2: Reinforced concrete beam section subjected to moment 'M' (*cracked–transformed section*)

The stresses in concrete and steel are computed as

$$f_c = \left[\frac{M}{0.5b(kd)(jd)} \right] \text{ and } f_{st} = \left[\frac{M}{A_{st}jd} \right]$$

2.5 BALANCED, UNDER-REINFORCED AND OVER-REINFORCED SECTIONS

In a given *reinforced concrete section*, analysis for the depth of the actual neutral axis and moment of resistance of the section can be determined as follows:

Let b = width of the section

d = effective depth

n_a = depth of actual neutral axis

n_c = depth of critical neutral axis

A_{st} = area of tension reinforcement

σ_{st} = permissible tensile stress in steel

σ_{cb} = permissible compressive bending stress in concrete

m = modular ratio

Then by equating the first moment of the areas above and below the neutral axis, we have

$$\frac{bn_a^2}{2} = mA_{st}(d - n_a)$$

Solving the equation, the depth of actual neutral axis n_s can be determined. The critical neutral axis depth depends on the permissible stresses and is computed as

$$n_c = \left[\frac{1}{1 + \dfrac{\sigma_{st}}{m \cdot \sigma_{cb}}} \right] d$$

If $n_a < n_c$, then the section is considered as *under-reinforced* and the moment of resistance of the section is compound from the tension side with the steel reaching maximum permissible stress σ_{st}. Hence the moment of resistance is given by the relation

$$M_r = \sigma_{st} A_{st} \left(d - \frac{n_a}{3} \right) \tag{2.3}$$

If $n_a > n_c$, then the section is considered as *over-reinforced* and the moment of resistance of the section is computed from the compression side since the concrete in the extreme fibre reaches the maximum permissible stress σ_{cb} first. Hence the moment of resistance is computed as

$$M_r = \frac{1}{2} \sigma_{cb} n_a b \left(d - \frac{n_a}{3} \right) \tag{2.4}$$

If $n_a = n_c$, then the section is designated as *balanced*. In this case, steel and concrete reach their maximum permissible stresses simultaneously and the moment of resistance of the section can be estimated by either of the Eqs (2.3) or (2.4) which yields the same value.

In practice, it is advisable to design the reinforced concrete sections as under-reinforced or balanced since there will be clear warning of failure of the member in the form of large deflections associated with well distributed cracks along the length of the member, before final failure. Balanced sections are the most economical but it may not be practicable in all the cases.

Over-reinforced sections are not preferred since they require larger quantities of steel and in addition, the members exhibit explosive failure without any warning with negligible deflections and very few cracks. Hence in practice, it is always economical to design balanced or under-reinforced sections.

The areas of round steel bars and their spacing for a given diameter of bars are given in Tables 2.4, 2.5(a) and 2.5(b).

Table 2.4: Area, perimeter and weight of round bars

Diameter of bar mm	Area cm^2	Perimeter cm	Weight kg/m
5	0.20	1.57	0.15
6	0.28	1.88	0.22
8	0.50	2.51	0.39
10	0.79	3.14	0.62
12	1.13	3.77	0.89
16	2.01	5.03	1.58
20	3.14	6.28	2.47
22	3.80	6.91	2.98
25	4.91	7.85	3.85
28	6.16	8.80	4.83
32	8.04	10.05	6.31
36	10.18	11.31	7.99
40	12.57	12.57	9.86
45	15.90	14.14	12.49
50	19.64	15.71	15.41

Table 2.5(a): Area, perimeter and spacing of round bars

Diameter of bar mm	Area of steel and perimeter per metre width for spacings of									
	6 cm	7 cm	8 cm	9 cm	10 cm	11 cm	12 cm	13 cm	14 cm	15 cm
5 area (cm²)	3.27	2.80	2.46	2.18	1.96	1.78	1.63	1.51	1.40	1.31
Perimeter (cm)	26.15	22.40	19.60	17.45	15.70	14.25	13.05	12.08	11.20	10.45
6 area (cm²)	4.72	4.04	3.54	3.14	2.83	2.57	2.36	2.17	2.02	1.89
Perimeter (cm)	31.35	26.85	23.50	20.85	18.80	17.10	15.65	14.45	13.45	12.55
8 area (cm²)	8.38	7.18	6.28	5.58	5.03	4.57	4.19	3.87	3.59	3.35
Perimeter (cm)	41.75	35.80	31.35	27.90	25.10	22.80	20.90	19.30	17.90	16.70
10 area (cm²)	13.08	11.22	9.82	8.72	7.85	7.14	6.54	6.04	5.61	5.24
Perimeter (cm)	52.30	44.90	39.25	35.00	31.40	28.55	26.15	24.20	22.50	20.95
12 area (cm²)	18.84	16.16	14.14	12.56	11.31	10.28	9.42	8.70	8.08	7.54
Perimeter (cm)	63.00	53.95	47.15	41.90	37.70	34.35	31.50	29.00	26.95	25.20
16 area (cm²)	33.52	28.72	25.14	22.34	20.11	18.28	16.76	15.47	14.36	13.41
Perimeter (cm)	84.00	72.00	63.00	56.00	50.30	45.75	42.00	38.75	36.00	33.55
20 area (cm²)	52.34	44.88	39.28	34.92	31.42	28.55	26.17	24.16	22.44	20.94
Perimeter (cm)	104.10	89.75	78.50	69.90	62.80	57.00	52.15	48.25	44.90	41.80
22 area (cm²)	63.34	54.30	47.52	42.24	38.01	34.55	31.67	29.24	27.15	25.34
Perimeter (cm)	115.00	98.90	86.50	77.00	69.10	63.00	57.85	53.10	49.50	46.10
25 area (cm²)	82.00	70.20	61.50	54.70	49.03	44.70	41.00	37.90	35.10	32.75
Perimeter (cm)	130.80	112.00	98.10	87.20	78.50	71.40	65.40	60.15	56.05	52.25

Table 2.5(b): Area, perimeter and spacing of round bars

Diameter of bar mm	Area of steel and perimeter per metre width for spacings of									
	16 cm	17 cm	18 cm	19 cm	20 cm	22.5 cm	25 cm	27.5 cm	30 cm	40 cm
5 area (cm²)	1.23	1.16	1.09	1.03	0.98	0.87	0.79	0.71	0.65	0.49
Perimeter (cm)	9.80	9.25	8.72	8.25	7.85	6.98	6.28	5.70	5.21	3.93
6 area (cm²)	1.77	1.66	1.57	1.49	1.41	1.25	1.13	1.02	0.94	0.71
Perimeter (cm)	11.75	11.05	10.45	9.90	9.40	8.35	7.52	6.82	6.25	4.70
8 area (cm²)	3.14	2.96	2.79	2.65	2.51	2.23	2.01	1.82	1.67	1.26
Perimeter (cm)	15.65	14.75	13.95	13.20	12.55	11.15	10.04	9.10	8.35	6.28
10 area (cm²)	4.91	4.62	4.36	4.13	3.93	3.50	3.15	2.85	2.62	1.97
Perimeter (cm)	19.65	18.50	17.45	16.55	15.70	14.00	12.56	11.40	10.50	7.85
12 area (cm²)	7.07	6.65	6.28	5.95	5.65	5.02	4.52	4.11	3.76	2.83
Perimeter (cm)	23.60	22.20	21.00	19.85	18.85	16.80	15.08	13.70	12.60	9.43
16 area (cm²)	12.57	11.83	11.17	10.58	10.05	8.90	8.02	7.30	6.70	5.03
Perimeter (cm)	31.50	29.60	28.00	26.50	25.15	22.40	20.12	18.35	16.80	12.58
20 area (cm²)	19.64	18.48	17.46	16.54	15.71	13.95	12.58	11.40	10.45	7.36
Perimeter (cm)	39.25	36.90	34.90	33.00	31.40	27.90	25.12	22.80	20.90	15.70
22 area (cm²)	23.76	22.36	21.12	20.01	19.01	16.85	15.22	13.85	12.70	9.51
Perimeter (cm)	43.25	40.65	38.50	36.45	34.55	30.75	27.64	25.20	23.10	17.28
25 area (cm²)	30.75	28.95	27.35	25.80	24.54	21.80	19.65	17.85	16.40	12.27
Perimeter (cm)	49.00	46.20	43.55	41.30	39.25	34.95	31.40	28.50	26.10	19.63

2.6 ANALYSIS EXAMPLES

1. A reinforced concrete beam of rectangular section 300 mm wide by 600 mm overall depth is reinforced with 4 bars of 25 mm diameter at an effective depth of 550 mm. Adopting M20 grade concrete and Fe 415 grade HYSD bars determine the following items:

   a. The cracking moment (assuming gross concrete section properties)

   b. The cracking moment (assuming transformed section properties)

   c. Compute stresses in steel and concrete due to an applied moment of 40 kN·m

   d. Compute stresses in steel and concrete under service load moment of 150 kN·m

   i. *Data*:

   Width of section $b = 300$ mm

   Overall weight $D = 600$ mm

   Effective depth $d = 550$ mm

   Area of tension steel $A_{st} = 1963$ mm²

   Modular ratio $m = 13$

   Grade of concrete M20

   Grade of steel Fe 415

   Modulus of rupture $f_{cr} = 0.7\sqrt{20} = 3.13$ N/mm²

ii. *Cracking moment (using gross concrete section)*:

Section modulus $Z = \left(\dfrac{bD^2}{6}\right) = \left(\dfrac{300 \times 600^2}{6}\right) = (18 \times 10^6)\ \text{mm}^3$

Approximate cracking moment $M_{cr} = (f_{cr}Z) = (3.13 \times 18 \times 10^6)/10^6 = 56\ \text{kN·m}$

iii. *Transformed section properties*:

Gross area of section $A_G = (300 \times 600) = (18 \times 10^4)\ \text{mm}^2$

Transformed area $A_T = [A_G + (m-1)A_{st} = [(18 \times 10^4)+(13-1)1963]$
$$= (20.35 \times 10^4)\ \text{mm}^2$$

iv. *Neutral axis depth*:

Depth of neutral axis (y) is determined by equating the moments of areas of the transformed section about the top edge

$$A_T y = (bD)\,(D/2) + (m-1)A_{st}(d)$$

$$y = \left[\frac{(300 \times 600)\,(300) + (13-1)\,1963 \times 550}{(20.35 \times 10^4)}\right] = 330\ \text{mm}$$

Distance from neutral axis to compression fiber $y_c = 330$ mm

Distance from neutral axis to extreme tension fiber $y_t = (600-330) = 270$ mm

Distance from neutral axis to tension steel $y_s = (550-330) = 220$ mm

Second moment of area of transformed section is computed as

$$I_T = \left\{\frac{by_c^3}{3} + \frac{by_t^3}{3} + (m-1)A_{st}y_s^2\right\}$$

$$= [300 \times (330^3 + 270^3)/3 + (12 \times 1963 \times 220^2]$$

$$= (6.73 \times 10^9)\ \text{mm}^4$$

v. *Cracking moment*:

$$M_{ct} = f_{cr}\left[\frac{I_T}{y_t}\right] = 3.13 \times \left(\frac{6.73 \times 10^9}{270}\right) = (78 \times 10^6)\ \text{N·mm} = 78\ \text{kN·m}$$

The estimate of cracking moment using the gross section is 56 kN·m and is underestimated by 28%.

vi. *Stresses in concrete and steel under applied moment*:

Applied moment $M = 40$ kN·m

Maximum compressive stress in concrete

$$f_c = \left(\frac{My_c}{I_T}\right) = \left(\frac{40 \times 10^6) \times 330}{6.73 \times 10^9}\right) = 1.96\ \text{N/mm}^2$$

Maximum tensile stress in concrete

$$f_{ct} = \left(\frac{My_t}{I_T}\right) = \left(\frac{40 \times 10^6) \times 270}{6.73 \times 10^9}\right) = 1.60\ \text{N/mm}^2$$

Tensile stress in steel $f_{st} = mf_c(y_s/y_t) = (13 \times 1.96)\,(200/330) = 17.0\ \text{mm}^2$

vii. *Stresses in concrete and steel under service load moment of 150 kN·m*:

Cracking moment M_{cr} = 78 kN·m

Under service load moment of 150 kN·m, the section would have cracked, hence the cracked transformed section properties should be used for evaluating the stresses in concrete and steel.

Transformed steel area (mA_{st}) = (13 × 1963) = 25519 mm^2.

Equating the first moment of areas about the neutral axis (n), we have

$$\left[\frac{300 \times n^2}{2}\right] = 25519\,(550 - n)$$

Solving the neutral axis depth n = 233 mm

Taking moments about the tension steel, we have

$$M = 0.5 f_c bn(d - n/3)$$

Hence concrete stress $f_c = \left[\dfrac{150 \times 10^6}{0.5 \times 300 \times 233\left(500 - \dfrac{233}{3}\right)}\right] = 9.08\ \text{N/mm}^2$

By taking moments about the compressive force in concrete

$$M = 0.5 f_{st} A_{st}(d - n/3)$$

Tensile stress in steel $f_{st} = \left[\dfrac{150 \times 10^6}{1963\left(500 - \dfrac{233}{3}\right)}\right] = 161.6\ \text{N/mm}^2$

2. A simply supported reinforced concrete one-way slab with an effective span of 4 m is reinforced with 10 mm diameter Fe 415 HYSD bars spaced at 200 mm centres at an effective depth of 180 mm. Using M20 grade concrete, calculate the permissible live load on the slab if the self-weight of slab and finishes are 5.5 kN/m^2.

i. *Data*:

Effective span of RC slab L = 4 m

Effective depth d = 180 mm

Spacing of 10 mm diameter bars = 200 mm

Materials: M20 grade concrete

Fe 415 HYSD bars

Thus $A_s = \left(\dfrac{100 \times 79}{200}\right) = 395\ \text{mm}^2/\text{m}$

ii. *Permissible stresses*:

σ_{cbc} = 7 N/mm^2 m = 13

σ_{st} = 230 N/mm^2 n_c = 0.284

iii. *Loads*:

Dead load and finishes g = 5.5 kN/m^2

$\therefore \qquad M_g = \left(\dfrac{5.5 \times 4^2}{8}\right) = 11\ \text{kN} \cdot \text{m}$

iv. *Neutral axis depth*:

Let n_a = depth of actual neutral axis

First moment of areas about neutral axis yields

$$0.5bn_a^2 = mA_{st}(d - n_a)$$

$$(0.5 \times 1000 \times n_a^2) = 13 \times 395 \times (180 - n_a)$$

Solving, n_a = 38.2 mm.

If n_c = critical neutral axis depth

$$n_c = \left[\frac{1}{1 + (\sigma_{st}/m\sigma_{cbc})}\right]d$$

$$n_c = \left[\frac{1}{1 + \left(\dfrac{230}{13 \times 7}\right)}\right]d = 0.284 \text{ or } d = (0.284 \times 180)$$

$$= 51.12 \text{ mm}$$

Since $n_a < n_c$, the section is under-reinforced.

v. *Moment of resistance*:

$$M_r = \sigma_{st}A_{st}(d - n/3)$$
$$= 230 \times 395 (180 - 38.2/3)$$
$$= 15.2 \times 10^6 \text{ N·mm}$$
$$M_q = \text{Live load moment} = (15.2 - M_g) = (15.2 - 11) = 4.2 \text{ kN·m}$$

If q = live load on the slab

$$q = \left(\frac{8M_q}{L^2}\right) = \left(\frac{8 \times 4.2}{4^2}\right)$$

$$= 2.1 \text{ kN/m}^2$$

∴ Permissible live load on slab q = 2.1 kN/m²

3. A simply supported RC slab having an overall thickness of 150 mm is reinforced with 12 mm diameter bars at an effective depth of 130 mm. The spacing of bars is 100 mm. The effective span of the slab is 4 m. If the self-weight of slab and finishes is 4.2 kN/m², estimate the maximum permissible live load on the slab. Adopt M15 grade concrete and Fe 250 grade I steel.

   i. *Data*:

   Effective span of slab = 4 m

   Overall thickness of slab = 150 mm

   Effective depth = 130 mm

   Spacing of 12 mm diameter bars = 100 mm

   $$\therefore \qquad A_{st} = \left(\frac{1000 \times 113}{100}\right) = 1130 \text{ mm}^2/\text{m}$$

   Materials: M15 grade concrete and Fe 250 grade I mild steel.

ii. *Permissible stresses*:

$$\sigma_{cb} = 5\,N/mm^2 \qquad m = 19$$
$$\sigma_{st} = 1400\,N/mm^2 \quad n_c = 0.4\,d$$

iii. *Loads*:

Dead load and finishes = 4.2 kN/m²

$$\therefore \qquad M_g = \left(\frac{4.2 \times 4^2}{8}\right) = 8.4\,kN\cdot m$$

iv. *Neutral axis depth*:

Let n_a = depth of actual neutral axis, then

$$\left(\frac{bn_a^2}{2}\right) = mA_{st}\,(d - n_a)$$

$$\left(\frac{1000n_a^2}{2}\right) = 19 \times 1130\,(130 - n_a)$$

$$\therefore\ n_a^2 + 43n^a - 5582 = 0$$
$$\therefore\ n_a = 56\,mm$$

If n_c = depth of critical neutral axis

$$n_c = \left[\frac{1}{1 + \dfrac{\sigma_{st}}{ms_{cb}}}\right]d$$

$$n_c = \left[\frac{1}{1 + \dfrac{\sigma_{st}}{19 \times 5}}\right]d$$

$$= 0.4d = (0.4 \times 130) = 52\,mm$$

Since $n_a > n_c$, the section is over-reinforced.

v. *Moment of resistance*:

$$M_r = \frac{1}{2}\sigma_{cb}n_ab\left(d - \frac{n_a}{3}\right)$$

$$= \frac{1}{2} \times 5 \times 56 \times 1000\left(30 - \frac{56}{3}\right)$$

$$= 15.58 \times 10^6\,N\cdot mm$$

$$= 15.58\,kN\cdot m$$

If M_q = Live load moment = $(M_r - M_g)$

$$= (15.58 - 8.4)$$
$$= 7.18\,kN\cdot m$$

Live load, $q = \left(\dfrac{8 \times 7.18}{4^2}\right) = 3.59\,kN/m^2$

4. A reinforced concrete beam having a rectangular section 300 mm wide is reinforced with 2 bars of 12 mm diameter at an effective depth of 550 mm. The section is subjected to a service load moment of 40 kN·m. Assuming M20 grade concrete and Fe 415 HYSD bars, estimate the stresses in concrete and steel.

i. *Data:*

$b = 300$ mm

$d = 550$ mm

$A_{st} = (2 \times 113) = 226$ mm^2

Materials: M20 grade concrete ($m = 13$)

Fe 415 HYSD bars

Service load moment $M = 40$ kN·m

ii. *Neutral axis depth:*

Let n_a = depth of actual neutral axis

Then, $0.5bn_a^2 = mA_{st}(d - n_a)$

$0.5 \times 300 \times n_a^2 = 13 \times 226(550 - n_a)$

Solving, $n_a = 94.5$ mm

Referring to Table 2.3, critical neutral axis depth for M20 grade concrete and Fe 415 HYSD bars is

$$n_c = 0.284d = (0.284 \times 550) = 156.2 \text{ mm}$$

Since $n_a < n_c$, the section is under-reinforced.

iii. *Stress in concrete and steel:*

Taking moments about the tension steel centroid,

$$M = 0.5\,\sigma_{cb}bn_a(d - n_a/3)$$

$$\therefore \quad \sigma_{cb} = \left[\frac{M}{0.5bn_a(d - n_a/3)}\right] = \left[\frac{40 \times 10^6}{0.5 \times 300 \times 94.5(550 \times 94.5/3)}\right]$$

$$= 5.44 \text{ N/mm}^2$$

Taking moments about the tension steel centroid,

$$M = \sigma_{st}A_{st}(d - n_a/3)$$

$$\therefore \quad \sigma_{st} = \left[\frac{M}{A_{st}(d - n_a/3)}\right] = \left[\frac{40 \times 10^6}{226(550 - 94.5/3)}\right]$$

$$= 340.8 \text{ N/mm}^2$$

5. A reinforced concrete beam of span 5 m has a rectangular section 250 mm wide by 500 mm depth. The beam is reinforced with 3 bars of 16 mm diameter on the tension side at an effective depth of 450 mm; and 2 bars of 16 mm diameter on the compression side at a cover of 50 mm from the compression face. If M15 grade concrete and Fe 250 grade mild steel is used, estimate the maximum permissible live load on the beam.

i. *Data:*

Effective span $L = 5$ m

Width of beam $b = 250$ mm

Overall depth $D = 500$ mm
Effective depth $d = 450$ mm
Area of tension steel $A_{st} = 603$ mm^2
Area of compression steel $A_{st} = 402$ mm^2
Materials: M15 grade concrete
Fe 250 grade mild steel

ii. *Permissible stresses*:

$\sigma_{cb} = 5$ N/mm^2 $m = 19$
$\sigma_{st} = 140$ N/mm^2 $n_c = 0.4\,d$

iii. *Depth of neutral axis*:

If n_a = depth of actual neutral axis

$$\frac{bn_a^2}{2} + (1.5\,m - 1)\,A_{sc}(n_a - d_c) = mA_{st}(d - n_a)$$

$$125 n_a^2 + (1.5 \times 19 - 1)\,402\,(n_a - 50) = 19 \times 603\,(450 - n_a)$$

$$n_a^2 + 180 n_a - 45667 = 0$$

Solving, $n_a = 142$ mm
Depth of critical neutral axis $= n_c$
or $n_c = 0.4d = (0.4 \times 450) = 180$ mm
Since $n_a < n_c$, the section is under-reinforced.

iv. *Moment of resistance*:

σ_{cc} = stress at level of compression steel
$\sigma_{st} = 140$ N/mm^2
σ_{cb} = stress in extreme fiber of concrete
$m = 19$

$$= \left[\frac{\sigma_{st}}{m(d - n_a)}\right] = \left[\frac{\sigma_{cb}}{n_a}\right]$$

$$\therefore \ \sigma_{cb} = \left[\frac{\sigma_{st} n_s}{m(d - n_a)}\right] = \left[\frac{140 \times 142}{19 \times (450 - 142)}\right]$$

$$= 3.39 \text{ N/mm}^2$$

$$\sigma_{cc} = \left[\frac{(n_a - d_c)\,\sigma_{cb}}{n_a)}\right] = \left[\frac{6(142 - 50)\,3.39}{142}\right]$$

$$= 2.19 \text{ N/mm}^2$$

Hence, the moment of resistance is given by

$$M_r = \frac{1}{2}\sigma_{cb} n_a b \left(d - \frac{n_a}{3}\right) + (1.5\,m - 1)\,A_{st}\sigma_{cb}(d - d_c)$$

$$= \frac{1}{2} \times 3.39 \times 142 \times 250 \left(450 - \frac{142}{3}\right)$$

$$+ (1.5 \times 19 - 1)\,402 \times 2.19 \times (450 - 50)$$

$$= 33.8 \times 10^6 \text{ N·mm}$$

$$= 33.8 \text{ kN·m}$$

v. *Permissible live load*:

Self-weight of beam = $(0.25 \times 0.5 \times 24) = 3$ kN/m

$M_g = (0.125 \times 3 \times 5^2) = 9.375$ kN·m

$M_q = (M_r - M_g)$

$= (33.8 - 9.375) = 24.4$ kN/m

Let q = Live load on the beam

or $q = \left(\dfrac{8M}{L^2}\right) = \left(\dfrac{8 \times 24.4}{25}\right) = 7.8$ kN/m

Maximum permissible live load on beam = 7.8 kN/m

REFERENCES

1. Coignet F, First Reinforced Concrete Three-Storeyed Building, Library archives, Paris, 1853 (Acquired a French patent for reinforced concrete in 1855).
2. Turneaure F, Cyclopedia of Civil Engineering, American Technical School, 1908.
3. Taylor WF and Thompson ES, *A Treatise on Concrete, Plain and Reinforced*, 1 edn 1905; 2 edn 1912; 3 edn 1916.
4. Faber O and Bowie PG, *Reinforced Concrete—Theory and Practice*, Vols 1 & 2, 1912–1920.
5. Adams H and Mathews RE, Reinforced Concrete Construction, 1911–1920.
6. Hool AG, *Reinforced Concrete Construction—Fundamental Principles*, Vol 1, 1912.
7. Park and Paulay T, *Reinforced Concrete Structures*, John Wiley & Sons, New York, 1975.
8. ACI:408R-03-2003, Bond and development of straight reinforcing bars in tension, American Concrete Institute, Farmington Hills, USA, p 49.
9. Lutz LA and Gergely P, Mechanics of bond and slip of deformed bars in concrete, *ACL Journal Proceedings*, Vol. 64, No. 11, pp 711–721 and Discussions, Vol. 65, pp 412–14.
10. IS:456-2000, Indian Standard Code of Practice for Plain and Reinforced Concrete (4th Revision), BIS, New Delhi, July 2000.

ASSIGNMENT

1. Compute the moment of resistance of a reinforced concrete section with a width of 250 mm, reinforced with 3 bars of 10 mm diameter at an effective depth of 500 mm. Assume M20 grade concrete and Fe 415 HYSD bars.

2. A reinforced concrete beam of rectangular section 300 mm wide by 650 mm deep is reinforced with 4 bars of 32 mm diameter at an effective depth of 600 mm using M20 grade concrete and Fe 415 HYSD bars, estimate the moment of resistance of the section.

3. A reinforced concrete slab of overall thickness 200 mm and effective span 4 m is provided with 10 mm diameter Fe 415 HYSD bars spaced at 100 mm centres at an effective depth of 170 mm. If the self-weight of slab and finishes is

6 kN/m^2, estimate the maximum permissible live load on the slab assuming M25 grade concrete.

4. A reinforced concrete beam of rectangular section is required to resist a service load moment of 120 kN·m. Design suitable breadth and depth of the section (assuming $b = 0.5d$) and reinforcements for the balanced section of the beam assuming M20 grade concrete and Fe 415 HYSD bars.

5. Calculate the moment of resistance of a reinforced concrete T-beam which has an effective flange width of 3 m and thickness 150 mm. The rib is 300 mm wide by 600 mm deep. The beam is reinforced with 6 bars of 25 mm diameter Fe 415 HYSD bars at an effective depth of 700 mm. Assume M20 grade concrete.

6. A doubly reinforced concrete beam of rectangular section 300 mm wide by 600 mm deep is reinforced with 2 bars of 20 mm diameter on the tension and compression side respectively at an effective cover of 50 mm. Adopt M20 grade concrete and Fe 415 HYSD bars and estimate the safe moment of resistance of the section.

7. A concrete beam of rectangular section 250 mm wide by 450 mm deep is reinforced with 3 bars of 20 mm diameter at an effective depth of 400 mm. Adopting M25 grade concrete and Fe 415 HYSD bars, determine (a) the cracking moment, and (b) the stresses in concrete and steel due to an applied moment of 40 kN·m.

8. A rectangular concrete beam with a width of 300 mm and an overall depth of 550 mm is reinforced with 4 bars of 25 mm diameter with an effective cover of 50 mm. If the concrete used is of grade M30 and reinforcements are Fe 500 HYSD bars, estimate the maximum stresses developed in concrete and steel under a working load moment of 200 kN·m.

9. A reinforced concrete beam is of trapezoidal section with widths of 400 mm at the top and 300 mm at the soffit. The overall depth of the beam is 650 mm. The beam is reinforced with 3 bars of 25 mm at an effective depth of 600 mm. Adopting M20 grade concrete and Fe 415 HYSD bars, estimate the maximum stresses developed when the section is subjected to a service load moment of 180 kN·m.

10. A reinforced concrete beam of symmetrical I-section having flanges 400 mm wide by 100 mm thick and a web depth and thickness of 600 and 150 mm respectively is reinforced with 4 bars of 20 mm diameter at an effective cover of 50 mm. Compute the stresses in the member when it is subjected to a moment of 50 kN·m. Adopt M20 grade concrete and Fe 415 HYSD bars.

REVIEW QUESTIONS

1. Mention the basic principles of elastic or working stress theory of reinforced concrete flexural members.

2. What are salient assumptions made in the elastic theory of reinforced concrete sections?

3. Explain the terms: neutral axis depth factor, modular ratio, lever arm coefficient and moment of resistance with respect to reinforced concrete sections.

4. Briefly outline the concept of cracked transformed section in the analysis of reinforced concrete sections.

5. Explain the terms: under-reinforced, balanced and over-reinforced sections as applied to reinforced concrete sections.

6. Why do you prefer to design under-reinforced or balanced sections instead of over-reinforced sections?

7. Explain the significance of the terms: (a) Actual neutral axis (b) Critical neutral axis with respect to reinforced concrete sections.

8. Discuss briefly the importance of modulus of rupture of concrete. In what way it influences the stresses developed in concrete and steel in a reinforced concrete structure?

9. What difference do you identify in the computation of moment of resistance of under-reinforced and over-reinforced sections according to the specifications of the IS code?

10. Why does the IS code specify an effectively higher modulus ratio for compression reinforcement as compared to tension reinforcement?

OBJECTIVE QUESTIONS

1. Elastic theory assumes that the stress–strain relationship of concrete and steel up to working load is
   a. non-linear
   b. linear
   c. linear with different slopes

2. The modular ratio depends upon the parameter
   a. permissible stress in steel
   b. modulus of elasticity of concrete
   c. permissible compressive stress in concrete

3. The effective depth of a reinforced concrete beam is inversely proportional to the
   a. load supported by the beam
   b. compressive stress in concrete
   c. width of the beam

4. The moment of resistance of a reinforced concrete beam depends upon
   a. tensile strength of concrete
   b. effective depth
   c. overall depth

5. Reinforced concrete beams designed with under-reinforced sections result in
   a. explosive failure
   b. progressive failure
   c. failure by yielding of steel

6. In case of reinforced concrete beams with balanced sections, the actual neutral axis lies
   a. above the critical neutral axis
   b. below the critical neutral axis
   c. coincides with the critical neutral axis

7. In reinforced concrete beams with under-reinforced sections, the amount of steel required is inversely proportional to
    a. the moment of resistance
    b. permissible stress in steel
    c. permissible stress in concrete

8. The modulus of rupture of concrete is directly proportional to the
    a. permissible stress in steel
    b. permissible stress in concrete
    c. characteristic compressive strength of concrete

9. In the case of over-reinforced sections, the actual neutral axis
    a. coincides with the actual neutral axis
    b. lies above the critical neutral axis
    c. below the critical neutral axis

10. In case of basement or cellar floors where the headroom is restricted, we can use
    a. singly reinforced sections
    b. doubly reinforced sections
    c. over-reinforced sections

Principles of Limit State Design

3.1 EVOLUTION OF LIMIT STATE DESIGN

The basic concepts of *limit state design*[1,2] emerged during the early part of 20th century as an advancement over the traditional design philosophies like the elastic theory and ultimate load methods of design. The main deficiency in the working stress or elastic method of design is the lack of information regarding the computation of collapse or ultimate load capacity of reinforced concrete structural elements. The ultimate load method[3,4] developed during the middle of 20th century predicted various procedures to evaluate the collapse or ultimate load capacity of structural elements. A structure designed solely on the basis of ultimate loads, although having a desirable margin of safety against collapse, may not be serviceable due to excessive deflections and/or development of objectionable cracks at working loads. This type of distress is particularly noticeable in structures designed by the ultimate load method using high strength materials.

The limit state design[5,6] concepts slowly emerged during this period as a solution to overcome the deficiencies of the elastic and ultimate load methods. The limit state design is a method of designing structures based on the statistical concept of safety and the associated probability of failure considering both the states of serviceability and strength. The limit state method of design ensures not only the serviceability of the structures at working loads, but also satisfies the requirement of desirable strength criterion by ensuring that the structure supports the ultimate loads.

The integrated limit state method of design has been accepted worldwide and incorporated in the national code of various countries like USA[7], UK[8], Germany[9], Canada[10] and India[11]. The limit state concepts have been examined thoroughly by various investigators and several research investigators have proved beyond doubt that the method can be universally adopted for the design of reinforced concrete structures. The evolution of the various design methods culminating in the limit state design is shown in Fig. 3.1.

1900–1930
Elastic theory or working stress method

i. Factor of safety applied to the yield or ultimate stress to get permissible stress
ii. Structure designed to support working loads without exceeding the permissible stresses in concrete and steel
Inadequacy of elastic theory: Actual safety against ultimate loads not known

1930–1960
Ultimate load or strength method

i. Load factors applied to working loads to estimate the ultimate loads
ii. Safety factors applied to characteristic strengths to estimate design strengths.
Inadequacy of ultimate load method: Serviceability aspects as deflection and cracking at service loads are not considered

1960 to date
Limit state method

i. Structures designed to satisfy the limit states of
  a. strength—collapse
  b. serviceability—deflection and cracking
ii. Statistical probability concepts incorporated for loads and strengths
  Characteristic loads and strengths obtained by applying partial safety factors for loads and material strengths
iii. Limit state method overcomes the inadequacies of working stress and ultimate load methods
iv. Limit state method incorporated in most of the national codes of various countries in American, European and Asian countries

Fig. 3.1: Evolution of limit state method

3.2 APPLICATION OF CLASSICAL RELIABILITY THEORY TO LIMIT STATE DESIGN

Probabilistic concepts are explicitly incorporated in the limit state design by considering the probability density of load and strength variables of the constituent materials. Applications of classical reliability theory[12,13] to structural design require comprehensive statistical data regarding loads and strengths and their exact shapes of normal distribution curves. At present, the probabilities of failure that are socially acceptable must be kept very low (1 in a million). At such low levels, the probability of failure is very sensitive to the exact shape of the normal distribution curves. To determine exact shapes of normal distribution curves, we require very large number of statistical data and such comprehensive data is not yet available. In particular, sufficient number of extreme values of the strengths of complete structures (to define accurately the shapes of the tails of the normal distribution curves) may never be available.

In a simple example, only one type of load and strength variables are used. For a real structure, there will in general be many types of loads and many modes of failure, normally with complex correlations between them making it very difficult to calculate the probability of failure. Hence in the limit state design, our engineering experience and judgment have been used to modify and to remedy the inadequacies of earlier design methods and partly use the probabilistic concepts. Hence, it is appropriate to designate the limit state design method currently practised as *semi probabilistic approach* to structural design as reported by Cornell[14].

The interaction between load effects and strength is shown in Fig. 3.2 where the normal distribution curves for loads and material strengths are superposed along with the characteristic loads and strengths.

For good design, the characteristic loads and strengths are expressed in terms of standard deviation, mean strength and probability factor as

$$F = (F_m + 1.65\sigma)$$

$$f_{ck} = (f_m + 1.65\sigma)$$

where,

F = characteristic load

F_m = mean load

f_{ck} = characteristic strength

σ = standard deviation

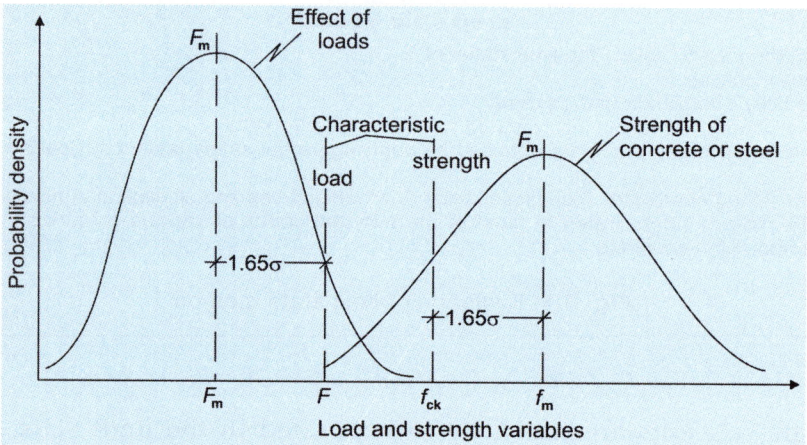

Fig. 3.2: Classical reliability model for strength design

3.3 PRINCIPLES OF LIMIT STATES

A safe and satisfactory design must ensure the achievement of an acceptable probability that the specified life of a structure is not curtailed prematurely due to the attainment of an unsatisfactory condition or *limit state* which covers the various forms of failure. The primary limit states are:

a. The limit state of collapse, and

b. The limit state of serviceability involving excessive cracking and deflection at service or working loads.

Each of these limit states may be attained due to different types of loading configurations. Some of the important limit states of failure or collapse are listed below:

1. Failure of one or more critical sections in flexure, shear, torsion or due to their combinations.
2. Failure due to fatigue under repeated loads.
3. Failure due to bond and anchorage failure of reinforcement.
4. Failure due to elastic instability of structural members.
5. Failure due to impact, earthquake, fire or frost.
6. Failure due to the destructive effects of chemicals, corrosion of reinforcements.

The structure may be rendered unfit for its intended purpose due to various serviceability limit states being reached, such as:

1. Excessive deflection or displacement, severely affecting the finishes and causing discomfort to the users of the structure.
2. Excessive local damage leading to cracking or spalling of concrete impairing the efficiency or appearance of the structure.

3.4 SAFETY FACTORS

In contrast to the factor of safety applied to the ultimate failure stress of concrete and steel in the working stress method, partial safety factors expressed in terms of the probability that the structure will not become unfit for its intended function during its useful life are applied for different limit states in the limit state design method.

Due to the large number of variables involved, a rational appraisal of the safety of a structure based on probability theory is not yet practical and hence, partial safety factors are applied for each limit state and these comprise of material strength (γ_m) reduction factors for characteristic strength of materials and partial safety factors for loads (γ_p) enhancement factors for characteristic loads on the structure.

The Indian standard code IS:456-2000 recommends the use of partial safety factors for loads (γ_f) as given in Table 3.1 and the partial safety factors for material strength (γ_m) are given in Table 3.2.

Table 3.1: Partial safety factors (γ_f) for loads (IS:456-2000, Table 18)

Load combination	Limit state of collapse			Limit states of serviceability		
	DL	IL	WL	DL	IL	WL
DL + IL	1.5		1.0	1.0	1.0	–
DL + WL	1.5 or 0.9^n	–	1.5	1.0	–	1.0
DL + IL + WL	1.2			1.0	0.8	0.8

Notes:
1. While considering earthquake effects, substitute EL for WL.
2. For the limit states of serviceability, the values of γ_f given in Table 3.1 are applicable for short term effects. While assessing the long term effects due to creep the dead load and that part of the live load likely to be permanent may only be considered.
[n] This value is to be considered when stability against overturning or stress reversal is critical.

Table 3.2: Partial safety factors for material strength (γ_m) (IS:456-2000)

Material	Limit state		
	Collapse	Deflection	Local damage
Concrete	1.50	1.00	1.00
Steel	1.15	1.00	1.00

3.5 CHARACTERISTIC AND DESIGN LOADS

The characteristic load acting on any structure can be generally expressed as

Characteristic load (F) = [Mean load + k (standard deviation)], where k is a factor, which ensures that the probability of the characteristic load being exceeded is small. Since exhaustive statistical information about the nature of loads and their variation is not available at present, the characteristic loads are assumed as equivalent to the values of loads specified in the loading standards given in IS:875 code. The revised Indian standard code, IS:456-2000, identifies three types of loads which are classified as dead, live (imposed) and wind load. In addition, loads resulting from the effects of creep, shrinkage and temperature are also considered wherever they are judged to be significant.

In general, the characteristic loads do not allow for lack of precision in design computation, inadequacies in the various methods of analysis and methods of construction. Hence, for the computation of design loads (F_d), the characteristic loads are enhanced by suitable partial safety factor (γ_f) for the various limit states.

The design load is given by the relation

$$F_d = F\gamma_f$$

where,

F = characteristic load

γ_f = partial safety factor appropriate to the nature of loading and the limit state being reached as shown in Table 3.1.

3.6 CHARACTERISTIC AND DESIGN STRENGTHS

The statistical variation in the strength properties of concrete and steel are expressed as characteristic strength related to the mean strength and standard deviation expressed as:

Characteristic strength = [Mean strength − k (standard deviation)],

where k is a factor chosen to ensure the probability of the characteristic strength not being exceeded is small. The Indian standard code, IS:456-2000 has recommended a value of 1.65 for k so that only 5% of the test results could have a strength less than that of the characteristic strength.

Due to lack of statistical data, the characteristic strength of concrete and steel may be assumed as the work cube strength and minimum proof or yield strength respectively as recommended in the current codes.

Since the strength of materials in a structure is likely to differ from those tested on standard specimens, the design strength is evaluated by dividing the characteristic strength by the appropriate partial safety factor (γ_m), for different limit states and expressed as:

Design strength (f_d) = (f/γ_m)

where,

f = characteristic strength of the material

γ_m = partial safety factor appropriate to the material and the limit state being considered as shown in Table 3.2.

3.7 GLOBAL FACTOR OF SAFETY

The global factor of safety concept has been proposed by Bill Moseley et al[15], in which a combined safety factor termed as global safety factor is computed by multiplying the appropriate partial safety factors for loads and materials. The use of partial safety factors on materials and load actions provides the considerable flexibility and this can be used for special situations such as very high standards of control and construction, e.g. construction of nuclear reactors or in structural elements where failure without warning may be very serious.

Thus, the global factor of safety in case of a beam failure caused by yielding of tensile reinforcement is computed as

$$(\gamma_f \times \gamma_m) = (1.50 \times 1.15)$$
$$= 1.725 \text{ [for permanent (dead) loads as per IS:456-2000 code]}$$

Alternatively, failure by crushing of concrete in the compression zone may have a global factor of safety of $(1.5 \times 1.5) = 2.25$ due to variable actions only emphasizing the fact that such failure is generally explosive without warning and may result in serious consequences.

REFERENCES

1. Rowe RE, Cranston WB and Best BC, New concepts in the design of concrete, *Structural Engineer*, Vol 43, pp 339–403.

2. Bate SCC, Why limit state designs, *Concrete*, March 1968, pp 103–108.

3. Evans RH, The plastic theories for the ultimate strength of reinforced concrete beams, *Journal of the Institution of Civil Engineers*, London, Vol 21, Dec. 1943, pp 98–121.

4. Hognestad E, Hansen NW and Henry D Mc, Concrete stress distribution in ultimate strength design, *Proc of the ACI Journal*, Vol 52, No 4, Dec 1955, pp 455–80.

5. CEB recommendations for International Code of Practice for Reinforced Concrete, American Concrete Institute and Cement & Concrete Association, London, 1964.

6. Krishna Raju N, Limit State Design for Structural Concrete, *Proc of the Institution of Engineers (India)*, Vol 51, No. 1, January 1971, pp 138–143.

7. ACI:318M-11, Building Code Requirements for Structural Concrete, American Concrete Institute, Farmington Hills, Michigan, 2005.

8. BS EN:1992-1-1-2004, Design of Concrete Structures, General Rules and Rules for Buildings, British Standard Institution, 2004.

9. DIN:1045-1988, Structural Use of Concrete, Design and Construction, Din Deutsches Institute, Fir Normung EV, 1988.

10. CSA Standard A23.3-94, Design of concrete structures, Canadian Standards Association, Rexdale, Ontario, 1994.

11. IS:456-2000, Indian Standard Code of Practice for Plain and Reinforced Concrete (4th Revision), BIS, 2000, p 100.

12. Ranganathan R, Reliability Analysis and Design of Structures, Tata McGraw-Hill, New Delhi, 1990.

13. Ellingwood B, Reliability Basis for Load and Resistance Factors for RC Design, NBS Building Science Series 110, National Bureau of Standards, Washington DC, 1978.

14. Cornell CA, A probability-based structural code, *Journal of ACI*, Vol 66, Dec 1969, pp 974–985.

15. Mosley B, Bungey J and Hulse R, *Reinforced Concrete Design to Euro Code-2*, Palgrave Macmillan, London, 2012.

REVIEW QUESTIONS

1. List the main reasons for the evolution of limit state method of design for RC structures.
2. Outline the difference between deterministic and probabilistic designs with reference to structural concrete members.
3. What are the various limit states to be considered in the design of structural concrete members?
4. Differentiate between safety and serviceability with respect to structural concrete members.
5. What are the various serviceability states and why they should be considered as design?
6. Explain the terms:(a) Characteristic load (b) Characteristic strength.
7. Explain clearly the concept of assigning different safety factors for different types of loads.
8. Differentiate between probabilistic design and semi-probabilistic design.
9. Explain the terms: standard deviation, mean strength and design strength.
10. What is global factor of safety? How do you compute the global factor of safety for flexural concrete members?

OBJECTIVE QUESTIONS

1. The inadequacy of the ultimate load/strength method is
   a. ultimate loads cannot be estimated
   b. collapse loads can be computed
   c. serviceability is not ensured
2. Probabilistic concepts are incorporated in
   a. ultimate load design
   b. limit state design
   c. working stress design
3. The first method to be used in the design of structural concrete members is
   a. the ultimate load method
   b. elastic method
   c. limit state method
4. The partial safety factor specified in the Indian standard code for the combination of live, dead and wind loads is
   a. 1.5
   b. 1.6
   c. 1.2

5. Excessive deflections in a reinforced concrete beam leads to
   a. sudden collapse
   b. damage for partitions
   c. instability of the structure
6. In the limit state design process, design loads are obtained by
   a. equating the characteristic loads
   b. enhancing the characteristic loads
   c. decreasing the characteristic loads
7. The partial safety factor used for material strength of steel at the limit state of collapse is
   a. 1.5
   b. 1.0
   c. 1.15
8. The design strength of material is
   a. directly proportional to partial safety factor
   b. nearly equal to the characteristic strength
   c. inversely proportional to the partial safety factor
9. The partial safety factor specified in the Indian standard code for evaluating the design strength of concrete for the limit state of deflection is
   a. 1.3
   b. 1.5
   c. 1.0
10. The global safety factor can be evaluated as the
    a. sum of partial safety factors
    b. product of partial safety factors
    c. product of safety factor for loads and strength

Flexural Strength of Reinforced Concrete Sections

4.1 INTRODUCTION

Reinforced concrete structural elements used in various types of structures like beams and slabs are invariably subjected to flexure or bending while resisting the working loads. In limit state design, the ultimate or collapse limit state is considered to be an important limit state and the structures designed should be able to support the desired ultimate loads. To achieve this goal, it is essential to ensure that the ultimate or collapse strength of the reinforced concrete sections are attained while supporting the designed ultimate loads.

The estimation of the flexural strength of reinforced concrete sections is based on the extensive experimental and research investigations by various investigators like Evans[1], Hognestad *et al*[2], Rusch[3], Jones[4], Reagen *et al*[5], Park and Paulay[6] and others. Based on the extensive research investigations on reinforced concrete flexural members, most of the national codes have used the experimental data and formulated simplified design procedures to estimate the ultimate flexural strength of rectangular and flanged sections with idealized stress blocks in the concrete compression zone.

In this chapter, the analysis of reinforced concrete sections subjected to flexure is examined in detail and the method specified in the Indian standard code IS:456-2000[7] for the computation of flexural strength of rectangular, T- and L-sections are presented with various types of practical examples. Structural engineers and students should be familiar with the various formulae given in the code and their implications which will facilitate them to produce safe and economical designs of flexural members. Alternatively, they can use the design tables and charts given in *design aids to IS:456-2000* published as SP:16[8] by the Bureau of Indian standards. Designers may also refer to the *Manual of Limit State Design of Reinforced Concrete Members* by Varyani and Radhaji[9] which contains exhaustive design tables and charts to facilitate faster design of structural concrete members. Reinforced concrete designers handbook by Reynolds *et al*[10] is also very useful for strength computations of flexural members.

4.2 FLEXURAL STRENGTH OF RECTANGULAR SECTIONS

4.2.1 Moment of Resistance Equations

The computation of ultimate flexural strength of reinforced concrete section subjected to flexure is based on the normal assumptions outlined in Section 38.1 of the IS code. The design stress block parameters recommended in IS:456-2000 code is shown in Fig. 4.1.

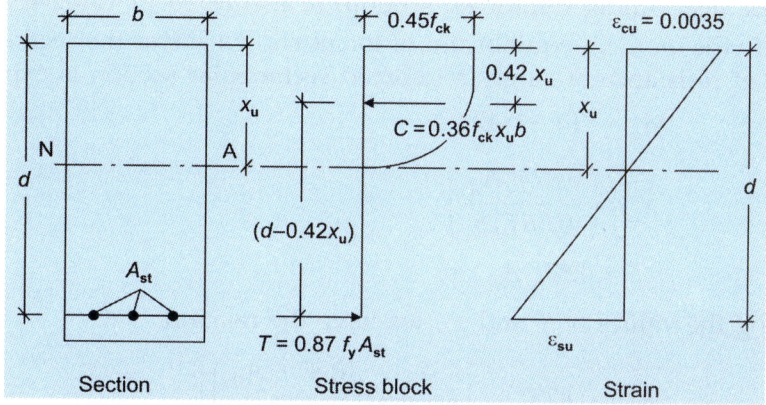

Fig. 4.1: Stress block parameters (rectangular section)

Equilibrium of forces across the section yields the following relations

$$C = T$$
$$0.36 f_{ck} x_u b = 0.87 f_y A_{st}$$
$$\left(\frac{x_u}{d}\right) = \left(\frac{0.87 A_{st} f_y}{0.36 f_{ck} bd}\right)$$

Limiting values of the ratio $(x_{u,max}/d)$ to prevent brittle failure (over-reinforced section) is determined from the condition that the strain in steel ε_{su} at failure should be more than the value given by the relation

$$\varepsilon_{su} = \left[\frac{0.87 f_y}{E_s} + 0.002\right]$$

The yield strain for design purposes for different grades of steel are given in Table 4.1, assuming the modulus of elasticity of steel $E_s = 2 \times 10^5$ N/mm^2.

Table 4.1: Limiting values of neutral axis depth ratio $(x_{u,max}/d)$			
Grade of steel	f_y (N/mm^2)	*Yield strain* (ε_{su})	$(x_{u,max}/d)$
Fe 250 mild steel	250	0.0031	0.53
Fe 415 HYSD bars	415	0.0038	0.48
Fe 500 HYSD bars	500	0.0042	0.46

From the strain diagram (Fig. 4.1), we have

Ultimate strain in concrete $\varepsilon_{cu} = 0.0035$

Hence

$$\left(\frac{x_u}{d}\right) = \left(\frac{\varepsilon_{cu}}{\varepsilon_{cu} + \varepsilon_{su}}\right) = \left(\frac{0.0035}{0.0035 + \varepsilon_{su}}\right)$$

Substituting the values of ε_{su} for different grades of steel, the maximum limiting values of the ratio $(x_{u,max}/d)$ for different grades of steel to ensure failure of suction by yielding of steel (under-reinforced section) is also listed in Table 4.1.

Referring to the moment equilibrium of forces shown across the section in Fig. 4.1, the moment of resistance of singly reinforced rectangular section is expressed as

$$M_u = T(d - 0.42x_u)$$

and

$$x_u = \left(\frac{0.87 f_y A_{st}}{0.36 f_{ck} b}\right)$$

and

$$T = 0.87 f_y A_{st}$$

Substituting the values of T and x_u, we have the relation

$$M_u = 0.87 f_y A_{st}\left[d - 0.42\left(\frac{0.87 f_y A_{st}}{0.36 f_{ck} b}\right)\right]$$

Rearranging the terms, the final expression for computing the moment of resistance of a singly reinforced rectangular section M_u (refer to Section G-1, IS:456-2000)

$$M_u = 0.87 f_y A_{st} d\left[1 - \left(\frac{A_{st} f_y}{bd f_{ck}}\right)\right] \tag{4.1}$$

Moment of resistance of balanced section using the limiting values of $(x_{u,max}/d)$ for different grades of steel are compiled in Table 4.2.

For under-reinforced sections, the area of tension reinforcement to resist the ultimate moment M_u can be estimated using Eq. (4.1). Also for a known limiting moment $M_{u,lim}$, the effective depth d (required), can be computed from the expressions given in Table 4.2.

Table 4.2: Moment of resistance for limiting values of $(x_{u,max}/d)$

Grade of steel	$(x_{u,max}/d)$	Expression for $M_{u,lim}$
Fe 250	0.53	$0.148 f_{ck} b d^2$
Fe 415	0.48	$0.138 f_{ck} b d^2$
Fe 500	0.46	$0.133 f_{ck} b d^2$

4.2.2 Analysis Examples

1. A rectangular reinforced concrete beam has a width of 200 mm and is reinforced with 2 bars of 20 mm diameter at an effective depth of 400 mm. If M20 grade concrete and Fe 415 HYSD bars are used, estimate the ultimate moment of resistance of the section.

  i. *Data*:

$b = 200$ mm

$d = 400$ mm

$A_{st} = (2 \times 314) = 628$ mm^2

$f_{ck} = 20$ N/mm^2

$f_y = 415$ N/mm^2

  ii. *Depth of neutral axis*:

If x_u = depth of neutral axis

$$\left(\frac{x_u}{d}\right) = \left(\frac{0.87 f_y A_{st}}{0.36 f_{ck} bd}\right) = \left(\frac{0.87 \times 415 \times 628}{0.36 \times 20 \times 200 \times 400}\right) = 0.39$$

Limiting value of (x_u/d) for Fe 415 grade steel is 0.48.

Since $(x_u/d) = 0.39 < 0.48$, the section is under-reinforced.

  iii. *Moment of resistance*:

$$M_u = 0.87 f_y A_{st} d \left[1 - \frac{A_{st} f_y}{bd f_{ck}}\right]$$

$$= (0.87 \times 415 \times 628 \times 400)\left[1 - \frac{(628 \times 415)}{(200 \times 400 \times 20)}\right]$$

$$= 76 \times 10^6 \text{ N·mm}$$

$$= 76 \text{ kN·m}$$

2. A reinforced concrete beam of rectangular section 200 mm wide by 500 mm deep is reinforced with 4 bars or 25 mm diameter at an effective depth of 500 mm. Using M20 grade concrete and Fe 415 HYSD bars, calculate the safe moment of resistance of the section.

  i. *Data*:

$b = 200$ mm

$d = 500$ mm

$D = 550$ mm

$A_{st} = (4 \times 491) = 1964$ mm^2

$f_{ck} = 20$ N/mm^2

$f_y = 415$ N/mm^2

  ii. *Depth of neutral axis*:

$$\left(\frac{x_u}{d}\right) = \left(\frac{0.87 f_y A_{st}}{0.36 f_{ck} bd}\right) = \left(\frac{0.87 \times 415 \times 1964}{(0.36 \times 20 \times 200 \times 500)}\right) = 0.984 > 0.48$$

Hence, the section is over-reinforced.

  iii. *Moment of resistance*:

The limiting moment of resistance of the over-reinforced section is computed as

$$M_{u,lim} = 0.138 f_{ck} b d^2$$

$$= (0.138 \times 20 \times 200 \times 500^2)$$

$$= 138 \times 10^6 \text{ N·mm}$$

$$= 138 \text{ kN·m}$$

3. A reinforced concrete beam, 300 mm wide is reinforced with 1436 mm² of Fe 415 HYSD bars at an effective depth of 500 mm. If M20 grade concrete is used, estimate the moment of resistance of the section.

   i. *Data*:

   $b = 300$ mm

   $d = 500$ mm

   $A_{st} = 1436$ mm²

   $f_{ck} = 20$ N/mm²

   $f_y = 415$ N/mm²

   ii. *Depth of neutral axis*:

   $$\left(\frac{x_u}{d}\right) = \left(\frac{0.87\,f_y\,A_{st}}{0.36\,f_{ck}\,bd}\right) = \left(\frac{0.87 \times 415 \times 1436}{(0.36 \times 20 \times 300 \times 500)}\right) = 0.48$$

   Since $(x_d/d) = (x_{u,max}/d) = 0.48$, the section is balanced.

   iii. *Moment of resistance*:

   $M_u = 0.138 f_{ck} bd^2$

   $= (0.138 \times 20 \times 300 \times 500^2)$

   $= 207 \times 10^6$ N·mm

   $= 207$ kN·m

4. Determine the area of reinforcement required for a singly reinforced concrete section having a breadth of 300 mm and an effective depth of 600 mm to resist a factored moment of 200 kN·m. Adopt $f_{ck} = 20$ N/mm² and $f_y = 415$ N/mm².

   i. *Data*:

   $b = 300$ mm

   $d = 600$ mm

   $M_u = 200$ mm²

   $f_{ck} = 20$ N/mm²

   $f_y = 415$ N/mm²

   ii. *Limiting moment of resistance*:

   For Fe 415 HYSD bars:

   $M_{u,lim} = 0.138 f_{ck} bd^2$

   $= (0.138 \times 20 \times 300 \times 600^2)$

   $= 298 \times 10^6$ N·mm

   $= 298$ kN·m > 200 kN·m

   iii. *Area of reinforcement*:

   $$M_u = 0.87\,f_y\,A_{st}\,d\left[1 - \left(\frac{A_{st}f_y}{bdf_{ck}}\right)\right]$$

   $$(200 \times 10^6) = (0.87 \times 415 \times A_{st} \times 600)\left[1 - \left(\frac{415 A_{st}}{(300 \times 600 \times 20)}\right)\right]$$

   Solving, $A_{st} = 1365$ mm².

5. Determine the minimum effective depth required and the corresponding area of tension reinforcement for a rectangular beam having a width of 200 mm to resist an ultimate moment of 200 kN·m, using M20 grade concrete and Fe 415 HYSD bars or ribbed tor steel.

i. *Data*:

$b = 200$ mm

$M_u = 200$ mm^2

$f_{ck} = 20$ N/mm^2

$f_y = 415$ N/mm^2

ii. *Minimum effective depth*:

For Fe 415 HYSD bars, limiting moment of resistance is expressed as

$$M_{u,lim} = 0.138 f_{ck} bd^2$$

$$\therefore \quad d = \sqrt{\frac{M_{u,lim}}{0.138 f_{ck} b}}$$

$$= \sqrt{\frac{200 \times 20^6}{0.138 \times 20 \times 200}}$$

$$= 601.9 \text{ mm}$$

iii. *Area of reinforcement*:

$$\left(\frac{x_{u,lim}}{d}\right) = 0.48 = \left(\frac{0.87 f_y A_{st}}{0.36 f_{ck} bd}\right)$$

$$A_{st} = \left(\frac{0.48 \times 0.36 f_{ck} bd}{0.87 f_y}\right)$$

$$= \left(\frac{0.48 \times 0.36 \times 20 \times 200 \times 601.9}{0.87 \times 415}\right)$$

$$= 1152.2 \text{ mm}^2$$

4.3 ULTIMATE FLEXURAL STRENGTH OF FLANGED SECTIONS

4.3.1 Moment of Resistance Equations

In case of flanged section (T- and L-sections), the ultimate flexural strength is influenced by the position of neutral axis which may lie in the flange or outside the flange depending upon the area of reinforcement on the tension face. The Indian standard code IS:456-2000 specifies equations for computing the moment of resistance of flanged sections by assuming the stress block and the following parameters:

Referring to Fig. 4.2:

Let b_w = width of rib

b_f = width of flange

d = effective depth

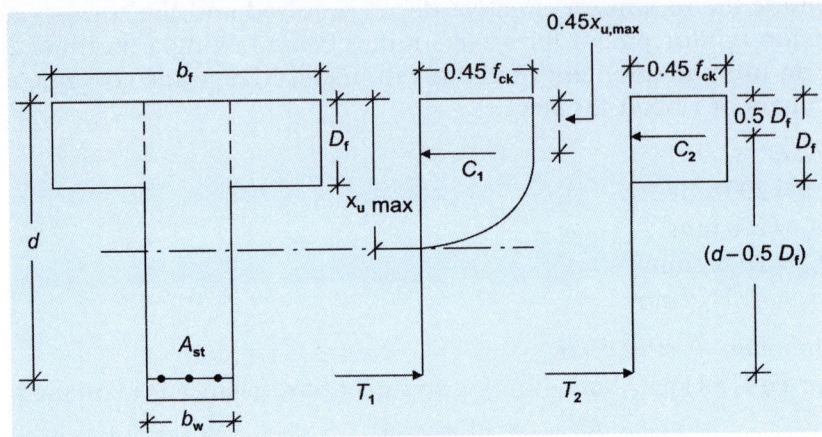

Fig. 4.2: Stress block parameters (flanged section)

D_f = depth of flange
x_u = neutral axis depth
A_{st} = area of tension reinforcement

The following two cases are considered:

(a) Neutral axis lies within the flange

In this case, since $x_u < D_f$, the section can be considered as rectangular with the width of compression flange

$$b = b_f$$

and

$$x_u < x_{u,lim}$$

The moment of resistance of the under-reinforced section is computed by the relation

$$M_u = 0.87 f_y A_{st} d \left[1 - \left(\frac{A_{st} f_y}{b_f d f_{ck}} \right) \right]$$

$$= 0.87 f_y A_{st} (d - 0.42 x_u)$$

In case of over-reinforced sections, where $x_u > x_{u,lim}$, the moment of resistance is computed by the equation

$$M_u = 0.36 f_{ck} x_{u,lim} b_f [d - 0.42 x_{u,lim}] \qquad (4.2)$$

(b) Neutral axis lies outside the flange and $(D_f/d) > 0.2$

When the neutral axis falls outside the flange ($x_u > D_f$) and the ratio $(D_f/d) \not> 0.2$ and $(D_f/x_u) < 0.43$, the flexural strength can be computed by using the stress block parameter shown in Fig. 4.2. The stress blocks are considered separately for the web portion and the flanges. Considering the tensile and compressive forces shown in Fig. 4.2, the moment of resistance of the flanged section is expressed as:

$$M_u = C_1 (d - 0.42 x_{u,max}) + C_2 (d - 0.5 D_f)$$

where,

$C_1 = 0.36 f_{ck} x_{u,max} b_w$

$C_2 = 0.45 f_{ck} D_f (b_f - b_w)$

The final relation is expressed as

$$M_u = 0.36\left(\frac{x_{u,max}}{d}\right)\left[1 - 0.42\left(\frac{x_{u,max}}{d}\right)\right]f_{ck}b_w d^2$$
$$+ 0.45 f_{ck}(b_f - b_w)D_f(d - 0.5D_f) \tag{4.3}$$

Equation (4.3) is valid for the case $(D_f/d) < 0.2$ and $(D_f/x_u) < 0.43$.

(c) Neutral axis lies outside the flange and $(D_f/d) > 0.2$

When the ratio $(D_f/d) > 0.2$, the moment equation is modified by substituting y_f for D_f in Eq. (4.3), where

$$y_f = (0.15x_u + 0.65D_f) \text{ but not greater than } D_f.$$

The modified equation for moment of resistance is expressed as

$$M_u = 0.36\left(\frac{x_{u,max}}{d}\right)\left[1 - 0.42\left(\frac{x_{u,max}}{d}\right)\right]f_{ck}b_w d^2$$
$$+ 0.45 f_{ck}(b_f - b_w)y_f(d - 0.5y_f) \tag{4.4}$$

The Indian standard code further stipulates that for $x_{u,max} > x_u > D_f$, the moment of resistance may be calculated by Eq. (4.3), when (D_f/x_u) does not exceed 0.43 and when (D_f/x_u) exceeds 0.43, the moment of resistance is computed by the Eq. (4.4) by substituting $x_{u,max}$ by x_u.

(d) Computation of tension steel in T-beam

Case 1$[x_u < D_f]$:

The section is considered as rectangular and area of reinforcement is computed by Eq. (4.2) for a known value of M_u.

$$M_u = 0.87 f_y A_{st}d\left[1 - \left(\frac{A_{st}f_y}{b_f d f_{ck}}\right)\right] \tag{4.4a}$$

Case 2$[(x_u > D_f)\ (D_f/d) \not> 0.2 \text{ and } (D_f/x_u) \not> 0.43]$:

For a known value of M_u, compute x_u using Eq. (4.2), force equilibrium results in the following relations:

$$T_1 = C_1$$
$$(A_{stw}\ 0.87f_y) = (0.36f_{ck}b_w x_u)$$

$\therefore$
$$A_{stw} = \left[\frac{0.36 f_{ck}b_w x_u}{0.87 f_y}\right] \tag{4.5}$$

Also
$$T_2 = C_2$$
$$(A_{stf}\ 0.87f_y) = 0.45f_{ck}(b_f - b_w)D_f$$

$\therefore$
$$A_{stw} = \left[\frac{0.45 f_{ck}(b_f - b_w)D_f}{0.87 f_y}\right] \tag{4.6}$$

Hence, the total tension reinforced in the T-section is obtained as

$$A_{st} = \left[\frac{0.36 f_{ck}b_w x_u}{0.87 f_y}\right] + \left[\frac{0.45 f_{ck}(b_f - b_w)D_f}{0.87 f_y}\right] \tag{4.7}$$

Case 3[$(x_u > D_f)(D_f/d) \not> 0.2$ and $(D_f/x_u) \not> 0.43$]:

For a known value of M_u, evaluate x_u by using Eq. (4.2) by replacing $x_{u,max}$ by x_u.

Since
$$y_f = (0.15x_u + 0.65D_f)$$

$$A_{stw} = \left[\frac{0.36 f_{ck} b_w x_u}{0.87 f_y}\right]$$

$$A_{stf} = \left[\frac{0.45 f_{ck}(b_f - b_w)y_f}{0.87 f_y}\right]$$

Hence
$$A_{st} = (A_{stw} + A_{stf})$$

Case 4[$(x_u > D_f)(D_f/d) > 0.2$ and $(D_f/x_u) > 0.43$]:

This is similar to Case 3.

4.3.2 Analysis Examples

1. Determine the moment of resistance of a T-section having the following section properties:

    Width of flange = 2500 mm
    Depth of flange = 150 mm
    Width of rib = 300 mm
    Effective depth = 800 mm
    Area of steel = 8 bars of 25 mm diameter

    Materials: M20 grade concrete
    Fe 415 HYSD bars

    i. *Data*:

    $b_f = 2500$ mm $A_{st} = (8 \times 491) = 3928$ mm^2
    $D_f = 150$ mm $f_{ck} = 20$ N/mm^2
    $b_w = 300$ mm $f_y = 415$ N/mm^2
    $d = 800$ mm

    ii. *Depth of neutral axis*:

    Assuming the neutral axis to lie within the flange:

    $$\left(\frac{x_u}{d}\right) = \left(\frac{0.87 f_y A_{st}}{0.36 f_{ck} b_f d}\right) = \left(\frac{0.87 \times 415 \times 628)}{(0.36 \times 20 \times 200 \times 400)}\right) = 0.098$$

    $\therefore$ $x_u = (0.098 \times 800) = 78.4$ mm $< D_f = 150$ mm

    iii. *Moment of resistance*:

    $$M_u = 0.87 f_y A_{st} d\left[1 - \frac{A_{st} f_y}{b_f d f_{ck}}\right]$$

    $$= (0.87 \times 415 \times 3928 \times 800)\left[1 - \frac{415 \times 3928}{2500 \times 800 \times 20}\right]$$

    $$= 1089 \times 10^6 \text{ N·mm}$$

    $$= 1089 \text{ kN·m}$$

2. Calculate the ultimate flexural strength of a T-beam section having the following section properties:

   Width of flange = 1200 mm
   Depth of flange = 120 mm
   Width of rib = 300 mm
   Effective depth = 600 mm

   Area of tensile steel = 8 bars of 25 mm diameter

   Materials: M20 grade concrete
   Fe 415 HYSD bars

   i. *Data*:

   $$b_f = 1200 \text{ mm} \qquad A_{st} = 4000 \text{ mm}^2$$
   $$D_f = 120 \text{ mm} \qquad f_{ck} = 20 \text{ N/mm}^2$$
   $$b_w = 300 \text{ mm} \qquad f_y = 415 \text{ N/mm}^2$$
   $$d = 600 \text{ mm}$$

   ii. *Depth of neutral axis*:

   Since the flange is thin and area of steel is large, assume that the neutral axis falls outside the flange and $(D_f/x_u) > 0.43$, using the force compatibility relations, we have

   $$C_1 + C_2 = T_1 + T_2 = T$$

   Let x_u = depth of neutral axis

   $$= (0.36 f_{ck} b_w x_u) + 0.45 f_{ck} (b_f - b_w)(0.15 x_u + 0.65 D_f)$$
   $$= (0.87 f_y A_{st})$$
   $$= (0.36 \times 20 \times 300 \times x_u) + 0.45 \times 20(1200 - 300)(0.15 x_u + 0.65 \times 120)$$
   $$= (0.87 \times 415 \times 4000$$

   Solving $x_u = 240.7$ mm

   ∴ $y_f = (0.15 x_u + 0.65 D_f)$
   $$= (0.15 \times 240.7) + (0.65 \times 120)$$
   $$= 114.1 D_f = 120 \text{ mm}$$

   $$\left(\frac{x_u}{d}\right) = \left(\frac{240.7}{600}\right) = 0.40$$

   $$\left(\frac{x_{u,max}}{d}\right) = 0.48$$

   ∴ $x_{u,max} > x_u > D_f$

   Also $\left(\dfrac{D_f}{x_u}\right) = \left(\dfrac{120}{240.7}\right) = 0.498 > 0.43$

   Hence, the moment of resistance is expressed as

   $$M_u = 0.36(x_d/d)[1 - 0.42(x_u/d)]f_{ck}b_w d^2$$
   $$+ 0.45 f_{ck}(b_f - b_w)] y_f (d - 0.5 y_f)$$

$$= (0.36 \times 0.40) [1 - (0.42 \times 0.40)] (20 \times 300 \times 600^2)$$
$$+ (0.45 \times 20) (1200 - 300) \, 114.1(600 - 0.15 \times 114.1)$$
$$= 760.5 \times 10^6 \text{ N·mm}$$
$$= 760.5 \text{ kN·m}$$

3. A T-beam has an effective flange width of 2500 mm and depth of flange = 150 mm, width of rib = 300 mm, effective depth = 800 mm. Using M20 grade concrete and Fe 415 HYSD bars, estimate the area of tension reinforcement required if the section has to resist a design ultimate moment of 1200 kN·m.

   i. *Data*:

$b_f = 2500$ mm	$f_y = 415$ N/mm^2
$D_f = 150$ mm	$b_w = 300$ mm^2
$d = 800$ mm	$M_u = 1200$ mm^2
$f_{ck} = 20$ N/mm^2	

   ii. *Moment of resistance*:

   Assuming the neutral axis depth $x_u = D_f$, the moment of resistance
   $$M_{uf} = 0.36f_{ck}b_fD_r(d - 0.42D_f)$$
   $$= (0.36 \times 20 \times 2500 \times 150) [800 - (0.42 \times 150)]$$
   $$= 1989.9 \times 10^6 \text{ N·mm}$$

   Since $M_u < M_{uf}$ ∴ $x_u < D_f$

   Considering the section as a rectangular width $b = b_f$

   $$M_u = 0.87 f_y A_{st} d \left[1 - \left(\frac{A_{st}f_y}{bd f_{ck}} \right) \right]$$

   $$(1200 \times 10^6) = (0.87 \times 415 \times A_{st} \times 800) \left[1 - \frac{415A_{st}}{(2500 \times 800 \times 20)} \right]$$

   Solving, $A_{st} = 4354$ mm^2

   iii. *Check for neutral axis depth*:

   $$x_u = \left(\frac{0.87 f_y A_{st}}{0.36f_{ck} \, bd} \right) = \left(\frac{0.87 \times 415 \times 1436}{0.36 \times 20 \times 300 \times 500} \right) = 0.48$$

   $$= 87.3 \text{ mm} < D_f = 150 \text{ mm}$$

4. A T-beam has the following cross-sectional details:

   Effective width of flange = 2000 mm
   Thickness of flange = 150 mm
   Width of rib = 300 mm
   Effective depth = 1000 mm

   Calculate the limiting or balanced moment capacity of the section and the corresponding area of tension reinforcement. Assume M20 grade concrete and Fe 415 HYSD bars.

i. *Data*:

$b_f = 2000$ mm $f_{ck} = 20$ N/mm^2

$b_w = 300$ mm $f_y = 415$ N/mm^2

$D_f = 150$ mm

$d = 1000$ mm

ii. *Moment of resistance*:

From Table 4.1, for Fe 415 HYSD bars,

limiting value of neutral axis depth $x_{u,max} = 0.48\,d$

$\therefore \ x_{u,max} = x_u = 0.48\,d = (0.48 \times 1000) = 480$ mm

Also, $(D_f/d) = (150/1000) = 0.15 < 0.2$

$(D_f/x_u) = (150/480) = 0.31 < 0.43$

Hence from Table 4.2 and Eq. (4.3), we have

$M_{u,lim} = 0.138 f_{ck} b_w d^2 + 0.45 f_{ck}(b_f b_w)\, D_f\,(d - 0.5 D_f)$

$\qquad = (0.138 \times 20 \times 300 \times 1000^2)$

$\qquad\qquad + (0.45 \times 20)\,(2000 - 300)\,150\,(1000 - 0.5 \times 150)$

$\qquad = 2950.5 \times 10^6$ N·mm

$\qquad = 2950$ kN·m

iii. *Tension reinforcement*:

$A_{st} = A_{stw} + A_{stf}$

$$= \left[\frac{0.36\,f_{ck} b_w x_u}{0.87 f_y} \right] + \left[\frac{0.45\,f_{ck}(b_f - b_w) D_f}{0.87 f_y} \right]$$

$$= \left[\frac{0.36 \times 20 \times 300 \times 480}{0.87 \times 415} \right] + \left[\frac{0.45 \times 20(2000 - 300)150}{0.87 \times 415} \right]$$

$= 9228$ mm^2.

4.4 ULTIMATE FLEXURAL STRENGTH OF DOUBLY REINFORCED CONCRETE SECTIONS

4.4.1 Necessity of Doubly Reinforced Sections

Doubly reinforced concrete sections are required in beams of restricted depth due to head room requirements. When the singly reinforced section is sufficient to resist the bending moment on the section, additional tension and compression reinforcements are designed based on steel beam theory.

The doubly reinforced section comprises two parts, outlined as:

1. Singly reinforced section with the restricted depth providing the limiting moment of resistance ($M_{u,lim}$) which is less than the design moment M_u.

2. Based on steel beam theory, a steel beam with tension and compression reinforcement providing the balance moment given by ($M_u - M_{u,lim}$).

4.4.2 Design Equation

Referring to Fig. 4.3:

Let M_u = ultimate flexural strength of doubly reinforced section

$M_{u1} = M_{u,lim}$ = limiting or maximum moment of resistance of the slightly reinforced section

M_{u2} = Moment of resistance of the steel beam neglecting the effect of concrete

$= f_{sc}A_{sc}(d - d')$

where, f_{sc} = stress in the compression steel corresponding to the strain reached by it when the extreme concrete fiber reaches a strain of 0.0035

A_{sc} = area of compression reinforcement

d = effective depth of tension steel

d' = depth of compression reinforcement from compression face

A_{st1} = area of tensile reinforcement for a singly reinforced section

A_{st2} = area of tensile reinforcement required to balance the compression reinforcement

$A_{st} = (A_{st1} + A_{st2})$

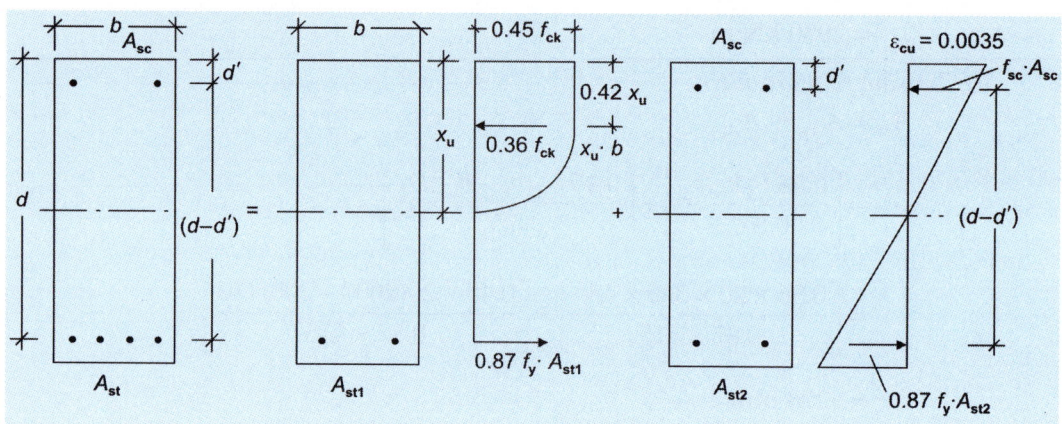

Fig. 4.3: Doubly reinforced section

4.4.3 Moment of Resistance

According to IS:456-2000 code, the following procedure is used to compute the ultimate flexural strength of the doubly reinforced section.

i. Limiting moment $M_{u,lim}$ of the singly reinforced section

$M_{u,lim} = 0.148 f_{ck}bd^2$ for Fe 250 grade steel

$= 0.138 f_{ck}bd^2$ for Fe 415 HYSD bars

$= 0.133 f_{ck}bd^2$ for Fe 450 grade steel

ii. Calculate $A_{st1} = \left[\dfrac{0.36 f_{ck}\, b\, x_{u,lim}}{0.87 f_y} \right]$

iii. Compute $A_{st2} = [A_{st} - A_{st1}]$

iv. Calculate $A_{sc} = \left[\dfrac{0.87 f_y A_{st2}}{f_{sc}} \right]$

$$f_{sc} = \left[\dfrac{0.0035 (x_{u,max} - d')}{x_{u,max}} \right] E_s$$

v. The ultimate moment capacity of the section is given by
$$M_u = M_{u,lim} + f_{sc} A_{sc} (d - d')$$

4.4.4 Analysis Examples

1. A doubly reinforced concrete beam having a rectangular section 250 mm wide and 540 mm overall depth is reinforced with 2 bars of 12 mm diameter in the compression side and 4 bars of 20 mm diameter in the tension side. The effective cover to bars is 40 mm. Using M20 grade concrete and Fe 415 HYSD bars, estimate the flexural strength of the section using IS:456-2000 code recommendations.

   i. *Data*:

$b = 250$ mm	$A_{st} = (4 \times 314) = 1256$ mm^2
$d = 500$ mm	$A_{sc} = (2 \times 113) = 226$ mm^2
$D = 540$ mm	$f_{ck} = 20$ N/mm^2
$d' = 40$ mm	$f_y = 415$ N/mm^2

   ii. *Limiting neutral axis depth*:
   $$x_{u,lim} = 0.48 \, d = (0.48 \times 500) = 240 \text{ mm}$$

   iii. *Stress in compression steel*:

   $$f_{sc} = \varepsilon_{sc} E_s = \left[\dfrac{0.0035 (x_{u,lim} - d')}{x_{u,lim}} \right] E_s$$

   $$= \left[\dfrac{0.0035 (240 - 40)}{240} \right] 2 \times 10^5$$

   $$= 583 \text{ N/mm}^2 \text{ but } \not> 0.87 f_y = (0.87 \times 415) = 361 \text{ N/mm}^2$$

   $$A_{st2} = \left[\dfrac{f_{sc} A_{sc}}{0.87 f_y} \right] = A_{sc} = 226 \text{ mm}^2$$

   $\therefore$ $A_{st1} = (A_{st} - A_{st2})$

   $$= (1256 - 226) = 1030 \text{ mm}^2$$

   iv. *Actual neutral axis depth*:
   $$0.36 f_{ck} b \, x_u = 0.87 f_y A_{st1}$$

   $\therefore$ $x_u = \left(\dfrac{0.87 \times 415 \times 1030}{0.36 \times 20 \times 250} \right) = 206.6 \text{ mm} < x_{u,lim} = 240 \text{ mm}$

   Hence, the section is under-reinforced.

v. *Moment of resistance*:

$$M_u = 0.87 f_y A_{st1} (d - 0.42 x_u + f_{ck} A_{sc}(d - d')$$
$$= 0.87 \times 415 \times 1030 \, (500 - 0.42 \times 206.6)$$
$$+ 0.87 \times 415 \times 226 \, (500 - 40)$$
$$= 191 \times 10^6 \, \text{N·mm}$$
$$= 191 \, \text{kN·m}$$

2. A doubly reinforced concrete section has a width of 300 mm is reinforced with tension steel of area 2455 mm^2 at an effective depth of 600 mm. Compression steel area of 982 mm^2 is provided at an effective cover of 60 mm. Using M20 grade concrete and Fe 415 HYSD bars, estimate the ultimate moment capacity of the section using the stress–strain curve of steel shown in Fig. 4.4.

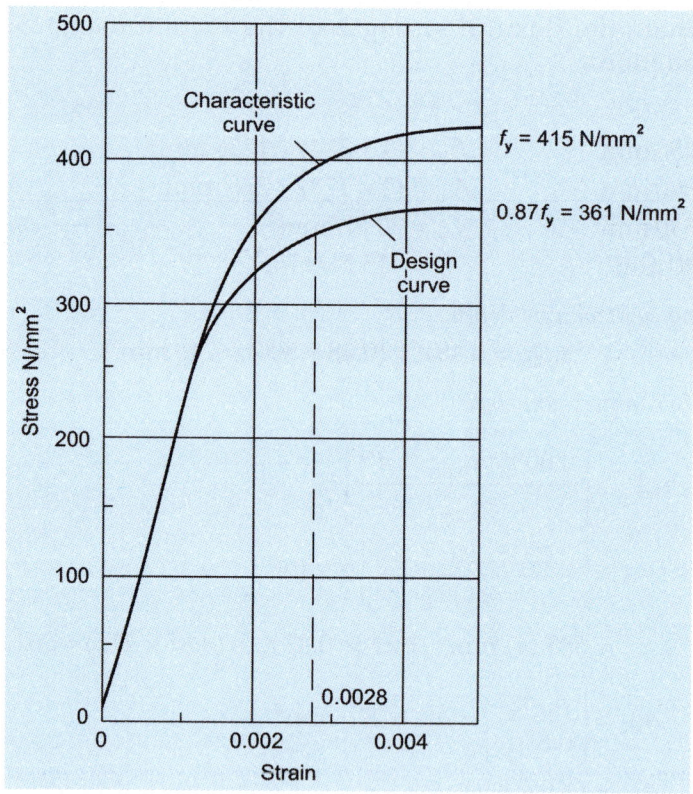

Fig. 4.4: Stress–strain curve for Fe 415 HYSD bars

i. *Data*:

$b = 300 \, \text{mm}$ $\qquad A_{sc} = 982 \, \text{mm}^2$

$d = 600 \, \text{mm}$ $\qquad f_{ck} = 20 \, \text{N/mm}^2$

$d' = 60 \, \text{mm}$ $\qquad f_y = 415 \, \text{N/mm}^2$

$A_{st} = 2455 \, \text{mm}^2$

ii. *Limiting neutral axis depth*:
$$x_{u,lim} = 0.48\,d = (0.48 \times 600) = 288 \text{ mm}$$

iii. *Stress in compression steel*:
$$= \left[\frac{0.0035\,(x_{u,lim} - d')}{x_{u,lim}}\right] = \left[\frac{0.0035\,(288 - 60)}{288}\right] = 0.0028$$

From stress–strain curve (Fig. 4.4) readout the stress in steel corresponding to the strain of 0.0028 as
$$f_{sc} = 350 \text{ N/mm}^2$$

iv. *Tension steel*:
$$A_{st2} = \left(\frac{f_{sc}A_{sc}}{0.87f_y}\right) = \left(\frac{350 \times 982}{0.87 \times 415}\right) = 952 \text{ mm}^2$$
$$\therefore \qquad A_{st1} = (A_{st} - A_{st2}) = (2455 - 952) = 1503 \text{ mm}^2$$

v. *Neutral axis depth*:
$$x_u = \left(\frac{0.87\,f_y A_{st1}}{0.36 f_{ck} b}\right) = \left(\frac{0.87 \times 415 \times 1503}{0.36 \times 20 \times 300}\right) = 252 \text{ mm}$$
$$= x_u < x_{u,lim}$$

Hence, the section is under-reinforced.

vi. *Moment of resistance*:
$$\begin{aligned}
M_u &= 0.87 f_y A_{st}\,(d - 0.42 x_u) + f_{ck} A_{sc}(d - d') \\
&= 0.87 \times 415 \times 1503\,(600 - 0.42 \times 252) + 350 \times 982\,(600 - 60) \\
&= 453 \times 10^6 \text{ N·mm} \\
&= 453 \text{ kN·m}
\end{aligned}$$

REFERENCES

1. Evans RH, The plastic theories for the ultimate strength of reinforced concrete beams, *Journal of the Institution of Civil Engineers*, London, Vol 21, Dec. 1943, pp 98–121.

2. Hognestad E, Hansen NW and Henry Mc D, Concrete stress distribution in ultimate strength design, *Proc of the ACI Journal*, Vol 52, No. 4, Dec. 1955, pp 455–480.

3. Rusch H, Researches toward a general flexural theory for structural concrete, *Journal of the ACI*, Vol 57, July 1960, pp 1–28.

4. Jones LL, *Ultimate Load Analysis of Reinforced and Prestressed Concrete Structures*, Pitman, London, 1969.

5. Reagen PE and Yu CW, *Limit State Design of Structural Concrete*, Chatto & Windus, London, 1973.

6. Park R and Paulay T, *Reinforced Concrete Structures*, John Wiley & Sons, New York, 1975.

7. IS:456-2000, Indian Standard Code of Practice for Plain and Reinforced Concrete (4th Revision), BIS, New Delhi, July 2000.

8. SP:16-1980, Design Aids for Reinforced Concrete to IS:456, 11th reprint, BIS, New Delhi, 1999.

9. Varyani UH and Radhaji A, *Manual of Limit State Design of Reinforced Concrete Members*, Khanna Technical Publications, New Delhi, 1984.

10. Reynolds CE, James C, Steedman and Anthony J Threlfall, *Reinforced Concrete Designers' Handbook*, 11 edn, London, 2007.

ASSIGNMENT

1. A reinforced concrete beam of rectangular section, 300 mm wide × 650 mm overall depth is reinforced with 4 bars of 25 mm diameter at an effective cover of 50 mm. Assuming the grade of concrete as M25 and HYSD reinforcement as Fe 500, estimate the ultimate flexural strength of the beam.

2. Determine the ultimate moment of resistance of a concrete beam of rectangular section, 350 mm wide × 750 mm deep is reinforced with 2 bars of 28 mm and 2 bars of 25 mm at an effective depth of 700 mm. Adopt M20 grade concrete and Fe 415 grade HYSD bars.

3. A reinforced concrete beam is required to resist an ultimate moment of 250 kN·m. The width of the beam is restricted to 250 mm. Effective cover is 50 mm. Adopting M20 grade concrete and Fe 415 grade HYSD bars, determine the overall depth of the beam.

4. A reinforced concrete beam of rectangular section 350 mm wide × 750 mm overall depth is reinforced with 4 bars of 25 mm diameter on the tension side at an effective cover of 50 mm and 3 bars of 20 mm on the compression side at an effective cover of 50 mm. If M250 grade concrete is used with Fe 415 HYSD bars, estimate the flexural strength of the beam using Indian standard code specifications.

5. A doubly reinforced concrete beam has a width 300 mm and overall depth 700 mm. The beam is reinforced with tension steel of area 1964 mm^2 at an effective cover of 45 mm. The beam has to be designed to have a flexural strength of 420 kN·m. Assuming M20 grade concrete and Fe 415 HYSD steel reinforcements, calculate the area of compression steel required at an effective cover of 45 mm.

6. Determine the moment of resistance of a T-beam having the following section properties:
   Effective width of flange = 2400 mm
   Depth of flange = 160 mm
   Width of rib = 300 mm
   Effective depth = 750 mm
   Area of steel: 6 bars of 25 mm diameter
   Materials: M20 grade concrete
   Fe 415 HYSD bars.

7. Calculate the ultimate flexural strength of a T-beam section having the following section properties:
   Width of flange = 1200 mm
   Depth of flange = 150 mm

Width of rib = 300 mm

Effective depth = 600 mm

Area of tension steel= 4000 mm^2

Materials: M20 grade concrete

Fe 415 HYSD bars.

8. A T-beam has an effective flange width 2500 mm, depth 150 mm, width of rib 300 mm, effective depth 800 mm. Using M20 grade concrete and Fe 415 HYSD bars, estimate the area of tension steel required if the section has to resist a factored moment of 1000 kN·m.

9. A doubly reinforced concrete beam having a rectangular section 300 mm wide and 750 mm overall depth is reinforced with 2 bars of 12 mm diameter on the compression face and 4 bars of 20 mm diameter on the tension side. The effective cover to the bars is 50 mm. Using M20 grade concrete and Fe 415 HYSD bars, estimate the flexural strength of the section using IS:456-2000 code specifications.

10. A doubly reinforced concrete section has a width 300 mm is reinforced with tension reinforcement of area 2465 mm^2 at an effective depth of 600 mm. Compression steel of area 942 mm^2 is provided at an effective cover of 50 mm. Using M25 grade concrete and Fe 500 HYSD bars, estimate the ultimate moment capacity of the section.

REVIEW QUESTIONS

1. What is the necessity of evaluating the ultimate flexural strength of reinforced concrete sections?

2. Explain briefly the basis of selecting stress block parameters for rectangular sections in the Indian standard code recommendations.

3. What is the purpose of limiting the neutral axis depth in the estimation of the flexural strength of reinforced concrete sections?

4. Explain the terms: (a) Moment of resistance (b) Balanced section (c) Limiting values of neutral axis depth.

5. Explain the significance of selecting separate stress block parameters for the rectangular and flanged portions for the computation of flexural strength of flanged sections in the IS code recommendations.

6. How do you compute the flexural strength of T-section in which the neutral axis lies outside the flange?

7. How do you compute the steel requirement in a T-beam in which the neutral axis depth is greater than the flange thickness, ratio of flange thickness to effective depth exceeds the value of 0.2 and ratio of flange thickness to the neutral axis depth is greater than 0.43?

8. What is the necessity for selecting doubly reinforced concrete sections while designing reinforced concrete beams?

9. How do you compute the ultimate flexural strength of a doubly reinforced concrete section using the IS code recommendations.

10. Explain the method of computing the ultimate moment capacity of a doubly reinforced concrete section using the stress–strain curve of steel.

OBJECTIVE QUESTIONS

1. In the computations of ultimate flexural strength of reinforced concrete sections, the Indian standard code assumes the shape of the stress block as
   a. rectangular
   b. parabolic
   c. rectangular-parabolic

2. To prevent brittle failure, the IS code prescribes that the failure strain in steel should be more than
   a. 0.0015
   b. 0.0005
   c. 0.002

3. The ultimate strain in concrete at the limit state of collapse is generally
   a. 0.0025
   b. 0.0035
   c. 0.0045

4. The moment of resistance of a reinforced concrete T-section is evaluated by assuming
   a. a single parabolic stress block
   b. separate stress blocks for the rib and flange portions
   c. a single rectangular stress block

5. The limiting value of the ratio of neutral axis to effective depth for beams reinforced with Fe 415 HYSD bars is
   a. 0.53
   b. 0.46
   c. 0.48

6. In the computation of flexural strength of reinforced concrete sections, the IS code limits the maximum stress in concrete to value of
   a. $0.55 f_{ck}$
   b. $0.35 f_{ck}$
   c. $0.45 f_{ck}$

7. If the neutral axis depth is less than the flange thickness while computing the flexural strength of flanged sections, then the flanged section can be considered as a rectangular section having a width equal to
   a. width of the rib
   b. width of the flange minus the rib
   c. width of flange

8. Flexural strength of doubly reinforced concrete sections can be determined by
   a. assuming the section as singly reinforced
   b. apportioning the steel areas as A_{st1} and A_{st2} and adding both the moments
   c. using steel beam theory only

9. In the estimation of ultimate flexural strength of doubly reinforced sections, it is necessary to
   a. check the stresses in concrete
   b. restrict the stress in compression steel to be within the characteristic tensile strength
   c. limit the neutral axis depth

10. Doubly reinforced concrete sections are generally preferred in
    a. single-storeyed domestic buildings
    b. industrial structures
    c. cellar floors of multistorey buildings

Shear Strength of Reinforced Concrete Sections

5.1 INTRODUCTION

In most of the structural elements like beams and slabs, bending is associated with transverse shear forces. Shear forces are generally maximum in the vicinity of supports and gradually reduce toward the mid span zones where bending moments are predominant. The shear developed in flexural members is termed *flexural shear*[1,2]. Shear can also develop in members such as circular girders supported on columns subjected to transverse loads. This type of shear is referred to as *torsional shear*[3,4]. In case of slab footings used as foundations for column, there can be one-way shear and also two-way shear (punching shear) associated with the possibility of punching through a relatively thin slab supporting concentrated loads from a column[5].

In limit state design[6], it is essential to design the structural elements to possess the required shear strength[7,8] when subjected to ultimate shear forces. This chapter deals with the Indian standard code methods recommended for the determination of the ultimate shear strength[9,10] of reinforced concrete structural elements.

5.2 SHEAR FAILURE MECHANISMS

Reinforced concrete members are generally subjected to maximum shear forces normally near the support sections of simply supported flexural members. In continuous beams, the support sections are subjected to shear coupled with moments. In case of corbels and brackets, large shear forces develop at the junction of the corbel and column.

The prominent types of shear failures observed in reinforced concrete members can be categorised under the following types:

a. Shear tension or diagonal tension failure.
b. Flexure–shear failure.
c. Shear–compression failure.
d. Shear–bond failure.

as *shear span to effective depth* (a/d).

The transverse shear force (V) is resisted by the following mechanisms:

1. Shear resistance of the uncracked portion of concrete (V_c).
2. Vertical component of interface shear (aggregate interlock force (V_a).
3. Dowel force (V_d) developed due to the tension steel.
4. The shear resistance (V_s) developed in shear reinforcement.

In case of I-beams having thin webs, failure is due to web crushing of concrete which can be prevented by suitable design of reinforcement in the web and using high strength concrete.

Table 5.1 shows the various types of shear failure modes associated with very deep, short and normal beams in terms of shear span/depth ratio (a/d). Figure 5.1 shows the various types of shear failures.

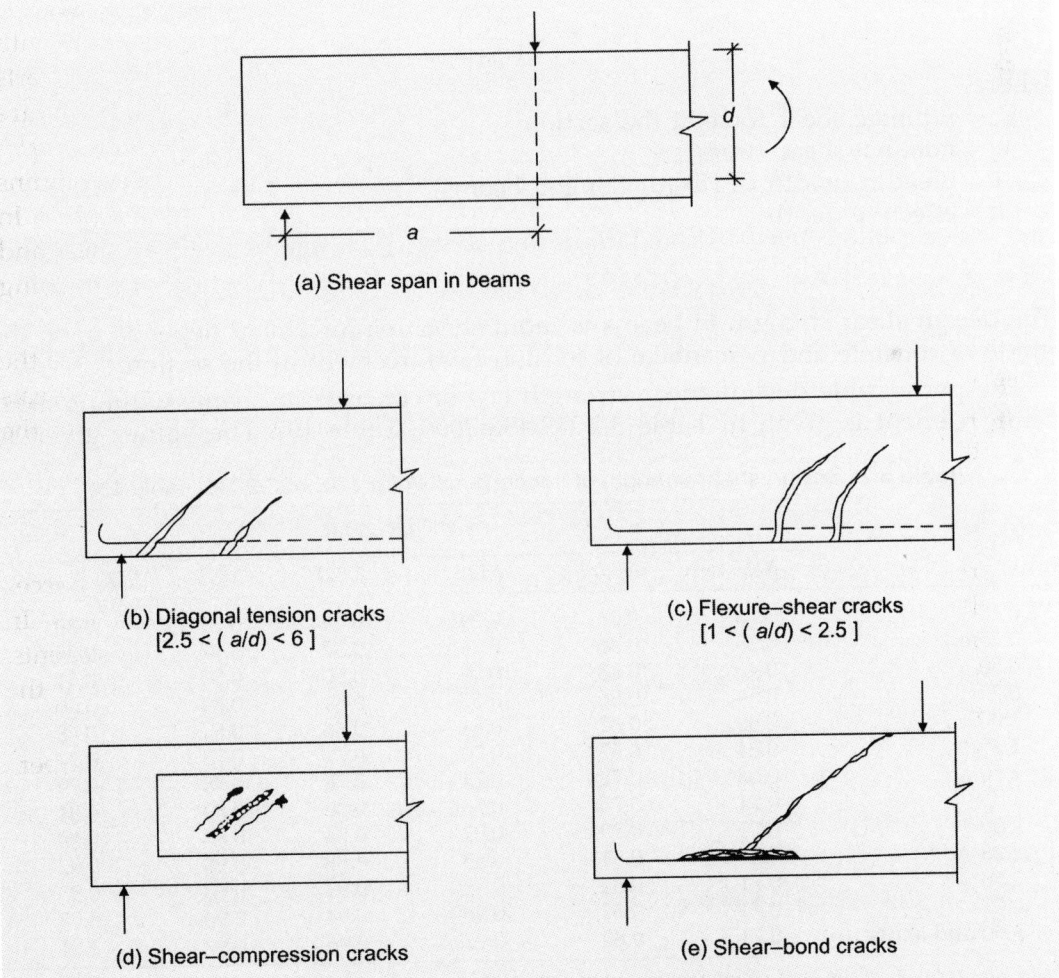

(a) Shear span in beams

(b) Diagonal tension cracks
[2.5 < (a/d) < 6]

(c) Flexure–shear cracks
[1 < (a/d) < 2.5]

(d) Shear–compression cracks

(e) Shear–bond cracks

Fig. 5.1: Various types of shear failures

Table 5.1: Shear failure modes

Sl. No.	Shear span to depth ratio (a/d)	Category of beam	Type of shear failure
1.	$(a/d) < 1$	Very deep	Inclined cracks from supports transforming the beam into an arch. Failure by yielding of steel
2.	$1 < (a/d) < 2.5$	Short	a. Combined shear and compression b. Shear–bond c. Flexure–shear cracking
3.	$2.5 < (a/d) < 6$	Normal	Flexure–shear cracking

5.3 NOMINAL SHEAR STRESS

The shear stress distribution in a reinforced concrete beam is influenced by the shear force acting on the section and the shape of cross-section in the elastic stage. At the ultimate stage, concrete below the neutral axis is ineffective due to cracking. Hence for simplicity, the nominal shear stress across the section is computed as average shear stress and expressed as

$$\tau_v = \left(\frac{V_u}{bd}\right)$$

where,

V_u = ultimate shear force at the section
τ_v = nominal shear stress
b = breadth (width of rib for flanged beams)
d = effective depth

5.4 DESIGN SHEAR STRENGTH OF CONCRETE

The design shear strength in beams without shear reinforcement depends upon the grade of concrete and percentage of tension reinforcement in the section.

The permissible design shear strength (τ_c) of concrete in beams without shear reinforcement is given in Table 5.2 (IS:456-2000, Table 19). The values given in

Table 5.2: Design shear strength of concrete τ_c (N/mm²) (IS:456-2000, Table 19)

$100\dfrac{A_s}{bd}$	Concrete grade					
	M15	M20	M25	M30	M35	M40 & above
≤ 0.15	0.28	0.28	0.29	0.29	0.29	0.30
0.25	0.35	0.36	0.36	0.37	0.37	0.38
0.50	0.46	0.48	0.49	0.50	0.50	0.51
0.75	0.54	0.56	0.57	0.59	0.59	0.60
1.00	0.60	0.62	0.64	0.66	0.67	0.68
1.25	0.64	0.67	0.70	0.71	0.73	0.74
1.50	0.68	0.72	0.74	0.76	0.78	0.79
1.75	0.71	0.75	0.78	0.80	0.82	0.84
2.00	0.71	0.79	0.82	0.84	0.86	0.88
2.25	0.71	0.81	0.85	0.88	0.90	0.92
2.50	0.71	0.82	0.88	0.91	0.93	0.95
2.75	0.71	0.82	0.90	0.94	0.96	0.98
3.00 and above	0.71	0.82	0.92	0.96	0.99	1.01

Note: The term A_s is the area of longitudinal tension reinforcement which continues at least one effective depth beyond the section being considered except at support where the full area of tension reinforcement may be used provided the detailing conforms to 26.2.2 and 26.2.3.

Table 5.2 are applicable for beams. In case of slabs having an overall thickness less than 300 mm, the shear strength being higher, the Indian standard code suggests an enhanced shear strength computed as $k\tau_c$, where k is a multiplying factor depending upon the overall depth of the slab as shown in Table 5.3.

Table 5.3: Maximum shear stress ($\tau_{c,max}$ N/mm²) (IS:456-2000, Table 20)

Concrete grade	M15	M20	M25	M30	M35	M40 and above
$\tau_{c,max}$ (N/mm²)	2.5	2.8	3.1	3.5	3.7	4.0

The code also specifies an upper limit for the design shear strength of concrete strengthened by shear reinforcements. Accordingly, the maximum shear stress in concrete ($\tau_{c,max}$) should not exceed the values specified in Table 5.4 (IS:456-2000, Table 20). If the values of the nominal shear stress (τ_v) exceeds the values of $\tau_{c,max}$, the section should be redesigned with increased cross-sectional dimensions.

Table 5.4: Shear strength factor k for slabs (IS:456-2000, Clause 40.2.1.1)

Overall depth of slab (mm)	300 or more	275	250	225	200	175	150 or less
k	1.00	1.05	1.10	1.15	1.20	1.25	1.30

5.5 DESIGN OF SHEAR REINFORCEMENTS

In case of reinforced concrete sections where the nominal shear stress (τ_v) exceeds the design shear strength of concrete (τ_c), shear reinforcements are to be designed comprising of:

1. Vertical stirrups
2. Tension reinforcement bent up near supports to resist the shear forces.

Let V_u = total design shear force

   V_c = shear resisted by concrete

   V_{us} = shear resisted by reinforcement in the form of links or bent up bars.

Then $V_{us} = (V_u - V_c)$

   $= (\tau_u - \tau_c)\, bd$

where, τ_v = nominal shear stress

   τ_c = design shear stress of concrete

If s_v = spacing of the stirrups

   A_{sv} = total area of the legs of shear reinforcements

   d = effective depth of section

Then $s_v = \left(\dfrac{0.87\, f_y A_{sv} d}{V_{us}} \right)$

The shear resisted by the bent up bars inclined at an angle 'α' to the horizontal is given by:

   $V_{us} = 0.87 f_y A_{sv} \sin \alpha$

The expressions for s_v and V_{us} are recommended for the design of shear reinforcements in IS:456-2000, Clause 40.4.

5.6 MINIMUM SHEAR REINFORCEMENTS

In designing reinforced concrete beams, the IS:456-2000 code stipulates that the minimum shear reinforcements are to be designed even if the design shear strength of concrete (τ_c) exceeds the nominal shear stress (τ_v) to safeguard against local cracking and nominal safety requirements.

The minimum shear reinforcements to be designed using the relation

$$\left(\frac{A_{sv}}{bs_v} \right) \geq \left(\frac{0.4}{0.87 f_y} \right)$$

Provision of nominal shear reinforcement is equivalent to designing the shear reinforcement for a shear stress of $(\tau_v - \tau_c) = 0.4$ N/mm² and it safeguards against spalling of concrete cover and bond failures.

5.7 ENHANCED SHEAR NEAR SUPPORTS

Investigations on shear failures of beams and cantilevers without shear reinforcement indicate that shear cracks develop on planes inclined at 30° angle as shown in Fig. 5.2a. Hence if a section is considered near the support, it is customary to enhance the shear strength capacity, the common examples being the design of brackets, corbels etc. Hence the design shear strength is different when beams are supported on members which are in Fig. 5.2b and when supported on members which are in tension as shown in Fig. 5.2c.

The following specifications of IS:456 code are useful in the design of shear reinforcements in the vicinity of supports:

1. In the simplified approach, the IS:456 code specifies that the critical section for shear is taken at a distance equal to the effective depth d from the face of support when the beam supports uniformly distributed load or a concentrated load farther than $2d$ from the face of support. The value of τ_c is calculated in accordance with Table 19 and appropriate shear reinforcements are designed at sections closer to the support without any further check for shear at sections closer to the support.

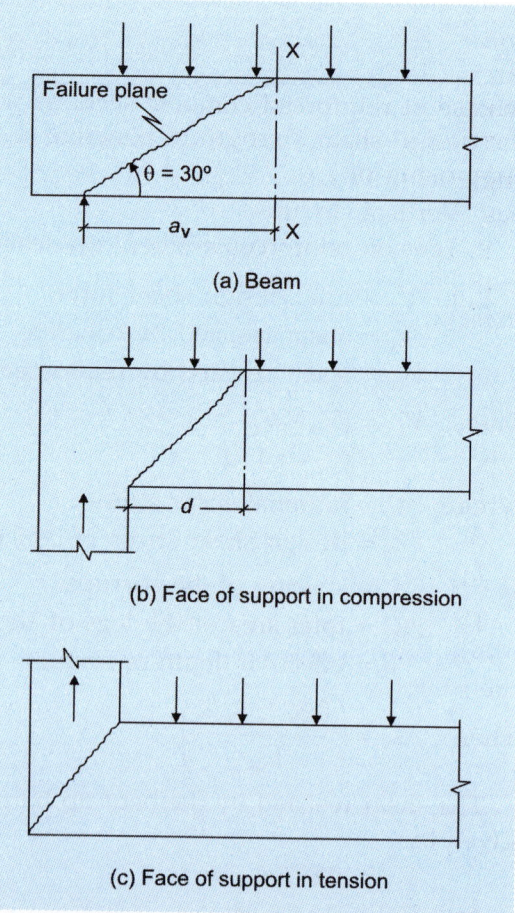

(a) Beam

(b) Face of support in compression

(c) Face of support in tension

Fig. 5.2: Critical sections for shear.

2. The enhancement of shear strength may be taken into consideration while designing sections near a support. The value of τ_c is enhanced by a factor and is given by the relation

$$[(\tau_c \cdot 2d)/a_v] \leq \tau_{c,max}$$

where, a_v = length of that part of a member traversed by a shear plane called the *shear span* as shown in Fig. 5.2(a).

If shear reinforcement is required, the total area of this reinforcement is given by

$$\left\{ \begin{array}{c} A_{sv} = a_v b\left(\tau_v - \dfrac{2d\tau_c}{a_v}\right)\Big/0.87 f_y \\[2mm] \geq (90.4 a_v b)/0.87 f_y \end{array} \right\}$$

This area is provided within a distance of $0.75a_v$. If a_v is less than the effective depth, horizontal shear reinforcement will be more effective than vertical steel since the action is similar to that of deep beams.

5.8 INFLUENCE OF AXIAL FORCE ON DESIGN SHEAR STRENGTH

In general, the actual shear strength of concrete is improved in the presence of uniaxial compression and weakened in the presence of uniaxial tension. The design shear strength of concrete is based on a safe estimate of the limiting nominal stress at which the first inclined crack develops.

The presence of tension accelerates the process of cracking and also increases the angle of inclination of the shear cracks while the presence of uniaxial compression has the opposite effect which is generally prevalent in prestressed concrete beams.

Hence, the IS:456-2000 code (Clause 40.2.2) specifies that the design shear strength in the presence of axial compression should be taken as $\delta\tau_c$, the multiplying factor δ is defined as

$$\delta = \left[1 + \frac{3P_u}{Ag f_{ck}}\right] \text{ or } 1.5, \text{ whichever is less}$$

where,

P_u = factored compressive force (N)
A_g = gross area of concrete section (mm^2)
f_{ck} = characteristic strength of concrete (N/mm^2)

The Indian standard code does not mention the case of axial tension which evidently reduces the design shear strength. However, the American code ACI:318-89 specifies the multiplying factor δ, as

$$\delta = \left[1 + \frac{P_u}{3.5 A_g}\right]$$

5.9 ANALYSIS EXAMPLES

1. A reinforced concrete beam has a support section with a width 250 mm and effective depth 500 mm. The support section is reinforced with 3 bars of

20 mm diameter on the tension side. 8 mm diameter 2 legged stirrups are provided at a spacing of 200 mm centres. Using M20 grade concrete and Fe 415 HYSD bars, calculate the shear strength of the support section.

i. *Data*:

$b = 250$ mm $f_{ck} = 20$ N/mm^2

$d = 500$ mm $f_y = 415$ N/mm^2

$A_{st} = (3 \times 314) = 942$ mm^2 $A_{sv} = (2 \times 50) = 100$ mm^2

$s_v = 200$ mm

ii. *Percentage tension reinforcement*:

$$p_t = \left(\frac{100 A_{st}}{bd}\right) = \left(\frac{100 \times 942}{250 \times 500}\right) = 0.75$$

Refer to Table 4.4 (IS:456-2000, Table 19) and readout the design shear strength of concrete τ_c corresponding to $f_{ck} = 20$ N/mm^2.

$$\tau_c = 0.56 \text{ N/mm}^2$$

iii. *Shear resisted by concrete*:

$$V_{uc} = (\tau_c b\, d) = (0.56 \times 250 \times 500)\, 10^{-3}$$
$$= 70 \text{ kN}$$

iv. *Shear resisted by vertical links*:

$$V_{us} = \left[\frac{A_{sv}(0.087 f_y)d}{s_v}\right] = \left[\frac{100 \times 0.87 \times 415 \times 500}{200}\right]$$
$$= 90.26 \text{ kN}$$

v. *Total shear resistance of support section*:

$$V_u = [V_{uc} + V_{us}]$$
$$= [70 + 90.26] = 160.26 \text{ kN}$$

2. A reinforced concrete beam of rectangular section with a width of 300 mm and effective depth 600 mm is reinforced with 4 bars of 25 mm diameter at tension reinforcement. Two of the tension bars are bent up at 45° near the support section. The beam is also provided with double legged vertical links of 8 mm diameter at 150 mm centres near supports. Using M25 grade concrete and Fe 415 HYSD bars, compute the ultimate shear strength of the support section using IS:456-2000 code specifications.

i. *Data*:

$b = 300$ mm $f_{ck} = 25$ N/mm^2

$d = 600$ mm $f_y = 415$ N/mm^2

$A_{st} = (2 \times 491) = 982$ mm^2 $s_v = 200$ mm

$A_{sv} = (2 \times 50) = 100$ mm^2

ii. *Percentage tension reinforcement*:

$$p_t = \left[\frac{100 A_{st}}{bd}\right] = \left[\frac{100 \times 982}{300 \times 600}\right] = 0.545$$

Refer to Table 4.4 (IS:456-2000, Table 19) and readout the permissible shear stress in concrete corresponding to $f_{ck} = 25$ N/mm^2 as $\tau_c = 0.50$ N/mm^2.

iii. *Shear resisted by concrete*:

$$V_{uc} = (\tau_c bd) = (0.50 \times 300 \times 600)\,10^{-3}$$
$$= 90 \text{ kN}$$

iv. *Shear resisted by vertical links*:

$$V_u = \left[\frac{A_{sv}(0.87 f_y)d}{s_v}\right] = \left[\frac{100 \times 0.87 \times 415 \times 600}{150}\right]$$
$$= 144.4 \text{ kN}$$

v. *Shear resistance of support section*:

$$V_{us} = A_{st}(0.87 f_y) \sin \alpha$$
$$= [982 \times 0.87 \times 415 \times \sin 45°]\,10^{-3}$$
$$= 250.6 \text{ kN}$$

vi. *Total shear resistance of support section*:

$$V_u = [V_{uc} + V_u + V_{us}]$$
$$= [90 + 144.4 + 250.6]$$
$$= 485 \text{ kN}$$

3. A reinforced concrete beam has a support section with a width 300 mm and effective depth 600 mm. The support section is reinforced with 3 bars of 20 mm diameter at an effective depth of 600 mm. 8 mm diameter 2 legged stirrups at a spacing of 200 mm is provided as shear reinforcement near supports using M20 grade concrete and Fe 415 HYSD bars, estimate the shear strength of the support section.

Method 1 (using IS:456-2000 code formulae)

i. *Data*:

$b = 300$ mm $\qquad\qquad f_{ck} = 20$ N/mm^2

$d = 600$ mm $\qquad\qquad f_y = 415$ N/mm^2

$A_{st} = (3 \times 314) = 942$ mm^2 $\quad A_{sv} = (2 \times 50) = 100$ mm^2

$s_v = 200$ mm

ii. *Percentage reinforcement*:

$$p_t = \left[\frac{100 A_{st}}{bd}\right] = \left[\frac{100 \times 942}{300 \times 600}\right] = 0.52$$

Refer to Table 19 (IS:456-2000 (Table 6.14 of text) and readout the design shear strength of concrete τ_c corresponding to $f_{ck} = 20$ N/mm^2.

$$\tau_c = 0.48 \text{ N/mm}^2$$

iii. *Shear resisted by concrete*:

$$V_{uc} = (\tau_c bd) = (0.48 \times 300 \times 600)\,10^{-3}$$
$$= 86.4 \text{ kN}$$

iv. *Shear resisted by stirrups:*

$$V_u = \left[\frac{A_{sv}(0.87\,f_y)d}{s_v}\right] = \left[\frac{100 \times 0.87 \times 415 \times 600}{200}\right]10^{-3}$$

$$= 108.3 \text{ kN}$$

v. *Total shear resistance of support section:*

$$V_u = [V_{uc} + V_{us}]$$
$$= [86.4 + 108.3] = 194.7 \text{ kN}$$

Method 2 (using SP:16 design tables)

i. *Shear resisted by concrete:*
Refer to Table 61, SP:16 (Table 6.11 of text) and readout the design strength of concrete as $\tau_c = 0.48 \text{ N/mm}^2$ for $p_t = 0.52$

$$V_{uc} = (\tau_c bd) = (0.48 \times 300 \times 600)\,10^{-3}$$
$$= 86.4 \text{ kN}$$

ii. *Shear resistance of two-legged vertical stirrups:*
Refer to Table 62, SP:16 (Table 6.14 of text) and readout the ratio (V_{us}/d) corresponding to $f_y = 415 \text{ N/mm}^2$ and diameter of stirrups as 8 mm and spacing $s_v = 200 \text{ mm}$ (20 cm).

$$\left(\frac{V_{us}}{d}\right) = 1.815 \text{ kN/cm}$$

$$\therefore \qquad V_{us} = (1.815 \times 60) = 108.9 \text{ kN}.$$

iii. *Total shear resistance:*

$$V_u = [V_{uc} + V_{us}]$$
$$= [86.4 + 108.9] = 195.3 \text{ kN}$$

4. A reinforced concrete beam of rectangular section has a width of 250 mm and an effective depth of 500 mm. The beam is reinforced with 4 bars of 25 mm diameter on the tension side. Two of the tension bars, one bent up at 45° near the support section in addition to the beam is provided with two-legged stirrups of 8 mm diameter at 150 mm centres near the suypports. If f_{ck} = 25 N/mm^2 and $f_y = 415 \text{ N/mm}^2$, estimate the ultimate shear strength of the support section.

Method 1 (using IS:456-2000 code formulae)

i. *Data:*

$$b = 250 \text{ mm} \qquad\qquad f_{ck} = 25 \text{ N/mm}^2$$
$$d = 500 \text{ mm} \qquad\qquad f_y = 415 \text{ N/mm}^2$$
$$A_{st} = (2 \times 491) = 982 \text{ mm}^2 \qquad s_v = 150 \text{ mm}$$
$$A_{sv} = (2 \times 50) = 100 \text{ mm}^2$$

ii. *Percentage reinforcement:*

$$p_t = \left[\frac{100 A_{st}}{bd}\right] = \left[\frac{100 \times 982}{250 \times 500}\right] = 0.78$$

Refer to Table 19, IS:456-2000 (Table 6.11 of text) and readout τ_c corresponding to $f_{ck} = 25 \text{ N/mm}^2$.

$$\tau_c = 0.584 \text{ N/mm}^2$$

iii. *Shear resisted by concrete*:

$$V_{uc} = (\tau_c bd) = (0.584 \times 250 \times 500) \, 10^{-3}$$
$$= 73.0 \text{ kN}$$

iv. *Shear resisted by stirrups*:

$$V_{us} = \left[\frac{A_{sv}(0.87 f_y)d}{s_v}\right] = \left[\frac{100 \times 0.87 \times 415 \times 500}{150}\right] 10^{-3}$$
$$= 120.3 \text{ kN}$$

v. *Shear resisted by bent up bars*:

$$V_{us\alpha} = A_s(0.87 f_y) \sin \alpha$$
$$= [982 \times 0.87 \times 415 \times \sin 45°] \, 10^{-3}$$
$$= 250.7 \text{ kN}$$

vi. *Total shear resistance of support section*:

$$V_u = [V_{uc} + V_{us} + V_{us\alpha}]$$
$$= [73.0 + 120.3 + 250.7]$$
$$= 444 \text{ kN}$$

Method 2 (using SP:16 design tables)

i. Refer to Table 61, SP:16 (Table 6.11 of text) and readout $\tau_c = 0.58 \text{ N/mm}^2$ for $p_t = 0.78$ and $f_{ck} = 25 \text{ N/mm}^2$

$$\therefore \quad V_{uc} = (\tau_c b \, d) = (0.584 \times 250 \times 500) \, 10^{-3}$$
$$= 73.0 \text{ kN}$$

ii. Refer to Table 62, SP:16 (Table 6.17 of text) and readout $(V_{us}/d) = 2.420$

$$\therefore \quad V_{us} = (2.42 \times 50) = 121 \text{ kN}.$$

iii. Refer to Table 63, SP:16 (Table 6.18 of text) and readout $V_{us\alpha}$ for single bar in kN

$$\therefore \quad V_{us\alpha} = (2 \times 125.32) = 250.64 \text{ kN}$$

iv. Total shear strength $V_u = [V_{uc} + V_{us} + V_{us\alpha}] = [73.0 + 121 + 250.64]$
$$= 444.64 \text{ kN}.$$

5.10 DESIGN EXAMPLES

1. A reinforced concrete beam of rectangular section, 300 mm wide is reinforced with four bars of 25 mm diameter at an effective depth of 600 mm. The beam has to resist a factored shear force of 400 kN at support section. Assuming $f_{ck} = 25 \text{ N/mm}^2$ and $f_y = 415 \text{ N/mm}^2$, design vertical stirrups for the section.

Method 1 (using IS:456-2000 code formulae)

i. *Data*:

$b = 300 \text{ mm}$	$f_{ck} = 25 \text{ N/mm}^2$
$d = 600 \text{ mm}$	$f_y = 415 \text{ N/mm}^2$
$A_s = (4 \times 491) = 1964 \text{ mm}^2$	$V_u = 400 \text{ kN}$

ii. *Nominal shear stress*:

$$V_u = 400 \text{ kN}$$

$$\tau_v = \left[\frac{V_u}{bd}\right] = \left[\frac{400 \times 10^3}{300 \times 600}\right] = 2.22 \text{ N/mm}^2 < \tau_{c,max} = 3.1 \text{ N/mm}^2$$

iii. *Shear resisted by concrete*:

$$p_t = \left(\frac{100 A_s}{bd}\right) = \left(\frac{100 \times 1964}{300 \times 600}\right) = 1.09$$

Refer to Table 19, IS: 456-2000 and readout τ_c, corresponding to p_t and f_{ck}.

$$\therefore \qquad \tau_c = 0.658 \text{ N/mm}^2 < \tau_v$$

Hence, stirrups are to be designed.

$$V_{uc} = (\tau_c b d)$$
$$= (0.658 \times 300 \times 600) \ 10^{-3} = 118.4 \text{ kN}$$

$\therefore$ Balance shear is given by

$$V_{us} = [V_u - V_{uc}] = [400 - 118.4]$$
$$= 281.6 \text{ kN}$$

iv. *Design of vertical stirrups*:

Using 10 mm diameter two-legged vertical stirrups, spacing is given by

$$S_v = \left[\frac{0.087 f_y A_{sv} d}{V_{us}}\right] = \left[\frac{0.87 \times 415 \times 2 \times 78.5 \times 600}{281.6 \times 10^3}\right] = 120.77 \text{ mm}$$

$$S_{v,max} = 0.75 \, d = (0.75 \times 600) = 450 \text{ mm}$$

Also $S_v \geq 300 \text{ mm}$

Provide 10 mm diameter two-legged vertical stirrups at 120 mm centres at support section.

Method 2 (using SP:16 design tables)

Compute the ratio $\left(\dfrac{V_{us}}{d}\right) \text{kN/cm} = \left(\dfrac{281.6}{60}\right) = 4.69 \text{ kN/cm}$.

Refer to Table 62, SP:16 (Table 6.14 of text) and readout the spacing corresponding to $f_y = 415 \text{ N/mm}^2$, 10 mm diameter stirrups and $(V_{us}/d) = 4.69$

$$S_v = 12 \text{ cm} = 120 \text{ mm}$$

Provide 10 mm diameter two-legged stirrups at 120 mm centres.

2. Design the shear reinforcements in a beam of rectangular section having a width 300 mm and effective depth 600 mm, the ultimate shear at the section is 100 kN. Use $f_{ck} = 20 \text{ N/mm}^2$ and $f_y = 415 \text{ N/mm}^2$. The beam is reinforced with 4 bars of 25 mm diameter in the tensile zone.

Method 1 (using IS:456-2000 code formulae)

i. *Data*:

$b = 300 \text{ mm}$ $\qquad\qquad$ $f_{ck} = 20 \text{ N/mm}^2$

$d = 600 \text{ mm}$ $\qquad\qquad$ $f_y = 415 \text{ N/mm}^2$

$A_s = (1964 \text{ mm}^2)$ $\qquad\quad$ $V_u = 100 \text{ kN}$

ii. *Nominal shear stress:*

$$V_u = 100 \text{ kN}$$

$$\tau_v = \left[\frac{V_u}{bd}\right] = \left[\frac{100 \times 10^3}{300 \times 600}\right] = 0.55 \text{ N/mm}^2$$

iii. *Shear strength of concrete:*

$$p_t = \left(\frac{100 A_s}{bd}\right) = \left(\frac{100 \times 1964}{300 \times 600}\right) = 1.09$$

Refer to Table 19, IS: 456-2000 and readout τ_c, corresponding to p_t and f_{ck}.

$$\therefore \qquad \tau_c = 0.658 \text{ N/mm}^2 > \tau_v$$

Hence, nominal shear reinforcements are to be designed.

iv. *Design of vertical stirrups:*

Using 8 mm diameter two-legged stirrups and the spacing is computed as

$$s_v = \left[\frac{A_{sv} \, 0.87 f_y}{0.4b}\right] = \left(\frac{2 \times 50 \times 0.87 \times 415}{0.4 \times 300}\right)$$

$$= 300.8 \text{ mm} < (0.75 \, d) = (0.75 \times 600) = 450 \text{ mm}.$$

Use 8 mm diameter two-legged stirrups at 300 mm centres.

Method 2 (using SP:16 design tables)

SP:16 requires the computation of the parameter $\left(\dfrac{V_{us}}{d}\right)$ kN/cm.

Design for nominal steel is equivalent to designing for a shear stress of 0.4 N/mm^2 (refer to Eq. (6.18))

$$\left(\frac{V_{us}}{d}\right) = \left(\frac{0.4 \times 300 \times 600}{10^3 \times 60}\right) = 1.2$$

where, V_{us} is expressed in kN and d in cm.

Refer to Table 62, SP:16 (Table 6.17 of text) and readout the spacing corresponding to $f_y = 415 \text{ N/mm}^2$ and diameter = 8 mm and $(V_{us}/d) = 1.20$

$$s_v = 30 \text{ cm} = 300 \text{ mm}$$

Adopt 8 mm diameter two-legged vertical stirrups at 300 mm centres.

3. A reinforced concrete beam of rectangular section 350 mm wide is reinforced with 4 bars of 20 mm diameter at an effective depth of 550 mm out of which 2 bars are bent up near the support section where a factored shear force of 400 kN is acting. Using M20 grade concrete and Fe 415 grade HYSD bars design suitable shear reinforcements at the support section.

Method 1 (using IS:456-2000 code formulae)

i. *Data:*

$b = 350 \text{ mm}$ $f_{ck} = 20 \text{ N/mm}^2$

$d = 550 \text{ mm}$ $f_y = 415 \text{ N/mm}^2$

$A_s = 628 \text{ mm}^2$ $V_u = 400 \text{ kN}$

ii. *Nominal shear stress*:

$$V_u = 400 \text{ kN}$$

$$\tau_v = \left[\frac{V_u}{bd}\right] = \left[\frac{400 \times 10^3}{350 \times 550}\right] = 2.07 \text{ N/mm}^2$$

$$p_t = \left(\frac{100 A_s}{bd}\right) = \left(\frac{100 \times 628}{350 \times 550}\right) = 0.32$$

Refer to Table 19, IS: 456-2000 and readout τ_c, corresponding to p_t and f_{ck}.

∴ $\tau_c = 0.40 \text{ N/mm}^2 < \tau_v < \tau_{c,max} = 2.8 \text{ N/mm}^2$

Hence, shear reinforcements have to be designed.

iii. *Shear resisted by concrete*:

$$V_{uc} = (\tau_c \, b \, d) = (0.40 \times 350 \times 550) \, 10^{-3} = 77 \text{ kN.}$$

∴ Shear to be carried by steel $V_{us} = [V_u - V_{uc}] = [400 - 77] = 323 \text{ kN.}$

iv. *Shear carried by bent up bars*:

$$V_{us\alpha} = A_s(0.87 f_y) \sin \alpha$$
$$= (628 \times 0.87 \times 415 \times \sin 45°) \, 10^{-3}$$
$$= 160.3 \text{ kN.}$$

∴ Shear to be carried by vertical stirrups = $[323 - 160.3] = 162.7 \text{ kN.}$

v. *Design of vertical stirrups*:

Using 10 mm diameter two-legged stirrups, the spacing s_v

$$= \left[\frac{0.87 f_y A_{sv} d}{V_s}\right] = \left[\frac{0.87 \times 415 \times 2 \times 78.5 \times 550}{162.7 \times 10^3}\right] = 192.8 \text{ mm}$$

$s_v < 0.75 \, d$ and $s_v < 300$ mm. Use 10 mm diameter two-legged stirrups at 190 mm centres.

Method 2 (using SP:16 design tables)

The shear to be carried by bent up bars and vertical stirrups = 323 kN.

i. *Shear resisted by bent up bars*:

2 bars of 20 mm diameter are bent up at an angle $\alpha = 45°$ near support section. Compute the shear taken by bent up bars using Table 63, SP:16 corresponding to $f_y = 415 \text{ N/mm}^2$, $\phi = 20$ mm and $\alpha = 45°$, $V_{us\alpha} = (2 \times 80.21)$ = 160.4 kN.

ii. *Design of vertical stirrups*:

Shear force to be resisted by vertical stirrups $V_{us} = [323 - 160] = 162.6 \text{ kN.}$

Compute the ratio $\left(\dfrac{V_{us}}{d}\right)$ kN/cm.

$$\left(\frac{V_{us}}{d}\right) = \left(\frac{162.6}{55}\right) = 2.96 \text{ kN/cm}$$

Using 10 mm diameter two-legged stirrups. Refer to Table 62, SP:16 and readout the spacing s_v corresponding to $\phi = 10$ mm and $f_y = 415$ N/mm^2 and the ratio of (V_{us}/d).

$\therefore \qquad s_v \cong 19$ cm $= 190$ mm.

$\therefore$ Adopt 10 mm diameter two-legged vertical stirrups at 190 mm centres.

REFERENCES

1. Ramakrishnan V and Arthur PD, *Ultimate Strength Design for Structural Concrete*, Pitman, London, 1969.

2. Park R and Paulay T, *Reinforced Concrete Structures*, John Wiley & Sons, New York, 1975.

3. Jones LL, *Ultimate Load Analysis of Reinforced and Prestressed Concrete Structures*, Pitman, London, 1969.

4. Winter G and Nilson SH, *Design of Concrete Structures*, McGraw-Hill Inc, New York, 1972.

5. Moseley WH, Bunget JH and Hulse R, *Reinforced Concrete Design to Euro Code-2*, 7 edn, Palgrave Macmillan, London, 2012, pp 218–224.

6. Rowe RE, Cranston WB and Best BC, New concepts in the design of concrete, *Structural Engineer*, Vol 43, 1965, pp 339–403.

7. ACI-ASCE-426:1974, The shear strength of reinforced concrete members—Slabs, *Proc of the ASCE*, Vol 100, No. ST8, pp 1543–91.

8. Warner RE, Rangan BV, Hall AS, *Reinforced Concrete*, Pitman, Australia, 1976.

9. IS:456-2000, Indian Standard Code of Practice for Plain and Reinforced Concrete (4th Revision), BIS, New Delhi, July 2000, pp 67–76.

10. SP:16-1980, Design Aids for Reinforced Concrete to IS:456, BIS, New Delhi, 1980.

ASSIGNMENT

1. A reinforced concrete beam has a support section with a width 250 mm and effective depth 550 mm. The support section is reinforced with 3 bars of 20 mm diameter on the tension side. 8 mm diameter two-legged stirrups are provided at a spacing of 200 mm centres. Using M25 grade concrete and Fe 500 HYSD bars, estimate the shear strength of the support section.

2. A reinforced concrete beam of rectangular section with a width 300 mm and an effective depth 600 mm is reinforced with 4 bars of 25 mm diameter as tension reinforcement. Two of the tension bars are bent up at 45° near the support section. The beam is also provided with two-legged vertical links of 8 mm diameter at 150 mm centres near supports. Using M20 grade concrete and Fe 415 HYSD bars, compute the ultimate shear strength of the support section using IS:456-2000 code specifications.

3. A reinforced concrete slab of a highway bridge culvert is 500 mm thick and is reinforced with 20 mm diameter Fe 415 HYSD bars at a spacing of 140 mm centres. If M25 grade concrete is used in the culvert, estimate the ultimate shear strength of the slab.

4. A T-beam girder of a national highway bridge is made of a top slab, 200 mm thick and a rib of width 300 mm and depth 1400 mm. The T-beam is reinforced with 16 bars of 32 mm at an effective cover of 150 mm. The beam is also provided with four-legged stirrups of 10 mm diameter at a spacing of 150 mm. If M25 grade concrete and Fe 415 HYSD bars are used in the beam, estimate the ultimate shear strength of the support section of the T-beam.

5. A reinforced concrete ring girder supporting the Intze type water tank is rectangular in section having a width 600 mm and overall depth 1200 mm. The girder is reinforced with 6 bars of 25 mm diameter at an effective cover of 60 mm, both on the compression and tension sides. The beam is also reinforced using four-legged vertical stirrups of 12 mm diameter spaced at 120 mm centres. If M20 grade concrete and Fe 415 HYSD bars are used, evaluate the shear strength of the section.

6. A reinforced concrete portal frame of an industrial workshop has a rectangular section 400 mm wide × 800 mm overall depth. The beam is reinforced with 4 bars of 22 mm diameter both on the compression and tension sides located at an effective cover of 50 mm. The stirrup reinforcement consists of 8 mm diameter bars at 150 mm centres. Adopting M25 grade concrete and Fe 500 HYSD bars, estimate the ultimate shear strength of the section.

7. A national highway reinforced concrete bridge girder is rectangular in section with a width 300 mm and overall depth 1600 mm. It is reinforced with 16 bars of 32 mm on the tension side at an effective cover of 150 mm. Design suitable shear reinforcements if the beam has to resist a factored shear of 1400 kN. Adopt M25 grade concrete and Fe 415 HYSD bars.

8. A reinforced concrete beam of rectangular section 350 mm wide is reinforced with 4 bars of 25 mm diameter at an effective depth of 600 mm out of which 2 bars are bent up near support section. The ultimate design shear force at the support section is 450 kN. Design suitable shear reinforcements using Fe 415 HYSD bars and M25 grade concrete.

9. A cantilever beam of a canopy is rectangular in section having a width 300 mm and overall depth 500 mm. The beam is reinforced with 2 bars of 25 mm on the tension side at the supports located at an effective depth of 450 mm. Design suitable shear reinforcements at the support section to resist a design ultimate shear force of 100 kN adopting M20 grade concrete and Fe 415 HYSD bars.

10. A continuous reinforced concrete beam of rectangular section with a width 350 mm and effective depth 650 mm, has to resist a design ultimate shear force of 250 kN. The beam is reinforced with 4 bars of 25 mm on the tension side. Design suitable shear reinforcements in the beam adopting M20 grade concrete and Fe 415 HYSD bars.

REVIEW QUESTIONS

1. What are shear failures? Under what conditions you would expect this type of failures in reinforced concrete structures?

2. Briefly outline the various types of sheer failure mechanisms encountered in reinforced concrete structures.

3. Explain the terms: (a) Nominal shear stress (b) Diagonal tension cracks (c) Design shear strength with respect to reinforced concrete members.

4. Explain briefly the various types of shear failure modes encountered in reinforced concrete structures subjected to loads.

5. Briefly explain the terms: (a) Shear strength factor (b) Maximum shear stress with respect to the design of reinforced concrete members subjected to shear.

6. What are the different types of reinforcements used to resist the shear forces developed in reinforced concrete members?

7. How do you design a reinforced concrete beam subjected to shear forces, when the nominal shear stress in concrete exceeds the maximum permissible shear stress?

8. What is the significance of providing minimum shear reinforcements in a reinforced concrete beam? Explain with reasons.

9. What is enhanced shear near supports? What are the Indian standard code recommendations for the design of shear reinforcements in such cases?

10. Briefly explain the effect of axial forces on the shear strength of reinforced concrete members with examples mentioning the code recommendations.

OBJECTIVE QUESTIONS

1. The critical section for shear in a reinforced concrete beam is located at
    a. quarter span
    b. mid span
    c. a distance equal to the effective depth from the support

2. In a beam subjected to live and dead loads, the shear cracks develop
    a. at the soffit of beam near mid span
    b. near the quarter span section
    c. in the inclined direction near the support

3. The various types of shear failure in reinforced concrete beams are inflammed by
    a. shear span to effective depth ratio
    b. neutral axis to effective depth ratio
    c. concrete to reinforcement strength ratio

4. Combined shear and compression mode of failure in a reinforced concrete beam develops when the ratio of shear span to depth is
    a. < 1
    b. < 6 but > 2.5
    c. < 2.5 but > 1

5. The design shear strength of concrete depends upon
    a. grade of concrete
    b. quantity of steel
    c. concrete grade and percentage reinforcement

6. The maximum spacing of shear reinforcement used as vertical stirrups expressed as a percentage of the effective depth should not exceed the value
    a. 0.95
    b. 0.65
    c. 0.75

7. The shear force resisted by concrete depends upon
    a. overall depth of beam
    b. percentage reinforcement
    c. shear strength of concrete
8. The shear force resisted by the vertical stirrups is inversely proportional to the
    a. effective depth of beam
    b. spacing of the stirrups
    c. area of reinforcement
9. The enhanced shear strength of concrete near the supports depends upon the
    a. percentage reinforcement in the section
    b. shear span length
    c. design shear stress of concrete
10. The presence of uniaxial compression in a reinforced concrete beam
    a. increases the shear cracks
    b. decreases the shear cracks
    c. has no effect on shear cracks

Torsional Strength of Reinforced Concrete Sections

6.1 INTRODUCTION

Design of reinforced concrete sections subjected to torsion requires a proper understanding of the behaviour of the members subjected to torsion[1,2]. Pure torsion is exceptional and almost not practicable in reinforced concrete members. Normally torsion is always associated with flexure and shear[3,4] in reinforced concrete members such as circular girders used in water tanks and towers, corner lintels where the load is eccentric to the line of reaction at supports. Primary torsion[5,6] is generally induced by eccentric loading and equilibrium conditions are sufficient to evaluate the torsional moments acting at critical sections.

Various research investigations[7,8,9] on the structural behaviour of members under combined flexure, shear and torsion has paved the way for formulation of specific design provisions in the Indian[10], British[11] and American[12] codes.

In reinforced concrete structures, torsion may develop at certain sections depending upon the type of loading. The resultant torsion is classified under two different categories:

   a. Primary or equilibrium torsion
   b. Secondary or compatibility torsion

6.2 PRIMARY AND SECONDARY TORSION

Primary torsion is the most common type in most of the structures like:

   a. Cantilever beam with slab
   b. Bow girder
   c. L-beams.

The type of loading on these structural members develops torsion as shown in Fig. 6.1. It is important to note that in all these cases, the loading is eccentric to the line of reaction at supports. The total torsion is equally distributed to the support sections. Primary or equilibrium torsion is induced by eccentric loading and equilibrium conditions are sufficient to evaluate the magnitude of torsional moments.

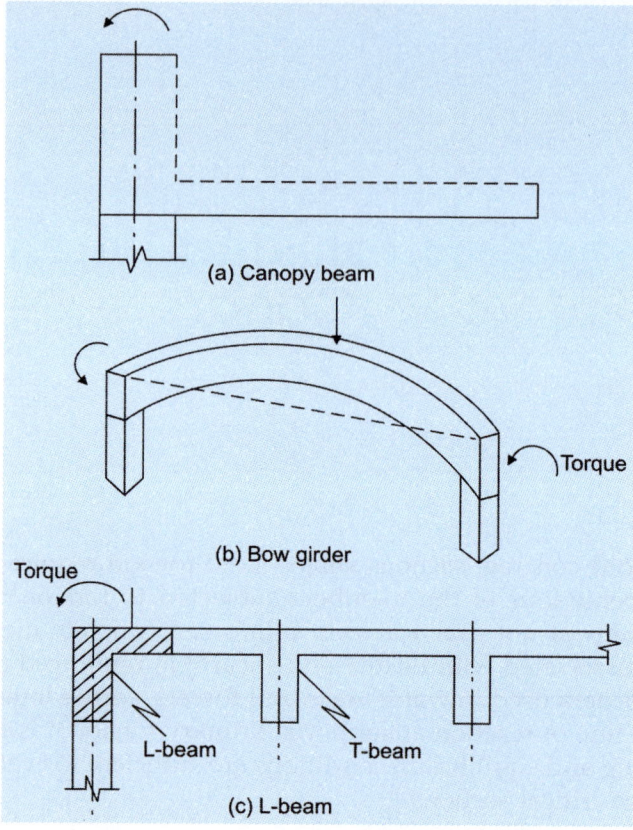

Fig. 6.1: Typical examples of primary torsion

Secondary or compatibility torsion develops in structural members as a secondary effect, due to rotations or twists applied at one or more points along the length of the member through the interconnected members rather than by direct application of loads. Figure 6.2 shows a typical example of a secondary beam monolithically connected to a primary beam rotating at the junction. The torsional moment will develop in the main beam due to the rotation at the junction and a bending moment in the secondary beam. The bending moment will be equal to and act in a direction opposite to the torsional moment to maintain static equilibrium.

6.3 DESIGN OF REINFORCEMENTS FOR TORSION, SHEAR AND FLEXURE

The design rules specified in IS code applies to beams of solid rectangular section and flanged sections in which the width of rib is considered for computations.

Sections subjected to torsion and shear are to be designed for an equivalent shear force computed as

$$V_e = V_u + 1.6(T_u/b)$$

where,

V_e = equivalent shear

V_u = transverse shear

T_u = torsional moment

b = breadth of beam

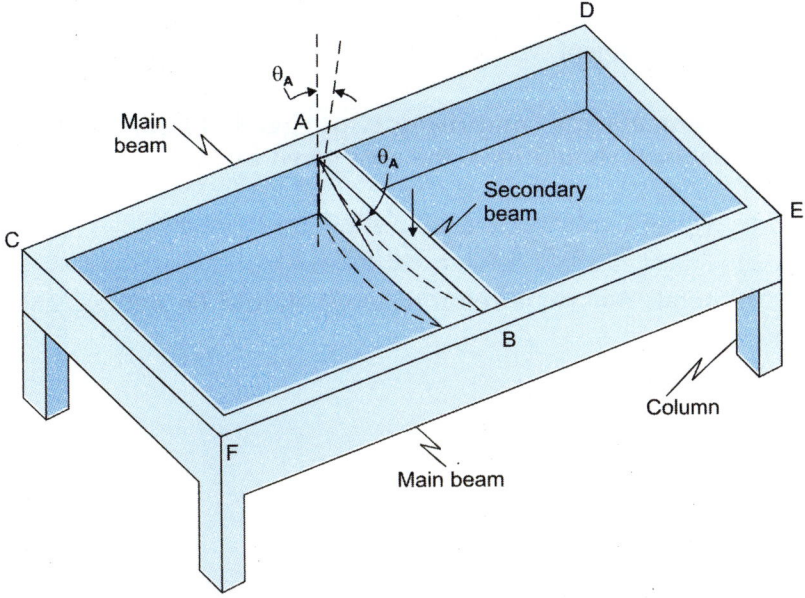

Fig. 6.2: Typical example of secondary torsion

The equivalent nominal shear stress is expressed as

$$\tau_{ve} = \left(\frac{V_u + 1.6(T_u / d)}{b\,d} \right)$$

The values of τ_{ve} should lie between τ_c, the permissible shear stress given in Table 5.2 and the maximum shear stress $\tau_{c,max}$ given in Table 5.3.

In cases where $\tau_{ve} > \tau_{c,max}$, the section has to be suitably redesigned by increasing the cross-sectional area and/or increasing the grade of concrete.

In $\tau_{ve} < \tau_c$, minimum shear reinforcements are designed according to that outlined in Section 4.5.4.

Longitudinal reinforcements are designed to resist an equivalent bending moment expressed as

$$M_{e1} = M_u + M_t$$

where,

M_{e1} = equivalent bending moment

M_u = design bending moment

M_t = bending moment developed due to torsion and expressed as

$$M_t = T_u \left(\frac{1 + (D/b)}{1.7} \right)$$

where,

T_u = torsional moment

D = overall depth

b = breadth of section

In cases where the numerical value of M_t exceeds the numerical value of M_u, longitudinal reinforcement should be provided on the flexural compression face such that the beam can also withstand an equivalent moment M_e computed as $M_{e2} = (M_t - M_u)$, the moment M_{e2} being taken as acting in the opposite sense to the moment M_u.

Transverse reinforcements comprising two-legged closed hoops enclosing the corner longitudinal bars should have the area given by

$$A_{sv} = \left[\frac{T_u s_v}{b_1 d_1 (0.87 f_y)} + \frac{V_u s_v}{2.5 d_1 (0.87 f_y)} \right]$$

However, the total transverse reinforcement should be not less than the value computed as

$$\left[\frac{(\tau_{ve} - \tau_c) b s_v}{0.87 f_y} \right]$$

where,

T_u = torsional moment
V_u = transverse shear force
s_v = spacing of vertical links
b_1 = centre to centre distance between corner bars in the direction of width
d_1 = centre to centre distance between corner bars in the direction of depth
f_y = characteristic strength of stirrup reinforcement
τ_{ve} = equivalent shear stress
τ_c = shear strength of concrete as given in Table 5.2.

6.4 ANALYSIS EXAMPLES

1. A reinforced concrete beam of rectangular section having a width 300 mm and an overall depth 500 mm is reinforced with bars of 16 mm diameter located at the corners at an effective cover of 48 mm. The single-legged stirrup reinforcement consists of 10 mm diameter bars spaced at 150 mm centres. If Fe 500 grade HYSD bars are used, estimate the ultimate torsional strength of the section.

   i. *Data*:

   $b = 300 \text{ mm}$
   $b_1 = [300 - (30 \times 2) - (10 \times 2) - 16] = 204 \text{ mm}$
   $D = 500 \text{ mm}$
   $d_1 = [500 - (30 \times (10 \times 2) \times 2) - (10 \times 2) - 16] = 404 \text{ mm}$
   $A_{sv} = (2 \times 78.5) = 157 \text{ mm}^2$
   $s_v = 150 \text{ mm}$

   Using the IS code formula with $V = 0$
   $T_u = [A_{sv} b_1 d_1 (0.87 f_y)] / s_v$
   $\quad = [157 \times 204 \times 404 \times (0.87 \times 500)] / 150$
   $\quad = [37.52 \times 10^6]$
   $\quad = 37.52 \text{ kN·m}$

2. A circular RCC girder has a rectangular section with a width 500 mm and overall depth 1000 mm. At a particular section, the factored values of bending and torsional moments are 150 and 30 kN·m respectively. The ultimate shear force at the section is 150 kN. Analyse the design moment and shear force for which the beam has to be designed.

  i. *Data*:

$$M_u = 150 \text{ kN·m} \qquad\qquad b = 500 \text{ mm}$$
$$T_u = 30 \text{ kN·m} \qquad\qquad D = 1000 \text{ mm}$$
$$V_u = 150 \text{ kN}$$

  ii. *Equivalent bending moment*:

$$M_{e1} = (M_u + M_t)$$

$$\text{where} \quad M_t = T_u\left[\frac{(1+D/b)}{1.7}\right] = 30\left[\frac{(1+1000/500)}{1.7}\right] = 53 \text{ kN·m}$$

$$\therefore \qquad M_{e1} = (150 + 53) = 203 \text{ kN·m}$$

  iii. *Equivalent shear force*:

$$V_e = V_u + 1.6\left(\frac{T}{b}\right) = \left[150 + 1.6\left(\frac{30}{0.5}\right)\right] = 246 \text{ kN}$$

  The cross-section has to be designed for an equivalent bending moment of 203 kN·m and an equivalent shear force of 246 kN.

3. A reinforced concrete beam of rectangular section has a width 350 mm and overall depth 700 mm. The beam is reinforced with 2 bars of 25 mm diameter both on the tension and compression faces at an effective cover of 50 mm. The side covers are 25 mm. 10 mm diameter two-legged stirrups are provided at 100 mm centres. The section is subjected to a factored shear force of 200 kN. If Fe 415 HYSD bars are used, estimate the torsional resistance of the beam using IS code provisions.

  i. *Data*:

$$b = 350 \text{ mm} \qquad\qquad b_1 = 300 \text{ mm}$$
$$D = 700 \text{ mm} \qquad\qquad d_1 = 600 \text{ mm}$$
$$V_u = 200 \text{ kN} \qquad\qquad A_{sv} = (2 \times 79) = 158 \text{ mm}^2$$
$$s_v = 100 \text{ mm}$$

  ii. *Torsional strength considering* $V_u = 0$:

$$T_u = \left(\frac{0.87 f_y A_{sv} b_1 d_1}{s_v}\right) = \left(\frac{0.87 \times 415 \times 158 \times 300 \times 600}{100}\right)$$

$$= (102.6 \times 10^6) \text{ N·mm}$$
$$= 102.6 \text{ kN·m}$$

  iii. *Torsional strength considering* $V_u = 200$ kN:

$$A_{sv} = \left[\left(\frac{T_u s_v}{b_1 d_1 \times 0.87 f_y}\right) + \left(\frac{V_u s_v}{2.5 d_1 \times 0.87 f_y}\right)\right]$$

$$158 = \left[\left(\frac{T_u \times 100}{300 \times 600 \times 0.87 \times 415} \right) + \left(\frac{200 \times 10^3 \times 100}{2.5 \times 600 \times 0.87 \times 415} \right) \right]$$

Solving, $T_u = (78.6 \times 10^6)$ N·mm = 78.6 kN·m. Hence, torsional strength is smaller of the two values.

4. A reinforced concrete rectangular beam has a breadth 400 mm and effective depth 800 mm. It has a factored shear force of 120 kN at a particular section. Assuming $f_{ck} = 25$ N/mm², $f_y = 415$ N/mm² and percentage of tensile steel at the section as 0.5%, determine the torsional moment which the section can resist for the following cases:

Case 1: If no additional reinforcement for torsion is provided.

Case 2: If the maximum steel for torsion is provided in the section.

Adopt IS:456 code provisions for the analysis.

  i. *Data:*

$$b = 400 \text{ mm} \qquad\qquad f_{ck} = 25 \text{ N/mm}^2$$
$$d = 800 \text{ mm} \qquad\qquad f_y = 415 \text{ N/mm}^2$$

 ii. *Permissible shear stress:*

For $p_t = 0.5\%$ and $f_{ck} = 25$ N/mm², refer to Table 19, IS:456 code

$$\tau_{c(min)} = 0.49 \text{ N/mm}^2$$

iii. *Allowable torsion:*

$$V_e = V_u + 1.6\,(T/b) \qquad \therefore \left(\frac{V_e}{bd} \right) = 0.49$$

$$\left[\frac{V_u + 1.6(T/b)}{bd} \right] = 0.49$$

$$\left[\frac{(120 \times 10^3) + 1.6(T/400)}{400 \times 800} \right] = 0.49$$

Solving, $T = (9.2 \times 10^6)$ N·mm = 9.2 kN·m.

iv. *Maximum shear capacity of the section (torsion + shear):*

Refer to Table 20 of IS:456 and readout the value of τ_{max} for $f_{ck} = 25$ N/mm²

$$\therefore \qquad \tau_{max} = 3.1 \text{ N/mm}^2$$

If T is maximum allowable torsion, then

$$\left[\frac{V_u + 1.6(T/b)}{bd} \right] = 3.1$$

$$\left[\frac{(120 \times 10^3) + 1.6(T/400)}{400 \times 800} \right] = 3.1$$

Solving, $T = (218 \times 10^6)$ N·mm = 218 kN·m.

5. A reinforced concrete beam of rectangular section with breadth 350 mm and overall depth 800 mm is reinforced with 4 bars of 20 mm diameter on the tension side at an effective depth of 750 mm. The section is subjected to an ultimate moment of 215 kN·m. If $f_{ck} = 30$ N/mm^2 and $f_y = 415$ N/mm^2, estimate the ultimate torsional moment that can be allowed on the section.

   i. *Data*:

$$b = 350\,\text{mm} \qquad\qquad f_{ck} = 30\,\text{N/mm}^2$$
$$d = 750\,\text{mm} \qquad\qquad f_y = 415\,\text{N/mm}^2$$
$$D = 800\,\text{mm} \qquad\qquad A_{st} = (4 \times 314) = 1256\,\text{mm}^2$$

   ii. *Neutral axis depth*:

$$x_{u,max} = (0.48d) = (0.48 \times 750) = 360\,\text{mm}$$

But $\qquad x_u = \left[\dfrac{0.87\,f_y A_{st}}{0.36\,f_{ck}b}\right] = \left[\dfrac{0.87 \times 415 \times 1256}{0.36 \times 30 \times 350}\right] = 120\,\text{mm} < x_{u,max}$

Hence, the beam is under-reinforced.

   iii. *Equivalent ultimate moment capacity of section*:

$$M_e = 0.87\,f_y A_{st} d\left[1 - \left(\frac{A_{st}f_y}{bdf_{ck}}\right)\right]$$

$$M_e = 0.87 \times 415 \times 1256 \times 750\left[1 - \left(\frac{1256 \times 415}{300 \times 750 \times 30}\right)\right]$$

$$= (317 \times 10^6)\,\text{N·mm} = 317\,\text{kN·m.}$$

   iv. *Allowable torsional moment*:

$$M_e = (M_u + M_t) = M_u + T_u\left(\frac{1 + (D/b)}{1.7}\right)$$

$$317 = 215 + T_u\left(\frac{1 + (800/350)}{1.7}\right)$$

Solving, $T_u = 62.2$ kN·m

6. A reinforced concrete beam of rectangular section with a width 300 mm and overall depth 600 mm is reinforced with 4 bars of 25 mm diameter distributed at each of the corners at an effective cover of 50 mm in the direction of depth and side covers of 25 mm in the direction of width. 8 mm diameter two-legged stirrups are provided at 100 mm centres. Estimate the torsional strength of the section adopting Fe 415 HYSD bars for the following cases:

   a. Torsional strength if $V_u = 0$

   b. Torsional strength if $V_u = 100$ kN

   i. *Data*:

$$b = 300\,\text{mm} \qquad\qquad b_1 = 250\,\text{mm}$$
$$D = 600\,\text{mm} \qquad\qquad d_1 = 500\,\text{mm}$$
$$s_v = 100\,\text{mm} \qquad\qquad A_{sv} = (2 \times 50) = 100\,\text{mm}^2$$

ii. *Case (a), torsional strength ($V_u = 0$):*

$$T_u = \left[\frac{A_{sv}b_1d_1(0.87f_y)}{s_v}\right] = \left[\frac{100 \times 250 \times 500 \times 0.87 \times 415}{100}\right]$$

$$= 45.13 \times 10^6 \text{ N·mm}$$
$$= 45.13 \text{ kN·m}$$

iii. *Case (b), torsional strength ($V_u = 100$ kN):*

$$A_{sv} = \left[\frac{T_u s_v}{b_1 d_1 (0.87 f_y)} + \frac{V_u s_v}{2.5 d_1 (0.87 f_y)}\right]$$

$$100 = \left[\frac{T_u \times 100}{(9250 \times 500 \times 0.87 \times 415)} + \frac{100 \times 10^3 \times 100}{(2.5 \times 500 \times 0.87 \times 415)}\right]$$

Solving, $T_u = 35.13 \times 10^6$ N·mm
$$= 35.13 \text{ kN·m}$$

Hence, the design ultimate torsional strength is the smaller of the two values.

7. A reinforced concrete beam of rectangular section with a breadth 300 mm and overall depth 850 mm is reinforced with 4 bars of 20 mm diameter on the tension side at an effective depth of 800 mm. The section is subjected to a factored bending moment of 200 kN·m. If $f_{ck} = 25$ N/mm², and $f_y = 415$ N/mm², calculate the ultimate torsional resistance that can be allowed on the section.

   i. *Data*:

$b = 300$ mm	$f_{ck} = 25$ N/mm²
$D = 850$ mm	$f_y = 415$ N/mm²
$d = 800$ mm	$A_{st} = (4 \times 314) = 1256$ mm²

   ii. *Neutral axis depth*:

   $$x_{u,max} = 0.48d = (0.48 \times 800) = 384 \text{ mm}$$

   $$x_u = \left[\frac{0.87 f_y A_{st}}{0.36 f_{ck} b}\right] = \left[\frac{0.87 \times 415 \times 1256}{0.36 \times 25 \times 300}\right] = 168 \text{ mm}$$

   Since $x_u < x_{u,max}$, section is under-reinforced.

   iii. *Equivalent ultimate moment capacity of section*:

   $$M_e = 0.87 f_y A_{st} d \left[1 - \left(\frac{A_{st} f_y}{b d f_{ck}}\right)\right]$$

   $$= (0.87 \times 415 \times 1256 \times 800)\left[1 - \frac{1256 \times 415}{(300 \times 800 \times 25)}\right]$$

   $$= 331 \times 10^6 \text{ N·mm}$$
   $$= 331 \text{ kN·m}$$

iv. *Allowable torsional moment*:

$$M_e = (M_u + M_t) = M_u + T_u \left[\frac{1 + (D/b)}{1.7} \right]$$

$$331 = 200 + T_u \left[\frac{1 + (850/300)}{1.7} \right]$$

Solving, $T_u = 58.1$ kN·m

6.5 DESIGN EXAMPLES

1. An RCC section 200 mm × 400 mm is subjected to a characteristic load torsional moment of 2.5 kN·m and a transverse shear of 60 kN. Assuming the use of M25 grade concrete and Fe 415 HYSD bars, determine the reinforcements required according to the IS:456 code provisions.

Method 1 (using IS:456 code formulae)

i. *Data*:

$b = 400$ mm	$f_{ck} = 25$ N/mm^2
$d = 350$ mm	$f_y = 415$ N/mm^2
$D = 400$ mm	$b_1 = 150$ mm
$T_u = 2.5$ kN·m	$d_1 = 300$ mm
$V_u = 60$ kN	

ii. *Equivalent shear force*:

$$V_e = V_u + 1.6 \left(\frac{T_u}{d} \right) = 60 + 1.6 \left(\frac{2.5}{0.2} \right) = 80 \text{ kN}$$

iii. *Equivalent bending moment*:

$$M_e = (M_u + M_t) = M_u + \frac{T_u(1 + D/b)}{1.7}$$

$$= 0 + \frac{2.5(1 + 400/200)}{1.7} = 4.41 \text{ kN} \cdot \text{m}$$

iv. *Longitudinal reinforcements*:

Since the equivalent bending moment $M_e = 4.41$ kN·m, design the longitudinal reinforcement for this moment.

The section is under-reinforced since the steel requirement to resist the small moment will be less than the minimum.

$$M_e = 0.87 f_y A_{st} d \left[1 - \left(\frac{A_{st} f_y}{b d f_{ck}} \right) \right]$$

$$M_e = 0.87 \times 415 \times A_{st} \times 350 \left[1 - \left(\frac{415 A_{st}}{200 \times 350 \times 25} \right) \right]$$

$[A_{st}^2 - 4218 A_{st} + (1.47 \times 10^5)] = 0$

∴ $A_{st} = 36$ mm^2

Providing minimum reinforcement of

$$A_s = \left(\frac{0.85bd}{f_y}\right) = \left(\frac{0.85 \times 200 \times 350}{415}\right) = 143 \text{ mm}^2$$

Provide 2 bars of 10 mm diameter as tension reinforcement and 2 hanger bars of 10 mm diameter on compression side at an effective cover of 50 mm.

v. *Permissible shear stress*:

$$\tau_{ve} = \left(\frac{V_e}{bd}\right) = \left(\frac{80 \times 10^3}{200 \times 350}\right) = 1.14 \text{ N/mm}^2$$

$$p_t = \left(\frac{100A_s}{bd}\right) = \left(\frac{80 \times 10^3}{200 \times 350}\right) = 0.225$$

Refer to Table 19, IS:456 and readout τ_c corresponding to f_{ck} = 25 N/mm²,

$$\tau_c = 0.34 \text{ N/mm}^2$$

$$\tau_{ve} > \tau_c \text{ and } \tau_{ve} < \tau_{c,max} = 3.1 \text{ N·mm}^2 \text{ (Table 20, IS:456)}$$

∴ Design transverse reinforcements using the IS code recommendations.

vi. *Transverse reinforcements*:

Using 8 mm diameter two-legged stirrups with side covers of 25 mm, the spacing is given by

$$S_v = \left[\frac{A_{sv} 0.87 f_y}{(\tau_{ve} - \tau_c)b}\right] = \left[\frac{0.87 \times 415 \times 2 \times 50}{(1.14 - 0.34)200}\right] = 225 \text{ mm}$$

Also $A_{sv} = \left[\left(\frac{T_u s_v}{(b_1 d_1 0.87 f_y)}\right) + \left(\frac{V_u s_v}{2.5 d_1 0.87 f_y}\right)\right]$

$$= \left[\left(\frac{2.5 \times 10^6}{150 \times 300 \times 0.87 \times 415}\right) + \left(\frac{60 \times 10^3}{2.5 \times 300 \times 0.87 \times 415}\right)\right] s_v$$

Solving, $s_v = 266$ mm

Adopting the smaller of the two values, use 8 mm diameter two-legged stirrups at a spacing of 225 mm.

Method 2 (using SP:16 design tables)

i. *Longitudinal reinforcements*:

Equivalent bending moment $M_e = M_u = 4.41$ kN·m

$$\left(\frac{M_u}{bd^2}\right) = \left(\frac{4.41 \times 10^6}{200 \times 350^2}\right) = 0.18$$

Referring to Table 3, SP:16 (f_{ck} = 25 N/mm²), the minimum value of the parameter (M_u/bd^2) is listed as 0.30. Hence, the table cannot be used. Provide minimum longitudinal reinforcement of p_t = 0.20% for f_y = 415 N/mm², as worked out in method 1.

ii. *Transverse reinforcements*:

Compute the parameter

$$\left[\frac{A_{sv}(0.87f_y)}{s_v}\right] = \left[\left(\frac{T_u}{b_1 d_1}\right) + \left(\frac{V_u}{2.5 d_1}\right)\right] = \left[\left(\frac{2.5 \times 10^6}{150 \times 300}\right) + \left(\frac{60 \times 10^3}{2.5 \times 300}\right)\right]$$

$$= 135.5 \text{ N/mm}$$

Refer to Table 62 and readout spacing of 8 mm diameter two-legged stirrups as 27 cm = 270 mm.

Also $\left(\frac{A_{sv} 0.87 f_y}{s_v}\right) = (\tau_{ve} - \tau_c) b = (1.14 - 0.34)200 = 160 \text{ N/mm} = 1.6 \text{ kN/cm}.$

Refer to Table 62 and readout spacing of 8 mm diameter two-legged stirrups as 22.5 cm = 225 mm. Provide the smaller spacing of the two values which is 225 mm.

2. A reinforced concrete beam of rectangular section with width 350 mm and overall depth 700 mm is subjected to an ultimate torsional moment of 100 kN·m together with an ultimate bending moment of 200 kN·m. Adopting M20 grade concrete and Fe 415 HYSD bars and assuming top and bottom covers of 50 mm and side covers of 25 mm, design suitable longitudinal and transverse reinforcement for the section.

Method 1 (using IS:456 code formulae)

i. *Data*:

$b = 350$ mm	$b_1 = 300$ mm
$d = 650$ mm	$d_1 = 600$ mm
$D = 700$ mm	$f_{ck} = 20 \text{ N/mm}^2$
$M_u = 200$ kN·m	$f_y = 415 \text{ N/mm}^2$
$T_u = 100$ kN·m	

ii. *Equivalent bending moment and shear force*:

$$M_e = M_u + T_u \left[\frac{1 + D/b}{1.7}\right] = 200 + 100 \left[\frac{1 + 700/350}{1.7}\right] = 376 \text{ kN} \cdot \text{m}$$

$$V_e = V_u + 1.6\left(\frac{T_u}{b}\right) = 0 + 1.6\left(\frac{100}{0.35}\right) = 457 \text{ kN}$$

The longitudinal reinforcement is designed for M_e and transverse reinforcement for V_e.

iii. *Longitudinal reinforcements*:

$$M_u = 0.138 f_{ck} b d^2 = (0.138 \times 20 \times 350 \times 650^2) = 408 \times 10^6 \text{ N·mm} > M_e$$

Hence, section is under-reinforced.

$$M_e = 0.87 f_y A_{st} d \left[1 - \left(\frac{A_{st} f_y}{b d f_{ck}}\right)\right]$$

$$(376 \times 10^6) = (0.87 \times 415 \times A_{st} \times 650)\left[1 - \left(\frac{415 A_{st}}{350 \times 650 \times 20}\right)\right]$$

Solving, $A_{st} = 1940$ mm²

Use 4 bars of 25 mm diameter ($A_{st} = 1964$ mm²) on the tension side and 2 hanger bars of 16 mm diameter on the compression side with effective covers of 50 mm.

iv. *Transverse reinforcements*:

$$\tau_{ve} = \left(\frac{V_e}{bd}\right) = \left(\frac{457 \times 10^3}{350 \times 650}\right) = 2\,\text{N/mm}^2$$

$$p_t = \left(\frac{100 A_s}{bd}\right) = \left(\frac{100 \times 1964}{350 \times 650}\right) = 0.86$$

Refer to Table 19, IS:456 and readout permissible shear stress as $\tau_c = 0.59$ N/mm² and $\tau_{ve} > \tau_c$ but less than $\tau_{c,max} = 2.8$ N/mm². Hence, shear reinforcements are required. Assuming 10 mm diameter two-legged stirrups, the area of shear reinforcement is computed by using the equation specified in IS:456 code, clause 41.4.3.

$$A_{sv} = \left[\left(\frac{T_u s_v}{b_1 d_1 0.87 f_y}\right) + \left(\frac{V_u s_v}{2.5 d_1 0.87 f_y}\right)\right]$$

$$158 = \left[\left(\frac{100 \times 10^6}{300}\right) + (0)\right]\left[\frac{s_v}{600 \times 0.87 \times 415}\right]$$

Solving, $s_v = 102.7$ mm.

Also $\quad s_v \ngtr \dfrac{A_{sv} \cdot 0.87 f_y}{(\tau_{ve} - \tau_c)b} = \ngtr 115.6$ mm

Hence, adopt 10 mm diameter two-legged stirrups at a spacing given by smaller of the above two equations which is 100 mm.

Method 2 (using SP:16 design tables)

i. *Longitudinal reinforcements*:

$M_e = M_u = 376$ kN·m $\qquad b = 350$ mm
$f_{ck} = 20$ N/mm² $\qquad\qquad b_1 = 300$ mm
$f_y = 415$ N/mm² $\qquad\qquad d_1 = 600$ mm
$d = 650$ mm $\qquad\qquad\quad V_u = 0$

Compute the parameter $\left(\dfrac{M_u}{bd^2}\right) = \left(\dfrac{376 \times 10^6}{350 \times 650^2}\right) = 2.54$

Refer to Table 2, SP:16 and readout the percentage reinforcement p_t for $f_{ck} = 20$ and $f_y = 415$ N/mm².

$$p_t = 0.857 \qquad \therefore \left(\frac{100 A_{st}}{bd}\right) = 0.857$$

$$\therefore \qquad A_{st} = \left(\frac{0.857 \times 350 \times 650}{100}\right) = 1950\,\text{mm}^2$$

ii. *Transverse reinforcements*:

Compute the parameter given by

$$\left[\frac{A_{sv}(0.87f_y)}{s_v}\right] = \left[\left(\frac{T_u}{b_1 d_1}\right) + \left(\frac{V_u}{2.5 d_1}\right)\right] = \left[\left(\frac{100 \times 10^6}{300 \times 600}\right) + (0)\right]$$

$$= 555.5 \text{ N/mm}$$
$$= 5.55 \text{ kN/cm}$$

Refer to Table 62, SP:16 and readout spacing of 10 mm diameter two-legged stirrups as $s_v = 10.2$ cm $= 102$ mm.

Also,

$$\left(\frac{A_{sv}(0.87f_y)}{s_v}\right) = (\tau_{ve} - \tau_c)b = (2 - 0.59)350 = 493.5 \text{ N/mm}$$

$$= 4.93 \text{ kN/cm}$$

Refer to Table 62, SP:16 and readout spacing of 10 mm diameter two-legged stirrups as $s_v = 11.5$ cm $= 115$ mm. Provide the smaller of the two spacings, $s_v = 100$ mm.

3. A reinforced concrete beam of rectangular section with width 350 mm and overall depth 800 mm is subjected to a factored bending moment of 215 kN·m and ultimate torsional moment of 105 kN·m. Using M20 grade concrete and Fe 415 HYSD bars and side, top and bottom covers of 50 mm, design suitable reinforcement in the section.

i. *Data*:

$b = 350$ mm	$f_{ck} = 20$ N/mm^2
$b_1 = 250$ mm	$M_u = 215$ kN·m
$D = 800$ mm	$f_y = 415$ N/mm^2
$d = 750$ mm	$T_u = 105$ kN·m
$d_1 = 700$	$V_u = 150$ kN

ii. *Equivalent bending moment and shear force*:

$$M_e = (M_u + T_u)$$

$$M_e = M_u + T_u\left[\frac{1+D/b}{1.7}\right] = 215 + 105\left[\frac{1+800/350}{1.7}\right] = 418 \text{ kN} \cdot \text{m}$$

$$V_e = V_u + 1.6\left(\frac{T_u}{b}\right) = 150 + 1.6\left(\frac{105}{0.35}\right) = 630 \text{ kN}$$

Longitudinal reinforcements are designed for the equivalent bending moment M_e.

iii. *Longitudinal reinforcements*:

Since $M_t > M_u$, design reinforcements for M_e only.

$$M_{u,lim} = 0.138 f_{ck} b d^2$$
$$= (0.138 \times 20 \times 350 \times 750^2)$$
$$= (543.3 \times 10^6) \, \text{N·mm}$$
$$= 543.3 \, \text{kN·m} > M_e$$

Hence, section is under-reinforced:

$$M_e = 0.87 f_y A_{st} d \left[1 - \left(\frac{A_{st} f_y}{b d f_{ck}} \right) \right]$$

$$(418 \times 10^6) = (0.87 \times 415 \times A_{st} \times 750) \left[1 - \left(\frac{415 A_{st}}{350 \times 750 \times 20} \right) \right]$$

Solving, $A_{st} = 1802 \, \text{mm}^2$

Provide 4 bars of 25 mm diameter ($A_{st} = 1964 \, \text{mm}^2$).

iv. *Transverse reinforcements*:

$$\tau_{ve} = \left(\frac{V_e}{bd} \right) = \left(\frac{630 \times 10^3}{350 \times 750} \right) = 2.4 \, \text{N/mm}^2$$

$$p_t = \left(\frac{100 A_s}{bd} \right) = \left(\frac{100 \times 1964}{350 \times 750} \right) = 0.75$$

Refer to Table 19, IS:456-2000 and readout τ_c for $f_{ck} = 20 \, \text{N/mm}^2$ as

$$\tau_c = 0.56 \, \text{N·mm}^2 < \tau_{ve}$$

and $\tau_{ve} < \tau_{c,max} = 2.8 \, \text{N/mm}^2$

Hence, transverse reinforcements are required.

Using 10 mm diameter two-legged stirrups, spacing is computed as

$$A_{sv} = \left[\left(\frac{T_u s_v}{b_1 d_1 0.87 f_y} \right) + \left(\frac{V_u s_v}{2.7 d_1 0.87 f_y} \right) \right]$$

$$(2 \times 79) = \left[\left(\frac{105 \times 10^6}{250} \right) + \left(\frac{150 \times 10^3}{2.5} \right) \right] \frac{s_v}{0.87 \times 415 \times 700}$$

Solving, $s_v = 83.2 \, \text{mm}$

Also the spacing should conform to the equation,

$$s_v \not> \left(\frac{A_{sv} 0.87 f_y}{(\tau_{ve} - \tau_c) b} \right) \not> \left[\frac{158 \times 0.87 \times 415}{(2.4 - 0.56) 350} \right] \not> 88.5 \, \text{mm}$$

Hence, provide 10 mm diameter two-legged stirrups at a spacing of 80 mm centres.

Method 2 (using SP:16 design tables)

i. *Longitudinal reinforcements*:

$M_e = M_u = 418 \, \text{kN·m}$ $b = 350 \, \text{mm}$
$f_{ck} = 20 \, \text{N/mm}^2$ $d = 750 \, \text{mm}$
$f_y = 415 \, \text{N/mm}^2$ $b_1 = 250 \, \text{mm}$
$d_1 = 700 \, \text{mm}$

Compute the parameter $\left(\dfrac{M_u}{bd^2}\right) = \left(\dfrac{418 \times 10^6}{350 \times 750^2}\right) = 2.12$

Refer to Table 2, SP:16 and readout the percentage reinforcement p_t for $f_y = 415 \text{ N/mm}^2$ as $p_t = 0.685$.

$$\therefore \qquad A_{st} = \left(\dfrac{0.685 \times 350 \times 750}{100}\right) = 1798 \text{ mm}^2$$

ii. *Transverse reinforcements*:

Compute the parameter given by

$$\left[\dfrac{A_{sv}(0.87 f_y)}{s_v}\right] = \left[\left(\dfrac{T_u}{b_1 d_1}\right) + \left(\dfrac{V_u}{2.5 d_1}\right)\right] = \left[\left(\dfrac{105 \times 10^6}{250 \times 700}\right) + \left(\dfrac{150 \times 10^3}{2.5 \times 700}\right)\right]$$

$$= 685.7 \text{ N/mm}$$

Refer to Table 62, SP:16 and using 10 mm diameter two-legged stirrups, readout spacing $s_v = 8.8 \text{ cm} = 88 \text{ mm}$.

Also

$$\left(\dfrac{A_{sv}(0.87 f_y)}{s_v}\right) = (\tau_{ve} - \tau_c)b = (2.4 - 0.56)350 = 644 \text{ N/mm} = 6.44 \text{ kN/cm}$$

Refer to Table 62, SP:16 and using 10 mm diameter two-legged stirrups, readout the spacing as $s_v = 8.5 \text{ cm} = 85 \text{ mm}$

Adopt smaller of the two spacings ($s_v = 85 \text{ mm}$).

REFERENCES

1. HSU, TTC, Plain Concrete Rectangular Sections, Torsion of Structural Concrete, ACI Publication (SP-18), American Concrete Institute, Detroit, 1968, pp 207–38.

2. Collins MP, The Torque-Twist Characteristics of Reinforced Concrete Beams, Inelasticity and Nonlinearity in Structural Concrete, SM Study No. 8, University of Waterloo Press, Waterloo, 1972, pp 211–232.

3. HSU, TTC, Ultimate torque of rectangular reinforced concrete beams, *ASCE Journal (Structural Division)*, Vol 94, February 1968, pp 448–510.

4. Purushothaman P, *Reinforced Concrete Structural Elements, Behaviour*, Analysis and Design, Tata McGraw-Hill, New Delhi, 1984.

5. Lampert P and Collins MP, Torsion, bending and confusion—an attempt to establish the facts, *Journal of the American Concrete Institute*, Vol 69, August 1972, pp 500–504.

6. Warner RF, Rangan BV and Hall AS, *Reinforced Concrete*, Pitman Publications, Australia, 1976.

7. Iyengar KTS and Ramprakash N, Recommendations for the design of reinforced concrete beams for torsion, bending and shear, bridge and structural engineer, March 1974.

8. Zia P, What do we know about torsion in concrete members, *Proc American Society of Civil Engineers*, Vol 96, ST6, June 1970, pp 1185–99.

9. Rausch E, *Berechung Des Eisenbentons Gegen Verdrehung and Abschheren* (*Design of Reinforced Concrete for Torsion and Shear*), Springer Verlag, Berlin, 1929.

10. IS:456-2000, Indian Standard Code of Practice for Plain and Reinforced Concrete (4th Revision), BIS, New Delhi, July 2000, pp 67–76.

11. BS EN:1992-1-1, Euro Code-2, Design of Concrete Structures, General Rules and Rules for Building, British Standards Institution, London, 2004.

12. ACI:318M-11, Building code requirements for structural concrete, *American Concrete Institute*, Farmington Hills, Michigan, 2005, p 43.

ASSIGNMENT

1. A canopy slab is cantilevered 3 m from a lintel beam of span 5 m. The beam has a size of 300 mm by 500 mm and is well anchored into the two reinforced concrete columns. Assuming the thickness of the canopy slab as 150 mm and a service live load of 1.5 kN/m², compute the design torsional moment to be resisted by the sections at the end of beam using the Indian standard code specifications.

2. A reinforced concrete beam of rectangular section having width 400 mm and overall depth 800 mm is reinforced with 4 bars of 25 mm diameter at the corners having a clear cover of 50 mm in the direction of depth and 30 mm in the direction of width. The stirrup reinforcement consists of 10 mm two-legged bars at 150 mm centres. Adopting M25 grade concrete and Fe 415 HYSD bars, estimate the torsional strength of the section for the following cases:

   a. Ultimate torsional strength when shear force is zero

   b. Ultimate torsional strength when ultimate shear force is 150 kN.

3. The cross-section of a circular girder of the foundation of a water tower is rectangular with width 500 mm and overall depth 1000 mm. The girder is reinforced with 4 bars of 25 mm diameter at the corners having a clear cover of 50 mm. 10 mm diameter four-legged stirrups at a spacing of 170 mm are used. Adopting M20 grade concrete and Fe 415 HYSD bars, evaluate the torsional strength of the section if the shear strength of the section is 900 kN.

4. A reinforced concrete beam of rectangular section with width 400 mm and overall depth 800 mm is reinforced with 4 bars of 25 mm diameter on the tension side at an effective depth of 750 mm. The section is subjected to a factored bending moment of 250 kN·m. If M25 grade concrete and Fe 500 HYSD bars are used, calculate the ultimate torsional strength of the section using Indian standard code specifications.

5. A reinforced concrete beam of rectangular section with width 350 mm and overall depth 750 mm is reinforced with 4 bars of 20 mm diameter at the corners such that the clear cover to the bars is 50 mm in the depth side and 25 mm in the width side. If 8 mm diameter two-legged stirrups are used and the grade of concrete is M25 and Fe 500 HYSD bars are used, estimate the ultimate torsional strength of the section assuming the shear strength as 120 kN.

6. A reinforced concrete beam of rectangular section with width 350 mm and overall depth 850 mm has to be designed to support factored bending and

torsional moments of 200 kN·m and 100 kN·m respectively along with a factored shear force of 150 kN. Using M20 grade concrete and Fe 415 HYSD reinforcements, design suitable reinforcements in the section conforming to the specifications of IS:456-2000.

7. A reinforced concrete beam of rectangular section 250 mm wide by 500 mm overall depth is reinforced with 4 bars of 20 mm diameter at the corners having an effective cover of 50 mm. The stirrup reinforcements consist of 10 mm diameter bars spaced at 150 mm centres. Using M25 grade concrete and Fe 415 HYSD bars, estimate the ultimate torsional strength of the section.

8. A reinforced concrete beam of rectangular section 300 mm wide by 600 mm overall depth is cast using M20 grade concrete. Estimate the cracking torque of the section using the Indian standard code specifications.

9. A reinforced concrete ring beam of an Intze type water tank is supported on eight equally spaced columns along its perimeter. The distance of the ring beam is 10 m. The beam has to be designed for an ultimate shear force of 220 kN and torsional moment of 150 kN·m at a critical section. The beam has a width of 300 mm and overall depth of 600 mm. Adopting M20 grade concrete and Fe 415 HYSD bars, design suitable reinforcements at the critical section.

10. A reinforced concrete beam of rectangular section 300 mm wide by 650 mm overall depth is to be designed to support ultimate bending moment of 250 kN·m, torsional moment of 20 kN·m together with a shear force of 200 kN. Adopting M20 grade concrete and Fe 415 HYSD bars, design suitable reinforcements in the section as per IS:456-2000 code.

REVIEW QUESTIONS

1. Bring out the differences between primary torsion and secondary torsion with examples.

2. What is the effect of torsion in reinforced concrete flexural members?

3. What is torsional shear stress? How do you estimate the torsional shear stress in concrete rectangular sections?

4. Explain the significance of the term equivalent shear in concrete members subjected to shear and torsion.

5. Explain the term equivalent bending moment. How do you compute the equivalent bending moment in reinforced concrete flexural members subjected to flexure and torsion?

6. How do you design transverse reinforcements in reinforced concrete members subjected to torsion and shear?

7. Explain the Indian standard code method of designing reinforcements in concrete members subjected to combined flexure and torsion?

8. Briefly explain the method of placing the longitudinal reinforcements in concrete members subjected to major torsional moments.

9. What are the advantages of using SP:16 design charts in designing concrete members subjected to flexure, shear and torsion?

10. Briefly explain the IS code specifications regarding the distribution of torsional reinforcements in concrete members.

OBJECTIVE QUESTIONS

1. Torsional shear stresses are likely to develop in reinforced concrete
   a. simply supported beams
   b. T-beams
   c. ring girders of water tanks

2. Torsional moments are maximum at beam sections
   a. where positive bending moments are maximum
   b. where negative bending moments are maximum
   c. in between the sections of maximum positive and negative bending moments

3. Torsion in a reinforced concrete beam develops cracks
   a. near the soffit of the beam
   b. diagonally due to torsional shear stress
   c. at the level of reinforcements

4. Torsion in beams is better resisted by selecting
   a. rectangular sections
   b. I-sections
   c. box sections

5. Torsional shear stress in a beam section is inversely proportional to
   a. shear force
   b. torsional moment
   c. sectional dimensions

6. The ultimate torsional strength of a rectangular reinforced concrete section is
   a. directly proportional to the spacings of stirrups
   b. inversely proportional to the strength of reinforcement
   c. inversely proportional to the spacing of stirrups

7. The spacing of stirrups reinforcement in a reinforced concrete beam section subjected to torsion and shear is
   a. directly proportional to the characteristic strength of stirrup reinforcement
   b. directly proportional to the cross-sectional dimensions
   c. inversely proportional to the cross-sectional dimensions

8. The equivalent nominal shear stress in a reinforced concrete beam section subjected to torsional moment and shear force depends upon the
   a. bending moment at the section
   b. ratio of torsional moment to effective depth
   c. the ratio of shear force to effective depth

9. If the equivalent shear stress is greater than the permissible shear stress in concrete in a reinforced concrete beam section subjected to torsional moment and shear, then

    a. the main reinforcements in the beam should be increased

    b. the stirrups reinforcement should be increased

    c. the section should be redesigned by increasing the cross-sectional area and the grade of concrete

10. The design shear strength of concrete depends upon

    a. characteristic strength of reinforcement

    b. torsional and bending moments acting on the beam

    c. percentage reinforcement ratio and the grade of concrete

Bond and Anchorage in Reinforced Concrete Members

7.1 INTRODUCTION

The term bond refers to the adhesion between the reinforcing steel bars and the surrounding concrete. The success of reinforced concrete as a composite material is mainly attributed to the excellent bond between steel bars and concrete. This basic factor was recognised by the early investigators like Frederick Turneaure[1], Morsch & Goodrich[2], Oscar Faber and Bowie[3] during the early 19th century. The assumption of plane sections remains plane ever after bending in the simple bending theory is valid only when there is effective bond between concrete and steel. Depending upon the bond resistance, the stresses in the reinforcing bar can vary from point to point along the length of the member. In the absence of bond, the stress in the steel reinforcement will be constant along its length as in the case of a straight cable used in prestressed concrete members[4].

The coefficients of thermal expansion for steel and concrete are of the order of $10 \times 10^{-6}/°C$ and 7 to $12 \times 10^{-6}/°C$ respectively. These values are sufficiently close that problems with bond seldom arise from differential expansion between the two materials over normal temperature ranges. Based on several investigations[5,6], it has been proved beyond doubt that bond between steel bars and surrounding concrete can be improved significantly by using superior quality of cements in the concrete and changing the surface characteristics of reinforcing bars by introducing ribs and indentations to improve the frictional resistance.

7.2 MECHANISMS OF BOND

The bond between concrete and steel reinforcement can be achieved by the following three distinct mechanisms:

1. The microscopic hydration products of cement in concrete having a gum like property sticks firmly to the surface of the reinforcement due to chemical adhesion resulting in bond between concrete and steel.

2. The microscopic surface roughness of the reinforcement develops grip with drying concrete due to shrinkage of the material resulting in frictional resistance against relative motion between concrete and reinforcement.
3. Shearing resistance or dilatancy due to the mechanical interlock developed as a consequence of surface protrusions or ribs provided in deformed bars.

In the case of plain steel bars, the resistance due to mechanical interlock is not available. Due to this reason, the American[7] and British[8] codes prohibit the use of plain bars in reinforced concrete structural members. However, the Indian standard code[9] does not impose such restriction and in the future revised editions, plain bars are likely to be prohibited for use in reinforced concrete members.

7.3 FLEXURAL BOND STRESS

In the case of beam subjected to flexure, bond stresses develop due to variation in bending moment and shear at various sections. Referring to Fig. 7.1, the differential moment dM from section X to Y causes the additional tension dT expressed as

$$dT = (dM/jd)$$

where, jd = lever arm

This unbalanced bar force is transferred to the surrounding concrete by means of flexural bond developed along the interface.

If τ_b is flexural bond stress, then the equilibrium of forces yields the relation

$$\tau_b(\Sigma O)dx = dT$$

where, (ΣO) = total perimeter of the bars

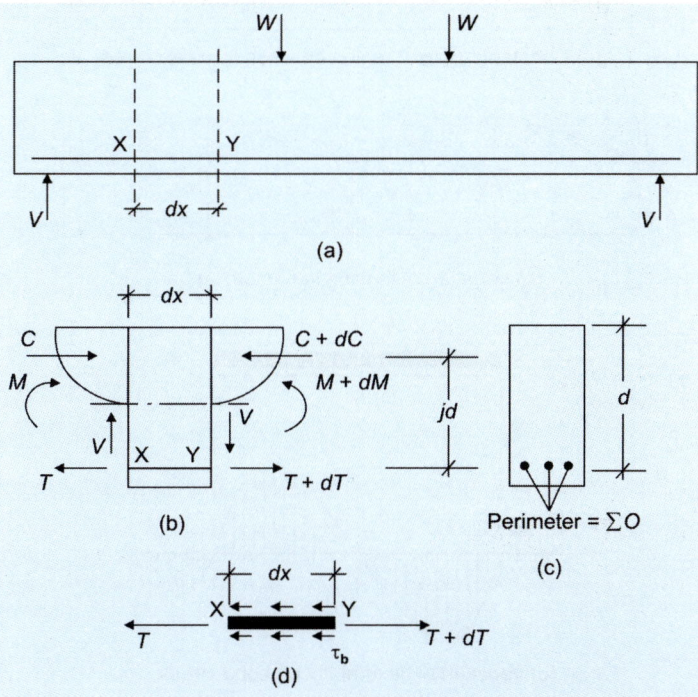

Fig. 7.1: Flexural bond stress

$\therefore$ $$\tau_b = \left(\frac{(dM/dx)}{(\Sigma O)jd}\right)$$

But $V = (dM/dx)$, hence we have

$$\tau_b = \left(\frac{V}{(\Sigma O)jd}\right)$$

The flexural bond stress is higher at locations of high shear force but it can be reduced by providing an increased number of bars of smaller diameter yielding the same equivalent area of reinforcement.

Figure 7.2a shows the effect of flexural cracks on flexural bond stresses in the constant moment region. The variation of tension in the reinforcing bar is shown in Fig. 7.2b and the bond stress variation in Fig. 7.2c. Experimental investigations have shown that splitting cracks develop in the vicinity of the flexural cracks where the local bond stresses are high. Hence, it is preferable to limit the magnitude of local bond stress by using a larger number of smaller diameter bars than using a few large diameter bars. However, with the development and wider use of deformed or high bond bars, more emphasis is laid on anchorage or development length requirements than the local bond stress. Hence, the Indian standard code IS:456-2000 does not specify any permissible values for the flexural or local bond stress.

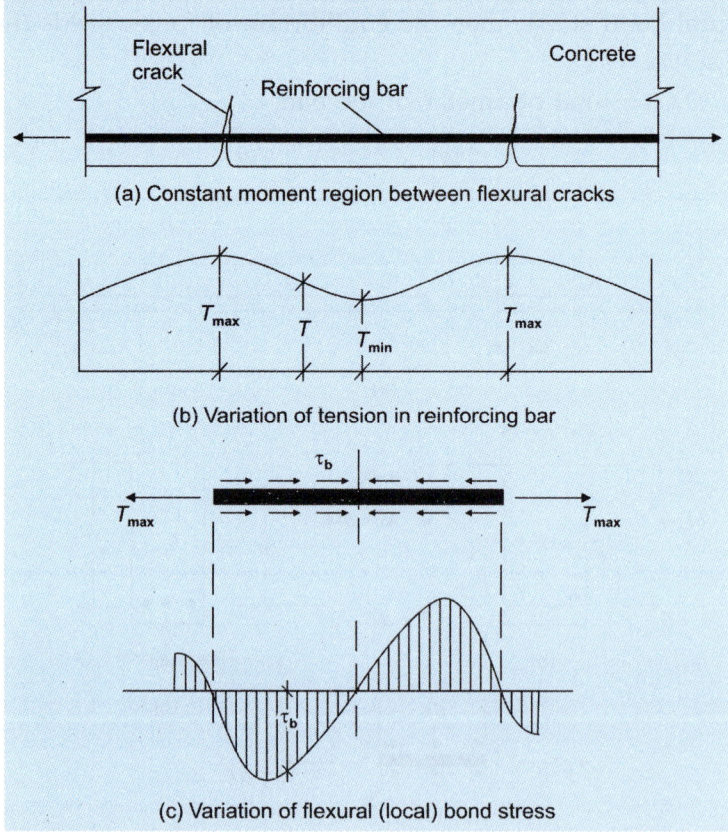

(a) Constant moment region between flexural cracks

(b) Variation of tension in reinforcing bar

(c) Variation of flexural (local) bond stress

Fig. 7.2: Variation of flexural bond stress between cracks

7.4 ANCHORAGE (DEVELOPMENT) BOND STRESS

The anchorage bond stress in a reinforcing bar develops from zero at the end of the bar to a maximum over a length of embedment where the bending moment is maximum as shown in Fig. 7.3a. The figure shows a cantilever slab in which the tensile stress in the bar varies from a maximum (σ_s) at the continuous end B to a value of zero at the end A. Since the moment is maximum at B, the tensile stress is also maximum at B and to develop the maximum stress σ_s, the bar requires a certain length (AB), which is termed *anchorage*, or *development length*. Also, the stress is zero at the discontinuous end A.

The variation of bond stress from A to B is shown in Fig. 7.3b. A similar situation exists in the bars terminated at the supports of a simply supported beam. For design purposes an average bond stress assuming uniform distribution over the length AB (Fig. 7.3c) is specified in the Indian standard code IS:456-2000. The average bond stress τ_{bd} can be expressed in terms of the diameter of the bar ϕ, the stress in steel σ_s and the anchorage length L_d, by considering the equilibrium of forces shown in Fig. 7.3c.

$$(\pi\phi L_d)\tau_{bd} = (\pi\phi^2/4)\sigma_s$$

The development length L_d required to develop the design stress σ_s in the bar is expressed as

$$L_d = \left(\frac{\phi\sigma_s}{4\tau_{bd}}\right)$$

The values of τ_{bd} depend upon the grade of concrete and in the limit state method for plain bars, the values of design bond stress specified in Clause 26.2.1.1, IS:456-2000 are given in Table 7.1.

The anchorage or development lengths of bends and hooks are shown in Fig. 7.4 as specified in SP:16.

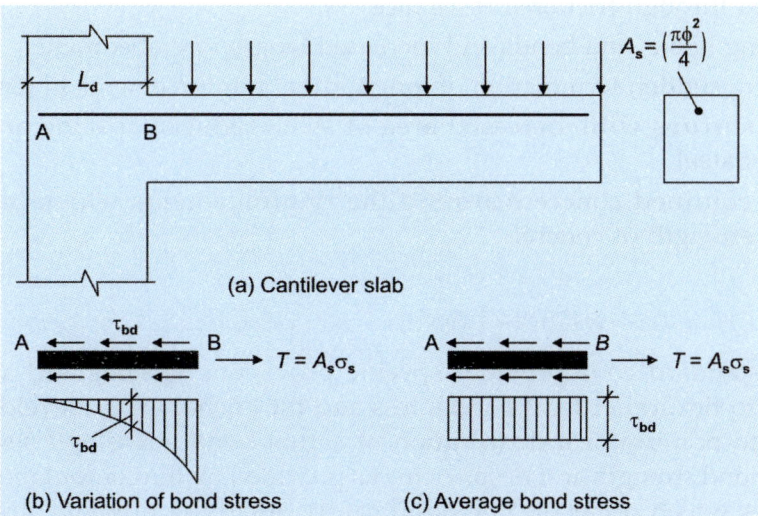

(a) Cantilever slab

(b) Variation of bond stress   (c) Average bond stress

Fig. 7.3: Anchorage bond stress

Table 7.1: Design bond stress in limit state method for plain bars in tension (Table 26.2.1.1 of IS:456-2000)

Grade of concrete	M20	M25	M30	M35	M40 & above
Design bond stress τ_{bd} (N/mm²)	1.2	1.4	1.5	1.7	1.9

Notes:
1. For deformed bars conforming to IS:1786, these values have to be increased by 60%.
2. For bars in compression, the value of bond stress in tension is increased by 25%.

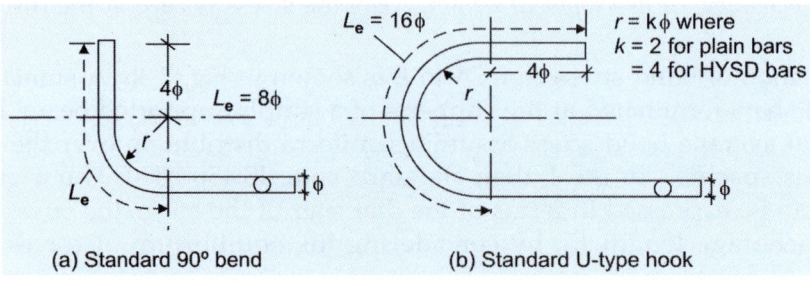

(a) Standard 90° bend   (b) Standard U-type hook

Fig. 7.4: Anchorage lengths of standard hooks and bends (SP:34)

7.5 FACTORS INFLUENCING BOND STRENGTH

The bond strength between steel reinforcements and the surrounding concrete can be enhanced by the following common measures:

1. Use of more smaller diameter bars instead of a few larger diameter bars, thus increasing the surface area of the bars leading to improved bond
2. Use of higher grade concrete resulting in improved tensile strength
3. Providing increased cover to reinforcements improves the bond between steel and concrete
4. Use of deformed bars with ribs and indentations on the surface will increase the bond through frictional resistance
5. Providing hooks and bends and increased length of embedment
6. Avoiding sudden termination of longitudinal reinforcements in tension zones
7. Use of stirrups with increased area of steel/reduced spacing and or higher grade of steel
8. Use of confined concrete around the reinforcements which improves the overall strength of concrete

7.6 CODE REQUIREMENTS FOR BOND

The Indian standard code provides specific provisions to safeguard against bond failures due to flexural or local bond stress and the anchorage or development bond stress. Due to non uniform distribution of actual bond stress and several factors influencing bond strength and despite checks provided by the computations, localised bond failures which occur do not significantly affect the ultimate strength of the member provided the reinforcement are properly anchored at their ends. In addition, the widespread use of deformed bars in place of plain bars, the design emphasis is centred around the anchorage or development bond stress rather than the flexural or local bond stress.

The code prescribes that deformed bars may be used without end anchorage provided development length requirements is satisfied. Hooks are normally provided for plain bars in tension. Bends and hooks shown in Fig. 7.4 should conform to the specifications of IS:2502-1963 and SP:16-1980[10].

7.7 SPLICING OF REINFORCEMENT

When the bars are to be extended beyond their available length as in the case of column bars in multistoreyed buildings, splicing of reinforcement is unavoidable to facilitate the extension of reinforcements in the subsequent floors. It is recommended that splices in flexural members should be kept away from sections with high bending stresses and shear stresses. Also the splices should be staggered in the individual bars of a group. The IS code recommends that splices in flexural members should not be at sections where the bending moment is more than 50% of the moment of resistance, and not more than half the bars shall be spliced at a section. When splicing in such situations becomes unavoidable, special precautions such as increasing lap length and using spirals or closely spaced stirrups around the length of splice, should be adopted.

The various types of splicing of reinforcements are:
- lapping of bars (lap splice)
- staggered splicing
- butt welding of bars
- stirrups at splice locations
- mechanical connections
- lap welding of bars

The different types of splicing of reinforcements are shown in Fig. 7.5.

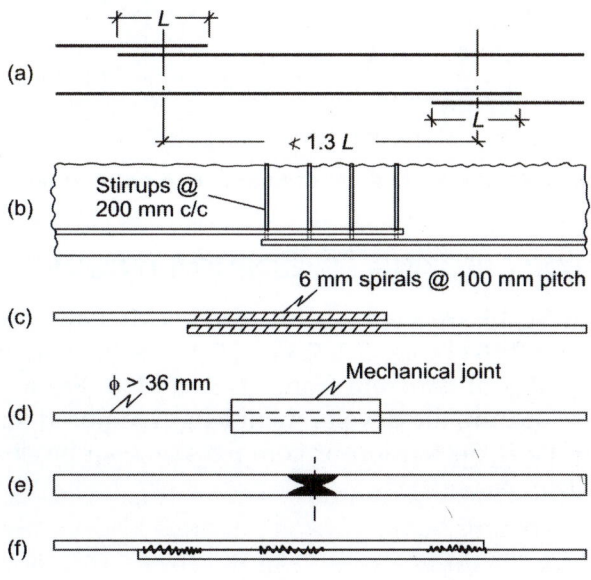

Fig. 7.5: Splicing of reinforcements

In the case of beams, particular care has to be taken in providing sufficient anchorage or development length near supports to limit the magnitude of bond stress. If the bond stresses are excessive, horizontal cracks at the level of reinforcement are formed at such locations. Typical horizontal cracks developed due to the failure of bond between concrete and steel bars in beams are shown in Figs 7.6 and 7.7.

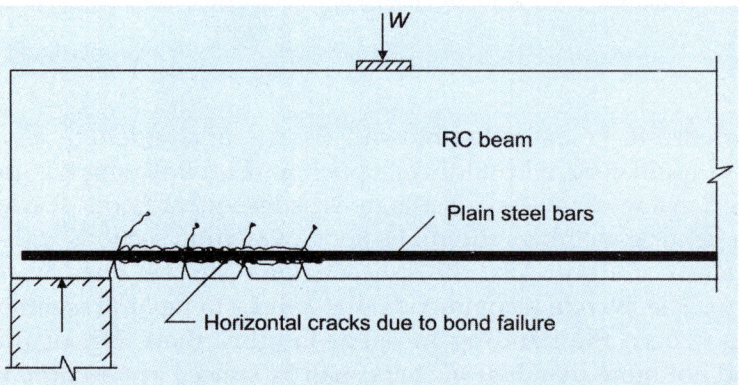

Fig. 7.6: Horizontal cracks due to bond failure

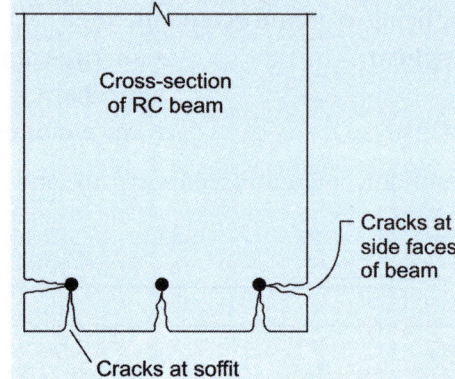

Fig. 7.7: Cracks at soffit of beams due to bond failure

7.8 USE OF SP:16 FOR CHECKING DEVELOPMENT LENGTH

The design tables of SP:16 are very useful for checking the development length. Tables 64, 65 and 66 of SP:16 (Tables 7.2, 7.3 and 7.4 of text) cover plain and deformed bars and different grades of concrete from M15 to M30. For a given bar diameter varying from 6 mm to 36 mm, the development length required for bars fully stressed to design strength of $0.87f_y$ in tension or compression, can be directly readout from the appropriate tables.

Table 7.2: Development length for fully stressed plain bars (Table 64, SP:16)
f_y = 250 N/mm² for bars up to 20 mm diameter
= 240 N/mm² for bars over 20 mm diameter
(Tabulated values are in centimetres)

Bar diameter (mm)	Tension bars Grade of concrete				Compression bars Grade of concrete			
	M15	M20	M25	M30	M15	M20	M25	M30
6	32.6	27.2	23.3	21.8	26.1	21.8	18.6	17.4
8	43.5	36.3	31.1	29.0	34.8	29.0	24.9	23.2
10	54.4	45.3	38.8	36.3	43.5	36.3	31.1	29.0
12	65.3	54.4	46.6	43.5	52.2	43.5	37.3	34.8
16	87.0	72.5	62.1	58.0	69.6	58.0	49.7	46.4
18	97.9	81.6	69.9	65.3	78.3	65.3	55.9	52.2
20	108.8	90.6	77.7	72.5	87.0	72.5	62.1	58.0
22	114.8	95.7	82.0	76.6	91.9	76.6	65.6	61.2
25	130.5	108.8	93.2	87.0	104.4	87.0	74.6	69.6
28	146.2	121.8	104.4	97.4	116.9	97.4	83.5	78.0
32	167.0	139.2	119.3	111.4	133.5	111.4	95.5	89.1
36	187.9	156.6	134.2	125.3	150.3	125.3	107.4	100.2

Note: The development lengths given above are for a stress of $0.87f_y$ in the bars.

Table 7.3: Development length for fully stressed deformed bars (Table 65, SP:16)
f_y = 415 N/mm²
(Tabulated values are in centimetres)

Bar diameter (mm)	Tension bars Grade of concrete				Compression bars Grade of concrete			
	M15	M20	M25	M30	M15	M20	M25	M30
6	33.8	28.2	24.2	22.6	27.1	22.6	19.3	18.1
8	43.1	37.6	32.2	30.1	36.1	30.1	25.8	24.1
10	56.4	47.0	40.3	37.6	45.1	37.6	32.2	30.1
12	67.7	56.4	48.4	45.1	54.2	45.1	38.7	36.1
16	90.3	75.2	64.5	60.2	72.2	60.2	51.6	48.1
18	101.5	84.6	72.5	67.7	81.2	67.7	58.0	54.2
20	112.8	94.0	80.6	75.2	90.3	75.2	64.5	60.2
22	124.1	103.4	88.7	82.7	99.3	82.7	70.9	66.2
25	141.0	117.5	100.7	94.0	112.8	94.0	80.6	75.2
28	158.0	131.6	112.8	105.3	126.4	105.3	90.3	84.2
32	180.5	150.4	128.9	120.3	144.4	120.3	103.2	96.3
36	203.1	169.3	145.0	135.4	162.5	135.4	116.1	108.3

Note: The development lengths given above are for a stress of $0.87f_y$ in the bars.

Table 7.4: Development length for fully stressed deformed bars (Table 66, SP:16)
$f_y = 500$ N/mm^2
(Tabulated values are in centimetres)

Bar diameter (mm)	Tension bars Grade of concrete				Compression bars Grade of concrete			
	M15	M20	M25	M30	M15	M20	M25	M30
6	40.8	34.0	29.1	27.2	32.6	27.2	23.3	21.8
8	54.4	45.3	38.8	36.3	43.5	36.3	31.1	29.0
10	68.0	56.6	48.5	45.3	54.4	45.3	38.8	36.3
12	81.6	68.0	58.3	54.4	65.3	54.4	46.6	43.5
16	108.8	90.6	77.7	72.5	87.0	72.5	62.1	58.0
18	122.3	102.0	87.4	81.6	97.9	81.6	69.9	65.3
20	135.9	113.3	97.1	90.6	108.8	90.6	77.7	72.5
22	149.5	124.6	106.8	99.7	119.6	99.7	85.4	79.8
25	169.9	141.6	121.4	113.3	135.9	113.3	97.1	90.6
28	190.3	158.6	135.9	126.9	152.3	126.9	108.8	101.5
32	217.5	181.3	155.4	145.0	174.0	145.0	124.3	116.0
36	244.7	203.9	174.8	163.1	195.8	163.1	139.8	130.5

Note: The development lengths given above are for a stress of $0.87f_y$ in the bars.

In general,

$$L_d(\text{compression}) = \frac{[L_d(\text{tension})]}{1.25}$$

For any other design stress level less than $0.87f_y$, the development length L'_d required is computed using the relation

$$L'_d = \left(\frac{\sigma_s}{0.87f_y}\right)L_d$$

where, σ_s is design stress in the bars.

Figure 7.4 gives the anchorage value of hooks and bends for tension reinforcement as specified in SP:16-1980. The effect of hooks and bends, if provided can also be considered as development length. However, in bars under compression, only the projected length of hooks or bends is considered as effective towards development length.

7.9 ANALYSIS EXAMPLES

1. A cantilever beam having width of 200 mm and effective depth 300 mm, supports a uniformly distributed load and is reinforced with 4 bars of 16 mm diameter on the tension side. If the factored total load on the cantilever is 80 kN, calculate:
   a. The maximum local bond stress
   b. The anchorage length required
   c. The average bond stress, if the anchorage length provided is 900 mm

   Assume M20 grade concrete and Fe 415 HYSD bars.

Method 1 (using IS:456-2000 code formulae)

i. *Data:*

$b = 200$ mm	$f_{ck} = 20$ N/mm^2
$d = 300$ mm	$f_y = 415$ N/mm^2
$A_{st} = 4$ bars of 16 mm diameter	$\tau_{bd} = (1.6 \times 1.2) = 1.92$ N/mm^2
$\Sigma O = 4(\pi \times 16) = 201$ mm	$j = 0.90$
$L_d = 900$ mm	

ii. *Maximum local bond stress:*

$$\tau_b = \left[\frac{V}{(\Sigma O)jd}\right] = \left[\frac{80 \times 10^3}{201 \times 0.9 \times 300}\right] = 1.47 \text{ N/mm}^2$$

iii. *Anchorage length:*

$$L_d = \left(\frac{\phi \sigma_s}{4\tau_{bd}}\right) = \left(\frac{16 \times 0.87 \times 415}{4 \times 1.92}\right) = 752 \text{ mm}$$

iv. *Average bond stress:*

$$\tau_{bd} = \left(\frac{\phi \sigma_s}{4L_d}\right) = \left(\frac{16 \times 0.87 \times 415}{4 \times 900}\right) = 1.6 \text{ N/mm}^2$$

Method 2 (using SP:16 design tables)

Refer to Table 65 of SP:16 (Table 7.3 of the text) and readout L_d for 16 mm bars (M20 concrete) as

$$L_d = 752 \text{ mm}$$

2. A reinforced concrete beam of 6 m span is uniformly loaded and is reinforced with 5 bars of 20 mm diameter on the tension side at an effective depth of 400 mm. Find the distance from the centre of the beam where one of the bars can be curtailed. Adopt M20 grade concrete and Fe 415 HYSD bars.

i. *Data:*

$L = 6$ m	$f_{ck} = 20$ N/mm^2
$d = 400$ mm	$f_y = 415$ N/mm^2
$\phi = 20$ mm	$\tau_{bd} = (1.6 \times 1.2) = 1.92$ N/mm^2

ii. *Theoretical cutoff point from bending moment considerations:*
Let x be the distance of cutoff point measured from centre of span.

Then $\dfrac{(wL^2/8) - (wx^2/2)}{(wL^2/8)} = \left(\dfrac{4}{5}\right).$

Solving, $x = 1.34$ m $= 1340$ mm.

iii. *Development length for maximum tension at centre in Fe 415 grade steel:*

$$L_d = \left(\frac{\phi \sigma_s}{4\tau_{bd}}\right) = \left(\frac{20(0.87 \times 415)}{4 \times 1.92}\right) = 940 \text{ mm}$$

iv. *Physical cutoff point (PCP):*

$\therefore$ PCP = (TCP + *d*) or 12

            = (1340 + 400) or (12 × 20) whichever is higher

            = 1740 mm

Hence, one bar can be curtailed at 1.74 m from centre of span.

3. A reinforced concrete column of a multistoreyed building is reinforced with 36 mm diameter longitudinal bars and tie at regular intervals. Assuming M25 grade concrete and Fe 415 HYSD bars: (i) calculate the lap length required, and (ii) specify the method of reducing the lap length to reduce the quantity of steel.

    i. *Data*:

       Diameter of bars $\phi = 36$ mm

                        $f_{ck} = 25$ N/mm^2

                        $f_y = 415$ N/mm^2

   ii. *Computation of lap length (L)*:

       L = development length of bars (only projected length of hooks and bends is considered)

       The value of design bond stress specified in limit state method for plain bars in Clause 26.2.1.1, IS:456-2000 are compiled in Table 7.1.

       Average bond stress τ_{bd} in compression (refer to Table 7.1)

             = (τ_{bd} in tension) × 1.25

             = (1.6 × 1.4) × 1.25 = 2.8 N/mm^2

$$L_d = \left(\frac{\phi\sigma_s}{4\tau_{bd}}\right) = \left(\frac{\phi \times 0.87 \times 415}{4 \times 2.8}\right) = 32.2\,\phi = (32.2)36 = 1160 \text{ mm.}$$

       From Table 65, SP:16, readout $L_d = 1161$ mm.

  iii. *Method of reducing lap length*:

       Shorter lap length can be used with welding. Using a lap length of 15ϕ together with lap welding at 5ϕ intervals, the welds are designed to resist the equivalent force (F) for a lap of (32.2 − 15)ϕ = 17ϕ

$$F = (0.87 \times 415)\left(\frac{\pi \times 36^2}{4}\right)\left(\frac{17.2}{32.2}\right) \text{N} = 196306\,\text{N} = 196.305 \text{ kN}$$

4. A simply supported beam of 8 m span is reinforced with 6 bars of 25 mm diameter at centre of span and 50% of the bars are continued into the supports. Check the length at supports. Check the development length at supports assuming M20 grade concrete and Fe 415 HYSD bars. The beam supports a characteristic total load of 50 kN/m.

    i. *Data*:

       $L = 8$ m $f_{ck} = 20$ N/mm^2

       $w = 50$ kN/m $f_y = 415$ N/mm^2

       $\phi = 25$ mm $\tau_{bd} = (1.6 \times 1.2) = 1.92$ N/mm^2

No. of bars at centre of span = 6

No. of bars at supports = 3

ii. *Bending moment and shear force:*

Design load $w_u = (1.5 \times 50) = 75$ kN/m

$$M_{max} = (0.125w_uL^2) = (0.125 \times 75 \times 8^2) = 600 \text{ kN·m}$$
$$V_{max} = (0.5w_uL) = (0.5 \times 75 \times 8) = 300 \text{ kN.}$$

iii. *Moment of resistance of bars continued into supports:*

$$M_1 = \frac{3}{6}(600) = 300 \text{ kN·m.}$$

iv. *Development length of 25 mm diameter bars:*

Use M20 grade concrete and Fe 415 grade HYSD bars.

$$L_d = \left(\frac{\phi\sigma_s}{4\tau_{bd}}\right) = \left(\frac{25 \times 0.87 \times 415}{4 \times 1.92}\right) = 1175 \text{ mm}$$

From Table 65 (SP:16), readout $L_d = 1175$ mm.

v. *Check for development length at supports:*

According to Clause 26.2.3.3 (IS:456) assuming 30% increase in development length computed as (M_1/V), we have

$$\left(\frac{1.3M_1}{V}\right) = \left(\frac{1.3 \times 300 \times 10^6}{300 \times 10^3}\right) = 1300 \text{ mm}$$

Condition to be satisfied is given by

$$\left[L_0 + \left(\frac{1.3M_1}{V}\right)\right] > L_d \text{, where } L_0 = \text{anchorage beyond support line}$$

$$[L_0 + 1300] > 1175$$

Hence, development length is satisfied without any anchorage value.

7.10 DESIGN EXAMPLES

1. A reinforced concrete section cantilever beam of rectangular section 300 mm wide by 600 mm deep is built into a column 500 mm wide as shown in Fig. 7.8. The cantilever beam is subjected to a hogging moment of 200 kN·m at the function of beam and column. Design suitable reinforcements in the beam and check for the required anchorage length. Adopt M20 grade concrete and Fe 415 HYSD bars (Fig. 7.8).

Method 1 (using IS:456 code formulae)

i. *Data:*

$b = 300$ mm $f_{ck} = 20$ N/mm^2

$d = 550$ mm $f_y = 415$ N/mm^2

$M_u = 200$ kN·m

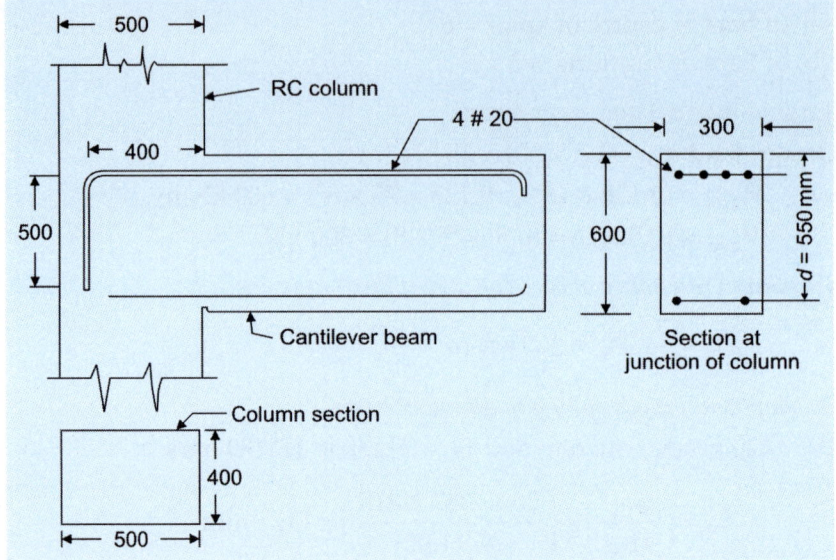

Fig. 7.8: Cantilever beam

ii. *Limiting moment of resistance*

$$M_{u,lim} = 0.138 f_{ck} bd^2 = (0.138 \times 20 \times 300 \times 550^2)$$
$$= (250 \times 10^6) \text{ N·mm}$$
$$= 250 \text{ kN·m}$$

Since $M_u < M_{u,lim}$, section is under-reinforced.

iii. *Reinforcements*:

$$M_u = (0.87 f_y A_{st} d)\left[1 - \left(\frac{A_{st} f_y}{bd f_{ck}}\right)\right]$$

$$(200 \times 10^6) = (0.87 \times 415 \times A_{st} \times 550)\left[1 - \left(\frac{415 A_{st}}{300 \times 550 \times 20}\right)\right]$$

Solving, $A_{st} = 1181 \text{ mm}^2$

∴ Provide 4 bars of 20 mm diameter ($A_{st} = 1256 \text{ mm}^2$).

iv. *Anchorage length*:

$$L_d = \left(\frac{0.87 f_y}{4\tau_{bd}}\right)\phi$$

Using 20 mm diameter bars, $\phi = 20$ mm.

For bars in tension, $\tau_{bd} = (1.6 \times 1.2) = 1.92 \text{ N/mm}^2$ and M20 grade concrete.

∴ $$L_d = \left[\frac{0.87 \times 415}{4 \times 1.92}\right]\phi = 47\phi = (47 \times 20) = 940 \text{ mm}$$

The bars are extended into the column to a length of 400 mm with a 90°
bend and 500 mm length as shown in Fig. 7.8.

Anchorage length provided = [400 + (8 × 20) + 500] = 1060 mm > 940 mm.

Method 2 (using SP:16 design tables)

Refer to Table 2 (SP:16) and readout the percentage steel corresponding to the
ratios

$$\left(\frac{M_u}{bd^2}\right) = \left(\frac{200 \times 10^6}{300 \times 550^2}\right) = 2.2$$

Hence, $p_t = \left(\frac{100 A_{st}}{bd}\right) = 0.717$

∴ $A_{st} = \left[\dfrac{0.717 \times 300 \times 550}{100}\right] = 1183 \text{ mm}^2$ (adopt 4 bars of 20ϕ)

From Table 65 (SP:16) for f_y = 415 N/mm² and 20 mm diameter, the required
anchorage length is L_d = 94 cm = 940 mm. Hence, the bars are built into the
column and bent at 90° for the required total anchorage length.

2. A reinforced column subjected to compression combined with flexure is shown
   in Fig. 7.9. It is required to reduce the longitudinal reinforcement diameter
   from 25 mm in the ground floor to 20 mm in the first floor. Design a suitable
   lap splice. Assume M20 grade concrete and Fe 415 HYSD bars.

   i. *Data*:

   Diameter of bars: 25ϕ at GF and 20ϕ at FF

   f_y = 415 N/mm² f_{ck} = 20 N/mm²

   τ_{bd} = (1.2 × 1.25) = 1.5 N/mm²

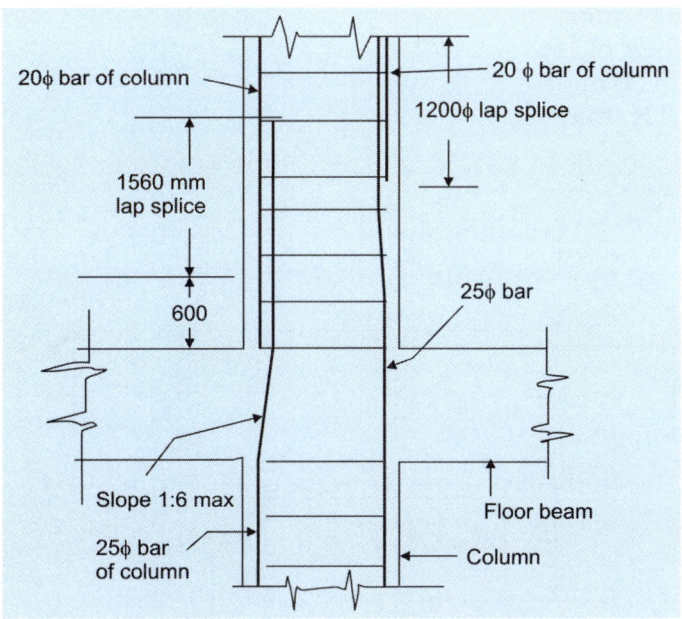

Fig. 7.9: Typical lap splice

ii. *Lap length*:

At location of splice, smaller diameter bars (20 mm diameter) are adequate in providing the desired strength. The lap length is based on smaller diameter bars.

$L = L_d$ or 30ϕ whichever is greater

$$L_d = \left[\frac{0.87 f_y}{4\tau_{bd}}\right]\phi = \left[\frac{0.87 \times 415}{4 \times 1.5}\right]\phi = 60\phi$$

$\therefore$ $L = L_d = (60 \times 20) = 1200$ mm.

iii. *Staggered splicing*:

According to IS:456 code (Clause 26.2.5.1), the splicing of bars should be ideally staggered with minimum centre-to-centre spacing of

$1.3L = (1.3 \times 1200) = 1560$ mm

The lap length and staggering of spacing is shown in detail in Fig. 7.9.

3. A doubly reinforced beam having width 300 mm and overall depth 500 mm is built into a column having width 600 mm as shown in Fig. 7.10. The section of the beam at supports is reinforced with 3 bars of 16 mm diameter to resist the hoging moment and 2 bars of 12 mm diameter on the compression side. Using $f_{ck} = 20$ N/mm² and $f_y = 415$ N/mm², design and detail the anchorage length required at the junction of column and beam.

   i. *Data*:

   $b = 300$ mm
   $d = 500$ mm
   $A_{sc} = 2$ bars of 12ϕ
   $f_{ck} = 20$ N/mm²
   $A_{st} = 3$ bars of 16ϕ
   $f_y = 415$ N/mm²
   $\tau_{bd} = 1.2$ N/mm²

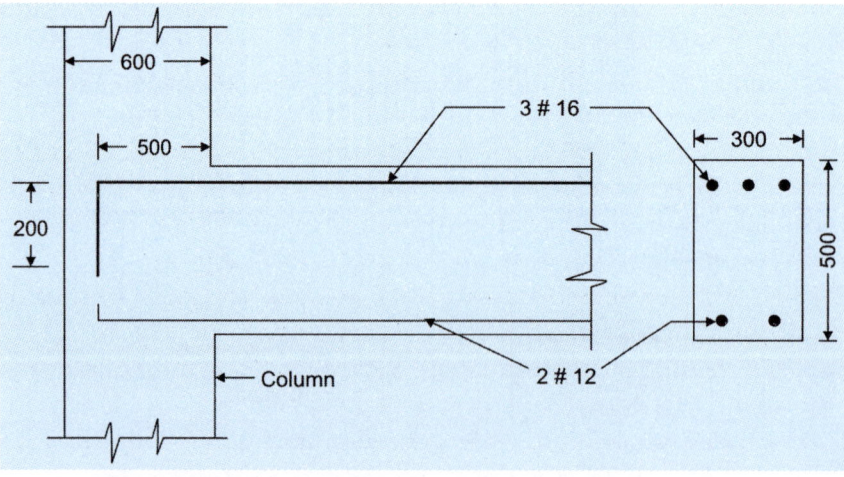

Fig. 7.10: Anchorage details in doubly reinforced beam.

ii. *Anchorage length for tension bars (top)*:

$$L_d = \left[\frac{0.87 f_y}{4\tau_{bd}}\right]\phi = \left[\frac{0.87 \times 415}{4 \times 1.2 \times 1.6}\right]\phi = 47\phi = (47 \times 16) = 752 \text{ mm}$$

Anchorage length provided = $[500 + (8 \times 16) + 200] = 764$ mm.

iii. *Anchorage length for compression bars (soffit)*:

τ_{bd} can be increased by 25%

∴ $L_d = (47\phi) \, 0.8 = 37.6\phi = (37.6 \times 12) = 451$ mm

Provide an anchorage length $L_d = 500$ mm as shown in Fig. 7.10.

4. A reinforced concrete column of a multistorey building, which is rectangular section 300 mm by 600 mm is reinforced with 6 bars of 28 mm diameter in the first floor and 6 bars of 25 mm diameter in the second floor. Adopting M25 grade concrete and Fe 415 HYSD bars, design suitable lap splice at the junction of the floors.

Columns in multistorey buildings are subjected to compression and bending and some of the bars may be under tension. Hence, the lap length should be calculated by taking into account flexural tension. In such cases, lap length $(L) = L_d$ or 30ϕ, whichever is greater.

In this example, $L_d = \left[\dfrac{0.87 \times 415 \times \phi}{4\tau_{bd}}\right] = \left[\dfrac{0.87 \times 415 \times \phi}{4 \times 1.4 \times 1.6}\right] = 40\phi > 30\phi$

Hence, the lap length = $(40 \times 25) = 1000$ mm.

REFERENCES

1. Turneaure F, Cyclopedia of Civil Engineering, American Technical School, 1908.
2. Morsch E and Goodrich EP, *Concrete and Steel Construction*, 3 edn, 1909–1910.
3. Faber O and Bowie PG, *Reinforced Concrete—Theory and Practice*, Vol 1 & 2, London, 1920.
4. Krishna Raju N, *Prestressed Concrete*, 5 edn, McGraw-Hill Education (India), New Delhi, 2002, pp 90–123.
5. Kemp EL and Wilhelm WJ, Investigation of the parameters influencing bond cracking, *Journal of the American Concrete Institute*, Vol 76, January 1979, pp 47–71.
6. Jimenez R, White RN and Gergely P, Bond and dowel capacities of reinforced concrete, *Journal of American Concrete Institute*, Vol 76, January 1979, pp 73–92.
7. ACI:318M-11, Building code requirements for structural concrete, *American Concrete Institute*, Farmington Hills, Michigan, 2005, p 443.
8. BS EN:1992-1-1, Euro Code-2, Design of Concrete Structures, General Rules and Rules for Buildings, British Standard Institution, London, 2004.
9. IS:456-2000, Indian Standard Code of Practice for Plain and Reinforced Concrete (4th Revision), BIS, New Delhi, July 2000, pp 1–100.
10. SP:16-1980, Design Aids for Reinforced Concrete, IS:456, BIS, New Delhi, 1980.

ASSIGNMENT

1. A 200 mm cantilever beam of rectangular section, 250 mm wide × 500 mm overall depth is reinforced with 2 bars of 25 mm diameter on the tension side and 2 bars of 12 mm diameter on the compression face. The length of the cantilever beam is 2.25 m. The total anchorage length provided for the tension and compression bars are 1200 mm and 700 mm. The beam has to support a total uniformly distributed ultimate load of 120 kN. Assuming M20 grade concrete and Fe 415 HYSD bars, check the adequacy of the anchorage provided for the longitudinal bars.

2. A cantilever beam having width 200 mm and effective depth 400 mm supports a uniformly distributed load and is reinforced with 4 bars of 16 mm diameter on the tension side. If the factored total load on the cantilever is 100 kN, calculate:
   a. the maximum local bond stress
   b. the anchorage length required
   c. The average bond stress, if the anchorage length provided is 1000 mm
   Assume M20 grade concrete and Fe 415 HYSD bars.

3. A reinforced concrete beam of 5 m span is uniformly loaded and reinforced with 4 bars of 25 mm diameter on the tension side at an effective depth of 400 mm. Find the distance from the centre of the beam where one of the bars can be curtailed. Adopt M30 grade concrete and Fe 415 HYSD bars.

4. A reinforced concrete column of a multistoreyed building is reinforced with 25 mm diameter longitudinal bars and tie at regular intervals. Assuming M20 grade concrete and Fe 415 HYSD bars (a) calculate the lap length required, and (b) specify the method of reducing the lap length to reduce the quantity of steel.

5. A simply supported beam of 6 m span is reinforced with 6 bars of 28 mm diameter at centre of span and 50% of the bars are continued into the supports. Check the length at supports. Check the development length at supports assuming M30 grade concrete and Fe 500 HYSD bars. The beam supports a characteristic total load of 80 kN/m.

6. A reinforced concrete column of a multistorey building, which is rectangular section 400 mm × 800 mm is reinforced with 8 bars of 36 mm diameter in the first floor and 6 bars of 25 mm diameter in the second floor. Adopting M20 grade concrete and Fe 415 HYSD bars, design suitable lap splice at the junction of the floors.

7. A tied reinforced concrete column (600 mm by 600 mm) of a multistorey building has sixteen 28 mm diameter bars arranged symmetrically with a clear cover of 70 mm tied with lateral ties of 12 mm at 300 mm centres. If Fe 500 HYSD bars are used, calculate the compression lap length required as per Indian standard code specifications.

8. A reinforced concrete beam of rectangular section 300 mm wide by 700 mm overall depth supports a factored load of 80 kN/m over a simply supported span of 6 m. The effective depth of the beam is 650 mm and the beam is reinforced with six 20 mm bars at the centre of span. If only 4 bars are continued into the support, check the development length at the support assuming M20 grade concrete and Fe 415 HYSD bars.

9. A cantilever beam of reinforced concrete is rectangular in section with 300 mm width and overall depth 600 mm. The beam is designed to support a factored moment of 360 kN·m at the critical support section. The section was designed as doubly reinforced and the area of steel required in the tension and compression sides are 2230 mm² and 272 mm² respectively. Five bars of 25 mm diameter at the tension side and 2 bars of 16 mm diameter on the compression side were provided. Calculate the actual anchorage length required for the tension and compression bars.

10. A reinforced concrete beam of rectangular section with width 300 mm and overall depth 700 mm was designed to resist a factored moment and shear force of 78 kN·m and 234 kN respectively. The beam was provided with 6 bars of 20 mm at the centre of span. If only 4 bars are continued into the support, check the development length at the support assuming M20 grade concrete and Fe 415 HYSD bars according to the Indian standard code specifications.

REVIEW QUESTIONS

1. What are the various mechanisms by which bond is achieved between concrete and steel in reinforced concrete structural elements?
2. What is development length in reinforced concrete members? Explain the Indian standard code clause pertaining to the development length.
3. What are the factors influencing the bond strength? What are the various measures used to enhance the bond strength between steel and concrete?
4. What are the devices that can be used to obtain the required development length if it is not achievable with straight bars?
5. What are the code specifications for the anchorage value of bends used to enhance the development length in reinforced concrete structural elements?
6. Explain clearly the code specifications for the use of splices in compression members.
7. What is the length of bars that are usually available in the market? When does a steel bar require splicing? Briefly explain the significance of splicing.
8. What is the anchorage value of a standard U-hook according to the specifications of the Indian standard code?
9. Write a brief note on: (a) Curtailment of reinforcements in flexural members (b) Mechanical splices.
10. Where do you locate the splices in flexural members? Mention the IS code specifications regarding the location of splices in reinforced concrete continuous beams.

OBJECTIVE QUESTIONS

1. The bond between reinforcement and surrounding concrete can be improved significantly by using
   a. large diameter bars
   b. bars with ribs or indentations on the surface
   c. plain bars

2. The anchorage bond stress is maximum at
   a. the end of the bar
   b. the centre of span
   c. a distance equal to the development length from the end
3. The bond developed due to shearing resistance or dilatancy is due to
   a. the chemical adhesion
   b. mechanical interlock due to ribs or indentations on bar
   c. frictional resistance
4. The design bond stress increase with the
   a. moment
   b. shear force
   c. grade of concrete
5. The development length required to develop the design bond stress is
   a. directly proportional to the design bond stress
   b. directly proportional to the characteristic strength of steel
   c. directly proportional to the grade of concrete
6. For bars in compression, the value of bond stress in tension is increased by
   a. 50%
   b. 10%
   c. 25%
7. The anchorage value of a standard U-hook is
   a. 16 times the diameter of the bar
   b. 8 times the diameter of the bar
   c. 4 times the diameter of the bar for each bend, which should not exceed 16 times the diameter of the bar
8. Lap splices should not be used for bars larger than
   a. 25 mm
   b. 16 mm
   c. 36 mm
9. When bars of different diameter are to be spliced, the lap length should be calculated on the basis of diameter of the
   a. larger bar
   b. average diameters of the two bars
   c. smaller bar
10. Splices are generally provided in reinforcing bars at location of
    a. maximum shear force
    b. maximum bending moment
    c. far away from sections of maximum stress

Serviceability Limit States

8.1 INTRODUCTION

The importance of the concept of serviceability limit states was recognised by several research investigators[1,2] during the middle of 20th century when reinforced concrete structural elements designed solely by the ultimate load method exhibited excessive local deflections at service loads causing damage to partitions and distress to the users of the structure. It was realised that the behaviour of the structure at service loads is as important as that of its ultimate strength. Many leading codes like the European Concrete Committee[3], the American Concrete Institute[4], the British[5], Canadian[6], German[7] and the Indian standard code[8] introduced the concept of limit state philosophy[9] in which both the serviceability and strength criterion were included in the design of reinforced concrete structural elements.

The primary serviceability limit states comprise the following criteria:

1. The member should not undergo excessive deformation under service loads. This limit state is generally referred to as the *limit state of deflection*.
2. The width of cracks developed on the surface of reinforced concrete members under service loads should be limited to the values prescribed in the codes of practice. This limit state is referred to as *limit state of cracking*.

Other limit states such as fatigue, durability and vibration, should also be considered depending upon the environmental conditions and type of structure. These limit states are also important for structures like bridges located in marine environment.

IS:456-2000 code has specified the partial safety factors for load combinations under which the deflection and cracking are to be checked as detailed in Chapter 3.

Table 18, IS:456-2000 code (Table 3.1 of text) outlines the combinations of loads for serviceability conditions. The largest value should be used for the computations. The load combinations are as follows:

1. 1.0 DL + 1.0 LL
2. 1.0 DL + 1.0 WL
3. 1.0 DL + 0.8 LL + 0.8 WL (EL)

Generally the codes specify the following two methods for control of deflection:

1. The empirical method in which the span/effective depth ratios of the structural members are limited to specified values in the codes.
2. The theoretical method in which the actual deflection is computed and checked with the codified permissible deflections.

For control of crack widths also, two methods are specified:

1. The empirical method which requires the detailing of reinforcements[10] according to the codified provisions, such as minimum percentage of steel in the section, spacing of bars, curtailment and anchorage bars, lapping bars, etc.
2. The theoretical method of computing the actual width of cracks and checking them with the codified requirements for the specified environmental conditions.

The widespread use of high grade steels like Fe 415 and Fe 500 with higher allowable stresses replacing the Fe 250 grade during the last few decades coupled with the use of high performance concrete generally results in slender structural elements necessitating greater attention to deflections and crack control in the modern methods of design of reinforced concrete structures.

8.2 IS:456 CODIFIED DEFLECTION LIMITS

Deflections of flexural members like beams and slabs, if excessive, causes distress to users of the structure and also likely to cause cracking of partitions. As given in IS:456 code, Clause 23.2, the accepted limits to permissible deflections are given below:

1. The final deflection including the effects of all loads, temperature, creep and shrinkage of horizontal structural members should not exceed the value of span/250.
2. The deflection including the effects of temperature, creep and shrinkage occurring after the erection of partitions and the application of finishes should not exceed span/350 or 20 mm whichever is less.

8.2.1 Basic Span/Depth Ratios

The following basic span/depth ratios (Table 8.1) are specified in the code for control of deflection limit state in structural concrete members.

Table 8.1: Basic span to effective depth ratios for beams and slabs
(Clause 23.2.1, IS:456-2000)

Type of support	Rectangular sections	Flanged sections
Cantilever	7	Multiply values for rectangular sections by factor K_f shown in Fig. 5.3
Simply supported	20	
Continuous	26	

8.2.2 Modification Factors for Basic Span/Depth Ratios

The basic span/depth ratios have to be modified to account for the percentage of tension reinforcement and compression reinforcement in the section and the type of section, i.e. rectangular or flanged.

The modified expression for the span/effective depth ratio is specified as

$$\left(\frac{L}{d}\right) = \left(\frac{L}{d}\right)_{\text{basic}} \times K_t \times K_c \times K_f$$

where; $\left(\dfrac{L}{d}\right)_{\text{basic}}$ values are shown in Table 8.1.

K_t = modification factor for tension steel (refer to Fig. 8.1)

For flanged sections, the percentage tension steel is based on $b_f d$.

K_c = modification factor for compression steel (refer to Fig. 8.2)

K_f = modification factor or reduction factor for flanged sections (refer to Fig. 8.3)

For spans exceeding 10 m, the (L/d) basic values have to be multiplied by (10/span), except for cantilevers where deflection calculations are required. In the case of flanged sections, the use of flange width b_f in computations yields anomalous results. Hence, it is better to consider the width of web b_w in place of b_f in calculations which yields conservative results.

For preliminary proportioning, the thickness of slabs using Fe 415 HYSD bars, it is preferable to use 0.4% of tension reinforcement (p_t) which corresponds to a value of K_t = 1.25 resulting in the span/depth ratio of 25.

In case of beams supporting heavy loads, the span/depth ratio of 10:12 is recommended from practical considerations. In case of two way slabs of spans below

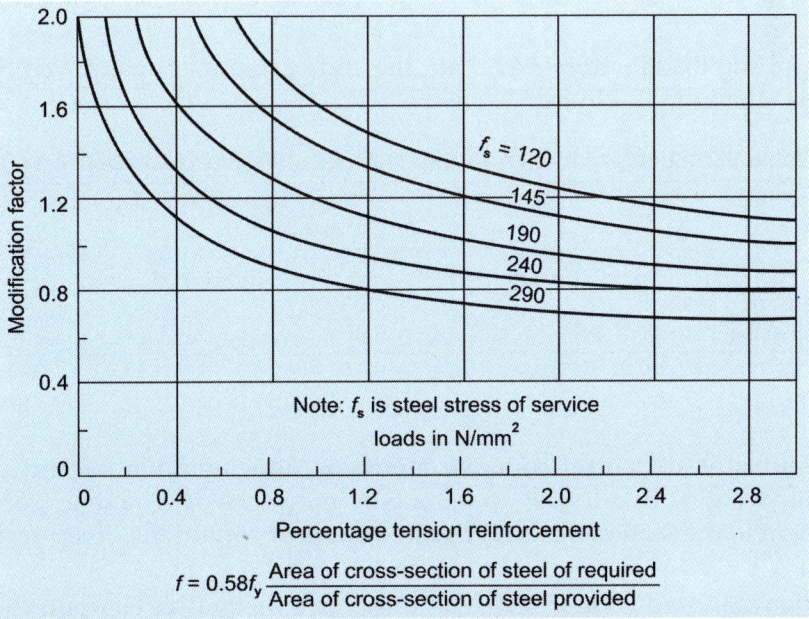

Fig. 8.1: Modification factor for tension reinforcement [K_t] (Fig. 4, IS:456-2000)

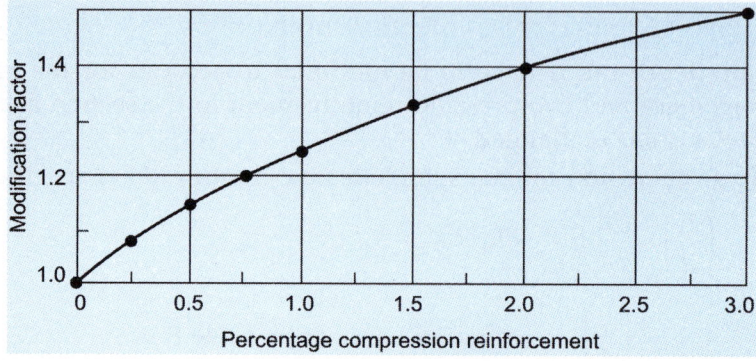

Fig. 8.2: Modification factor for compression reinforcement $[K_c]$ (Fig. 5, IS:456-2000)

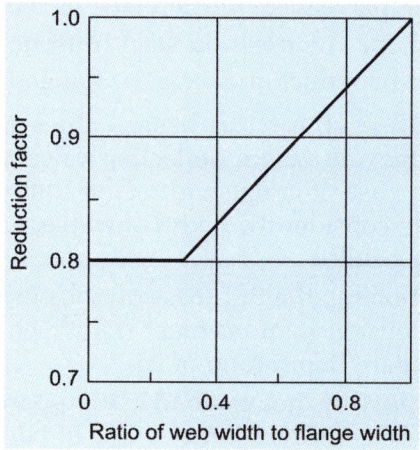

Fig. 8.3: Reduction factor for flanged beams $[K_f]$ (Fig.6, IS:456-2000)

3.5 m supporting loads within 3 kN/m², the code prescribes span/overall depth (L/D) ratios as shown in Table 8.2.

Table 8.2: Span/depth ratios for two way slabs (IS:456-2000, Clause 24.1)		
Support conditions	*Span/overall depth ratio*	
	Fe 250 grade steel bars	*Fe 415 HYSD bars*
Simply supported slabs	35	28
Continuous slabs	40	32

8.3 CALCULATION OF DEFLECTIONS (THEORETICAL METHOD)

The deflections of structural concrete members depend upon several parameters such as self-weight, live loads, span, elastic modulus of concrete, percentage of reinforcement in the section, flexural rigidity, support conditions, creep and shrinkage of concrete.

The Indian standard code IS:456-2000 presents a method of computing short term and long term deflections in Annexure-C of the code.

1. *Short term deflection*: The short term or instantaneous deflection is calculated using the elastic theory using the short term modulus of elasticity of concrete (E_c) and the effective second moment of area of the cracked concrete section (I_{eff}) expressed as

$$I_{eff} = \cfrac{I_r}{1.2 - \left(\dfrac{M_r}{M}\right)\left(\dfrac{Z}{d}\right)\left[1 - \left(\dfrac{x}{d}\right)\right]\left(\dfrac{b_w}{b}\right)}$$

but $I_r \le I_{eff} \le I_{gr}$

where,

I_r = second moment of area of the cracked section

I_{gr} = second moment of area of the gross cross-section about centroidal axis neglecting the reinforcement

M_r = cracking moment = $(f_{cr}I_{gr})/y_t$

f_{cr} = modulus of rupture (flexural strength) of concrete = $0.8\sqrt{f_{ck}}$

y_t = distance of extreme fiber in tension from centroidal axis of gross section neglecting reinforcement

M = maximum service load moment

Z = lever arm

x = depth of neutral axis

d = effective depth

b_w = breadth of web

b = breadth of compression face

The short term deflection is computed as

$$a_{i\ (perm)} = K_w \left(\frac{WL^3}{E_c\, I_{eff}}\right)$$

where,

K_w = constant depending upon the type of load and support conditions

$E_c = 5000\sqrt{f_{ck}}$

W = total load on the beam

L = span of the beam

2. *Deflection due to shrinkage*: Shrinkage deflection (a_{cs}) is expressed as

$$a_{cs} = K_3\psi_{cs}L^2$$

where,

K_3 = a constant depending upon the support conditions

= 0.5 for cantilevers

= 0.125 for simply supported members

= 0.086 for members continuous at one end

= 0.063 for fully continuous members

ψ_{cs} = shrinkage curvature = $K_4(\varepsilon_{cs}/D)$

where,

ε_{cs} = ultimate shrinkage strain of concrete (Clause 6.2.4.1, IS:456-2000)

D = overall depth of section

$$K_4 = 0.72 \left[\frac{P_t - P_c}{\sqrt{P_t}} \right] \le 1.0 \text{ for } 0.25 \le (P_t - P_c) < 1.0$$

$$= 0.65 \left[\frac{P_t - P_c}{\sqrt{P_t}} \right] \le 1.0 \text{ for } (P_t - P_c) \ge 1.0$$

where,

$P_t = (100 A_{st}/bd)$ and $P_c = (100 A_{sc}/bd)$

and L = span of the member.

3. *Deflection due to creep*: The creep deflection due to permanent load ($a_{cc(perm)}$) is expressed as

$$a_{cc(perm)} = [a_{icc(perm)} - a_{i(perm)}]$$

where,

$a_{icc(perm)}$ = initial plus creep deflection due to permanent loads obtained using an elastic analysis with an effective modulus of elasticity $E_{ce} = [E_c/(1 + \theta)$ and θ = creep coefficient

$a_{i(perm)}$ = short term deflection due to permanent load using E_c

4. *Long term deflection*: Hence, long term deflection (a_{Ld}) is evaluated using the relation

$$a_{Ld} = [\text{short term deflection}] + [\text{shrinkage deflection}] + [\text{creep deflection}]$$

$$= [a_{i(perm)} + a_{cs} + a_{cc(perm)}]$$

8.4 CRACKING IN STRUCTURAL CONCRETE

8.4.1 Reasons for Cracks

Cracks in structural concrete members may develop due to flexural tensile stress, shear and diagonal tension, differential shrinkage, creep, thermal and aggressive environmental conditions and also due to bond and anchorage failures.

The Indian standard code IS:456-2000, Clause 35.3.2 specifies that cracks in concrete should not adversely affect the appearance or durability of the structure.

8.4.2 Permissible Crack Width

The IS:456-2000 code prescribes the following limiting crack widths in structural concrete members depending upon the environmental exposure conditions as shown in Table 8.3.

Table 8.3: Limiting width of cracks (IS:456-2000, Clause 35.3.2)

Sl. No.	Type of exposure	Permissible width of cracks at surface
1.	Protected and not exposed to aggressive environmental conditions	0.3 mm
2.	Moderate environmental conditions	0.2 mm
3.	Aggressive environmental conditions	0.1 mm

8.4.3 Control of Cracking by Empirical Method

Proper detailing of reinforcements in structural concrete members results in cracks well within the permissible limits. The IS code specifies maximum and minimum spacing of reinforcing bars for control of cracking of reinforced concrete members. Clauses 26.3.2 and 26.3.3 of IS:456-2000 specifies the horizontal and vertical spacing of bars. The maximum spacing in the tension zone is a function of stress level in the steel and redistribution of moments to and from that section. Table 15, IS:456 code specifies the percentage redistribution as a function of the clear distance between bars for different grades of steel used in structural concrete members.

In case of slabs, the horizontal distance between parallel main reinforcement should not exceed three times the effective depth of the slab or 300 mm whichever is smaller. For distribution reinforcement, the horizontal distance is limited to 5 times the effective depth or 450 mm whichever is smaller (refer to Fig. 1.1 for details).

In case of beams of overall depth exceeding 750 mm, side face reinforcements of area not less than 0.1% of web area should be distributed equally on the side faces with a spacing not exceeding 300 mm or web thickness whichever is smaller.

The minimum tension reinforcement in beams to prevent failure in the tension zone by cracking of concrete is prescribed as

$$A_s = \left(\frac{0.85\,bd}{f_y} \right)$$

However, for slabs the minimum reinforcement is specified as 0.15% for Fe 250 grade steel and 0.12% for Fe 415 HYSD bars based on shrinkage and temperature effects rather than the strength considerations.

8.4.4 Crack Width Computations

In case of special structures and aggressive environmental conditions, it is preferred to compute the width of cracks and compare them with the permissible crack width to ensure the safety of the structure at the limit state of serviceability. The Indian standard code IS:456-2000 has specified an analytical method for the estimation of design surface crack width in Annexure-F which is based on the British code (BS:8110) specifications. The design surface crack width is expressed as

$$W_{cr} = \left[\frac{3\,a_{cr}\varepsilon_m}{1 + 2\left\{ \dfrac{a_{cr} - C_{min}}{h - x} \right\}} \right]$$

where,

a_{cr} = distance from the point considered to the surface of the nearest longitudinal bar (Fig. 8.4)

$a_{cr} = [(0.5S)^2 + C_{min}^2]^{1/2}$

s = spacing between bars

C_{min} = minimum cover to the longitudinal bars

x = depth of neutral axis

h = overall depth of the member

m = average strain at the level of steel where cracking is being considered, calculated by allowing for the stiffening effect of concrete in the tension zone and obtained from the equation

$$\varepsilon_m = \varepsilon_1 - \left\{ \frac{b(h-x)(a-x)}{3 E_s A_s (d-x)} \right\}$$

where,

ε_1 = strain at the level considered considering a cracked section

b = width of section at the centroid of tension steel

a = distance from the compression face to the point at which crack width is being calculated

d = effective depth

E_s = modulus of elasticity of steel (N/mm²)

A_s = area of tension reinforcement

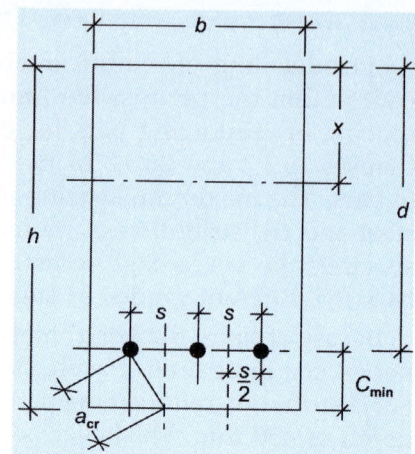

Fig. 8.4: Parameters for crack width computations

If the crack width is measured at the soffit of the beam, then $a = h$ and hence

$$\varepsilon_1 = \left(\frac{f_s}{E_s} \right) \left(\frac{h-x}{d-x} \right)$$

and

$$\varepsilon_m = \frac{1}{E_s} \left(\frac{h-x}{d-x} \right) \left[f_s - \frac{b(h-x)}{3 A_s} \right]$$

where, f_s = stress at the centroid of the tension reinforcement expressed in N/mm².

Generally, the maximum width of cracks are observed at the soffit of the beams and at sections of maximum moment and at point midway at soffit between the reinforcements and at corners.

8.5 ANALYSIS EXAMPLES

1. A simply supported beam of rectangular section spans over 10 m and has an effective depth of 700 mm. The beam is reinforced with 1% reinforcement on the tension side. Check for the deflection control of the beam by empirical method if:

   a. Fe 415 grade HYSD bars are used

   b. Fe 500 grade bars are used

   i. *Data:*

$L = 10$ m	$f_y = 415$ N/mm²
$d = 700$ mm $= 0.7$ m	$f_y = 500$ N/mm²
$A_{st} = 1\%$	

   ii. *Actual span/depth ratio*
   $$(L/d) = (10/0.7) = 14.28$$

iii. *Allowable span/depth ratio*:

$$(L/d) = (L/d)_{basic} \times K_t \times K_c \times K_f$$

Since $A_{sc} = 0, K_c = 1$

The beam is rectangular in section and hence $K_f = 1$.

From Table 8.1, $(L/d)_{basic} = 20$

From Fig. 8.1, $K_t = 1.0$ for Fe 415 steel bars

$\qquad\qquad\qquad = 0.85$ for Fe 450 steel bars

Case (a): $(L/d)_{max} = [20 \times 1.0 \times 1 \times 1] = 20 > 14.28$

Case (b): $(L/d)_{max} = [20 \times 0.85 \times 1 \times 1] = 17 > 14.28$

Since $(L/d)_{actual}$ is less than $(L/d)_{max}$, the deflection control is satisfactory.

2. A simply supported beam of the effective span 6 m is rectangular in section 250 mm wide with an effective depth of 500 mm. The beam is reinforced with 4 bars of 20 mm diameter on the tensile face and 2 bars of 16 mm diameter at the compression face. The effective cover is 50 mm. Using Fe 415 HYSD bars, check the limit state of deflection using the empirical method.

   i. *Data*:

   $L = 25\,m$ $\qquad\qquad\qquad A_{st} = (4 \times 314) = 1256\,mm^2$

   $d = 500\,mm$ $\qquad\qquad\quad A_{sc} = (2 \times 201) = 402\,mm^2$

   Fe 415 HYSD bars

   ii. *Actual span/depth ratio*:

   $$(L/d) = (6000/500) = 12$$

   iii. *Allowable or maximum span/depth ratio*:

   $$(L/d)_{max} = (L/d)_{basic} \times K_t \times K_c \times K_f$$

   From Table 8.1, $(L/d)_{basic} = 20$.

   iv. *Modification factors*:

   $$p_t = (100A_{st}/bd) = (100 \times 1256)/(250 \times 500) = 1.00\%$$
   $$p_c = (100A_{sc}/bd) = (100 \times 402)/(250 \times 500) = 0.32\%$$

   From Fig. 8.1, $K_t = 0.9$

   $\qquad$ Fig. 8.2, $K_c = 1.1$

   $\qquad$ Fig. 8.3, $K_f = 1$

   $\therefore (L/d)_{max} = (20 \times 0.9 \times 1.1 \times 1) = 19.8 > 12$

   Hence, the beam is safe with regard to serviceability limit state of deflection.

3. A T-beam continuous over 8 m spans has a flange width 900 mm and web width 300 mm. The effective depth is 400 mm, area of tension reinforcement is 1600 mm² and area of compression reinforcement = 900 mm². Adopting Fe 415 HYSD bars, estimate the safety of the beam for deflection control using the empirical method.

i. *Data*:

$L = 8$ m (continuous) $A_{sc} = 900$ mm^2

$b_f = 900$ mm $\dfrac{b_w}{b_f} = \dfrac{300}{900} = 0.33$

$b_w = 300$ mm

$A_{st} = 1600$ mm^2

ii. *Actual span/depth ratio*:

$$p_t = \left(\frac{100\,A_{st}}{b_f\,d}\right) = \left(\frac{100 \times 1600}{900 \times 400}\right) = 0.44\%$$

$$p_c = \left(\frac{100\,A_{st}}{b_f\,d}\right) = \left(\frac{100 \times 900}{900 \times 400}\right) = 0.25\%$$

iii. *Actual span/depth ratio*:

$(L/d)_{actual} = (8000/400) = 20$

iv. *Modification factors*:

From Fig. 8.1, $K_t = 1.30$

Fig. 8.2, $K_c = 1.08$

Fig. 8.3, $K_f = 0.8$

v. *Maximum permissible span/depth ratio*:

$(L/d)_{max} = [(L/d)_{basic} \times K_t \times K_c \times K_f]$

$= [(26 \times 1.30 \times 1.08 \times 0.8)]$

$= 29.2 > 20$

Hence, the T-beam is safe with regard to limit state of deflection.

4. A simply supported beam of rectangular section 250 mm wide by 450 mm overall depth is used over an effective span of 4 m. The beam is reinforced with 3 bars of 20 mm diameter Fe 415 HYSD bars at an effective depth of 400 mm. Two hanger bars of 10 mm diameter are provided. The self-weight of the beam together with the dead load on the beam is 4 kN/m. The service live load is 10 kN/m. Using M20 grade concrete, compute:

a. the short term deflection

b. the long term deflection according to the provisions of the Indian standard code IS:456-2000.

i. *Data*:

$b = 250$ m $f_{ck} = 20$ N/mm^2

$D = 450$ mm $f_y = 415$ N/mm^2

$d = 400$ mm $f_{cr} = 0.7\sqrt{f_{ck}}$

$L = 4$ m $= 0.7\sqrt{20} = 3.13$ N/mm^2

$g = 4$ kN/m $E_c = 5000\sqrt{f_{ck}}$

$q = 10$ kN/m $= 5000\sqrt{20}$

$A_{st} = (3 \times 314) = 942$ mm^2 $= 22360$ N/mm^2

$A_{sc} = (2 \times 79) = 158$ mm^2 $m = 13$

ii. *Second moment of area of cracked and gross cross-section*:

Let x be the depth of neutral axis, then

$$0.5bx^2 = m\,A_{st}(d-x)$$
$$0.5 \times 250x^2 = 13 \times 942\,(400-x)$$

Solving, $x = 155$ mm

Distance of centroid of steel from neutral axis is

$$r = (d-x) = (400-155) = 245 \text{ mm}$$

Second moment of area of cracked section is

$$I_r = [(bx^3/3) + mA_{st}r^2]$$

$$= \left(\frac{250 \times 155^3}{3}\right) + (13 \times 942 \times 245^2)$$

$$= 10.45 \times 10^8 \text{ mm}^4$$

Gross second moment of area is computed as

$$I_{gr} = \left(\frac{bD^3}{12}\right) = \left(\frac{250 \times 450^3}{12}\right) = 18.98 \times 10^8 \text{ mm}^4$$

iii. *Service load and cracking moment*:

$$M = 0.125\,(g+q)L^2$$
$$= 0.125\,(4+10)\,4^2$$
$$= 28 \text{ kN·m}$$
$$= 28 \times 10^6 \text{ N·mm}$$

$$M_r = \left(\frac{f_{cr}I_{gr}}{y_t}\right) = \left(\frac{3.13 \times 18.98 \times 10^8}{0.5 \times 450}\right) = 26 \times 10^6 \text{ N·mm}$$

Lever arm $z = \left(d - \dfrac{x}{3}\right) = \left(400 - \dfrac{155}{3}\right) = 348.34$ mm

iv. *Effective second moment of area* (I_{eff}):

$$I_{eff} = \left[\frac{I_r}{1.2 - \left(\dfrac{M_r}{M}\right)\left(\dfrac{z}{d}\right)\left(1-\dfrac{x}{d}\right)\left(\dfrac{b_w}{b}\right)}\right]$$

$$= \left[\frac{(10.45 \times 10^8)}{1.2 - \left(\dfrac{26 \times 10^6}{28 \times 10^6}\right)\left(\dfrac{348.34}{400}\right)\left(1-\dfrac{155}{400}\right)1}\right]$$

$$= 14.93 \times 10^8 \text{ mm}^4$$

∴ $I_r < I_{eff} < I_{gr}$

$$(10.45 \times 10^8) < (14.93 \times 10^8) < (18.98 \times 10^8)$$

v. *Maximum short term deflection*

$$a_{i(perm)} = \left[\frac{5(g+q)L^4}{384\,E_c\,I_{eff}} \right]$$

$$= \left[\frac{5 \times 14 \times (4000)^4}{384 \times 22360 \times 14.93 \times 10^8} \right] = 1.39 \text{ mm}$$

vi. *Long term deflection*:

$$a_{Ld} = [a_{i(perm)} + a_{cs} + a_{cc\,(perm)}]$$

Shrinkage deflection:

$$a_{cs} = k_3 \psi_{cs} L^2$$
$$= 29.2 > 20$$

where, k_3 = a constant = 0.125 for simply supported beam

ψ_{cs} = shrinkage curvature = $k_4(\varepsilon_{cs}/D)$

ε_{cs} = ultimate shrinkage strain of concrete = 0.0003

$$k_4 = \left\{ \frac{0.72(p_t - p_c)}{\sqrt{p_t}} \right\} \le 1.0 \text{ for } 0.25) \le (p_t - p_c) < 1.0$$

$$p_t = \left(\frac{100 \times 942}{250 \times 400} \right) = 0.942$$

$$p_c = \left(\frac{100 \times 158}{250 \times 400} \right) = 0.158$$

$$\therefore (p_t - p_c) = 0.942 - 0.158$$
$$= 0.784 \ge 0.25 \text{ and} < 1.00$$

$$k_4 = \left(\frac{0.72 \times 0.784}{0.97} \right) = 0.58 < 1.0$$

$$\psi_{cs} = k_4 \left(\frac{\varepsilon_{cs}}{D} \right) = 0.58 \left(\frac{0.0003}{400} \right) = 4.35 \times 10^{-7}$$
$$a_{csd} = k_3 \psi_{cs} L^2$$
$$= (0.125 \times 4.35 \times 10^{-7} \times 4000^2)$$
$$= 0.87 \text{ mm}$$

Creep deflection:

$$a_{cc(perm)} = [a_{i,\,cc(perm)} - a_{i(perm)}]$$

where,

$a_{i,\,cc(perm)}$ = initial + creep deflection due to permanent loads obtained by using the effective loads and by using the effective modulus of elasticity as $E_{ce} = [E_c/(1 + \theta)]$

where, θ = creep coefficient = 1.6 (for 28 days loading)

$$a_{1,\,cc(perm)} = \left[\frac{5(g+q)L^4}{384\,E_{ce}\,I_{eff}} \right]$$

$$E_{ce} = \left(\frac{E_c}{1+1.6}\right) = \left(\frac{E_c}{2.6}\right)$$

$\therefore a_{1,cc(perm)}$ = 2.6 (short term deflection)

$\qquad\qquad$ = (2.6 × 1.39)

$\qquad\qquad$ = 3.614

Hence,

$\qquad a_{cc(perm)} = a_{1,cc(perm)} - a_{1(perm)}$

$\qquad\qquad$ = (3.614 − 1.39)

$\qquad\qquad$ = 2.224 mm

Total long term deflection is given by

$\qquad a_{Ld}$ = (short term + (shrinkage + (creep
$\qquad\qquad$ deflection) deflection) deflection)

$\qquad\qquad = a_{1(perm)} + a_{cs} + a_{cc(perm)}$

$\qquad\qquad$ = 1.39 + 0.87 + 2.224

$\qquad\qquad$ = 4.484 mm

According to IS:456-2000 code, maximum permissible long term deflection should not exceed (span/250).

Permissible deflection = (span/250)

$\qquad\qquad\qquad\qquad$ = (4000/250)

$\qquad\qquad\qquad\qquad$ = 16 mm

Actual deflection $\qquad$ = 4.484 mm

Hence, the beam is safe with regard to the limit state of deflection.

5. A simply supported beam of rectangular section spanning over 6 m has a width 300 mm and overall depth 600 mm. The beam is reinforced with 4 bars of 25 mm diameter on the tension side at an effective depth of 550 mm spaced at 50 mm centres. The beam is subjected to a working load moment of 160 kN·m at the centre of span section. Using M25 grade concrete and Fe 415 HYSD bars, check the beam for the serviceability limit state of cracking according to IS:456-2000 code method.

   i. *Data*:

$b = 300$ mm	$f_{ck} = 25$ N/mm^2
$h = 600$ mm	$f_y = 415$ N/mm^2
$d = 550$ mm	$m = 11$
$M = 160$ kN·m	$s = 50$ mm
$A_{st} = 1963$ mm^2	$E_s = 2 \times 10^5$ N/mm^2

   ii. *Neutral axis depth*:

   Let x be the depth of neutral axis.

   Then we have $0.5bx^2 = m A_{st}(d - x)$

   $\qquad (0.5 \times 300x^2) = 11 \times 1963 (550 - x)$

   Solving, $x = 220$ mm

iii. *Second moment of area* (I_r):
$$I_r = (bx^3/3) + (mA_{st}r^2)$$
where, $r = (d - x) = (550 - 220) = 330$ mm

$$I_r = \left(\frac{300 \times 220^3}{3}\right) + (11 \times 1963 \times 330^2)$$
$$= 34.1 \times 10^8 \text{ mm}^4$$

iv. *Maximum width of cracks*:

Referring to Fig. 8.5,
$$\text{Cover} = C_{min} = (50 - 12.5) = 37.5$$
$$a_{cr} = [(0.5S)^2 + C_{min}^2)]^{1/2}$$
$$= [(0.5 \times 50)^2 + 37.5^2]^{1/2}$$
$$= 45$$

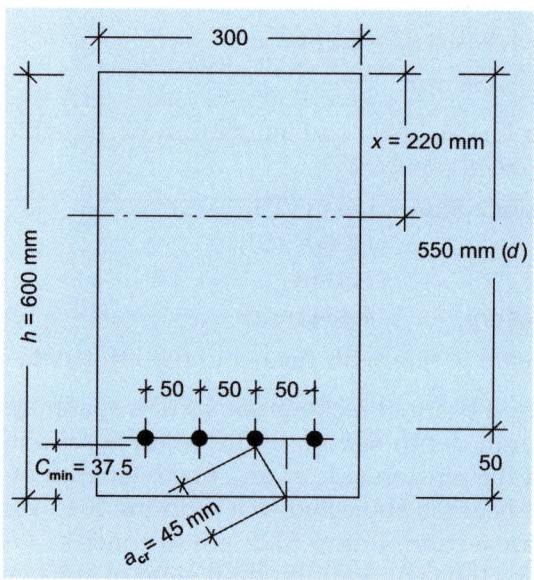

Fig. 8.5: Crack width computations

Crack width will be maximum at the soffit of the beam. Distance of centroid of steel from neutral axis is
$$r = (d - x) = (550 - 220) = 330 \text{ mm}$$

∴ $$\varepsilon_1 = \left(\frac{f_s}{E_s}\right)\left[\frac{h - x}{d - x}\right]$$

where, $$f_s = m\left[\frac{My}{I_r}\right] = 11\left[\frac{160 \times 10^6 \times 330}{34.1 \times 10^8}\right] = 170 \text{ N/mm}^2$$

∴ $$\varepsilon_1 = \left(\frac{170}{2 \times 10^5}\right)\left[\frac{600 - 220}{550 - 220}\right] = 9.78 \times 10^{-4}$$

$$\varepsilon_m = \varepsilon_1 - \left[\frac{b_t(h-x)(a'-x)}{3E_s A_s (d-x)} \right] \text{ and } a' = h$$

$$= (9.78 \times 10^{-4}) - \left[\frac{300(600-220)(600-220)}{3 \times 2 \times 10^5 \times 1963(550-220)} \right]$$

$$= 8.67 \times 10^{-4}$$

Maximum width of cracks is expressed as

$$W_{cr} = \left[\frac{3 a_{cr}\, \varepsilon_m}{1 + 2 \left\{ \dfrac{a_{cr} - C_{min}}{h - x} \right\}} \right]$$

$$W_{cr} = \left[\frac{3 \times 45 \times 8.67 \times 10^{-4}}{1 + 2 \left\{ \dfrac{45 - 37.5}{600 - 220} \right\}} \right] = 0.113 \text{ mm}$$

v. *Comparison with code limits*:

Under normal environmental conditions,

maximum crack width > 0.3 mm

Under aggressive environment,

$$W_{cr} \not> (0.004 \times C_{min}) \not> (0.004 \times 37.5) \not> 0.15 \text{ mm}$$

Since $W_{cr} = 0.113$ mm < 0.15 mm, the serviceability limit state of cracking is satisfied.

6. A doubly reinforced concrete beam of rectangular section with an overall depth 400 mm is reinforced with 3 bars of 28 mm on the tension side at an effective depth of 348 mm and with 3 bars of 20 mm diameter on the compression face at an effective cover of 48 mm. The beam has been designed to support a total service load moment of 124 kN·m. If M25 grade concrete and Fe 415 HYSD bars are used, estimate the maximum possible crack width and check the safety of the beam for the serviceability limit state of cracking according to the Indian standard code specifications. Assume spacing of tension bars as 73 mm.

i. *Data*:

$D = 400$ mm $A_{st} = 1848$ mm^2

$b = 250$ mm $A_{sc} = 942$ mm^2

$d = 348$ mm $f_{ck} = 25$ N/mm^2

$d' = 48$ mm $f_y = 415$ N/mm^2

$C_{min} = 38$ mm $M = 124$ kN·m

$s = 73$ mm $d_c = 52$ mm

ii. *Depth of neutral axis*:

Let x be the depth of neutral axis.

Modular ratio $m = [280/(3 \times 8.5)] \simeq 11$

Equating the first moment of the areas on either side of the neutral axis

$$0.5b(x^2) + (1.5m - 1)\, A_{sc}(x - d') = mA_{st}(d - x)$$

Substituting the numerical values and solving, $x = 146$ mm.

iii. *Second moment of area of cracked transformed section (I_{cr}):*

$$I_{cr} = \frac{bx^3}{3} + mA_{st}(d - x)^2 + [1.5m - 1](x - d')^2$$

Substituting the numerical values, the value of I_{cr} is computed as

$$I_{cr} = (1229 \times 10^6)\ \text{mm}^4$$

iv. *Tensile stress in steel at working loads*

$$f_{st} = \left\{ \frac{11 \times (124 \times 10^6) \times (348 - 146)}{(1229 \times 10^6)} \right\} = 224\ \text{N/mm}^2$$

v. *Calculation of crack width using IS code formula:*

For the corner bar, side face effective cover = $[250 - 73 - 73]0.5 = 52$ mm

a_{cr} = distance of corner of beam from the surface of corner bar

$$a_{cr} = \sqrt{\left(\frac{s}{2}\right)^2 + d_c^2} - \frac{(d_b)}{2}$$ Substituting the numerical values, $a_{cr} = 60$ mm

Strain in tension steel $\varepsilon_{st} = \left[\dfrac{D - x}{E_s(d - x)} \right]\left[f_{st} - \dfrac{b(D - x)}{3A_{st}} \right]$

Substituting the numerical values, $\varepsilon_{st} = 1.336 \times 10^{-3}$

The maximum probable crack width is given by the relation,

$$W_{cr} = \left[\frac{3a_{cr}\varepsilon_m}{1 + 2\left\{ \dfrac{a_{cr} - C_{min}}{h - x} \right\}} \right]$$

Substituting the numerical values, $W_{cr} = 0.204$ mm

Under normal environmental conditions,

Maximum permissible crack width = 0.3 mm

$$W_{cr} = 0.204\ \text{mm} < 0.3\ \text{mm}$$

The serviceability limit state of cracking is satisfied.

REFERENCES

1. Rowe RE, Cranston WB and Best BC, New concepts in the design of concrete, *Structural Engineer*, Vol. 43, 1965, pp 339-403.

2. Bate SCC, Why limit state design? *Concrete*, March 1968, pp 103–108.

3. CEB recommendations for International Code of Practice for Reinforced Concrete, American Concrete Institute and Cement & Concrete Association, London, 1964.

4. ACI:318-M11, Building code requirements for structural concrete, American Concrete Institute, Farmington Hills, Michigan, 2005.

5. BS EN:1992-1-12004, Design of concrete structures, General Rules & Rules for Buildings, British Standards Institution, 2004.

6. CSA Standard A23.3-94, Design of concrete structures, Canadian Standards Association, Rexdale, Ontario, 1994.

7. DIN:1045-1988, Structural use of concrete, design & construction, Din Deutsches Institute Fir Normung EV, 1988.

8. IS:456-2000, Indian Standard Code of Practice for Plain and Reinforced Concrete (4th Revision), BIS, 2000, p 100.

9. Krishna Raju N, Limit state design for structural concrete, Proc Institution of *Engineers* (*India*), January 1971, pp 138–143.

10. SP:34, Handbook of Concrete Reinforcement and Detailing, BIS, New Delhi, 1987.

ASSIGNMENT

1. A rectangular reinforced concrete beam 250 mm wide by 650 mm overall depth is reinforced with 1.5% reinforcement on the tension side at an effective cover of 50 mm. Check the adequacy of the span/depth ratio of the beam for deflection control using the empirical method for the following two cases:

   a. Fe 415 grade HYSD bars are used

   b. Fe 500 grade HYSD bars are used.

2. A simply supported beam of effective span 6 m is rectangular in section 300 mm wide by 650 mm deep. The beam is reinforced with 4 bars of 22 mm diameter on the tension face and 2 bars of 16 mm diameter on the compression side. The effective cover is 50 mm. Using Fe 500 HYSD bars, check the adequacy of the span/depth ratio of the beam for deflection control using the empirical method.

3. A T-beam, continuous over 12 m span, has a flange width and thickness of 1200 mm and 150 mm respectively. The rib is 300 mm wide. The beam is reinforced with 4 bars of 20 mm diameter at an effective depth of 800 mm. Adopting Fe 415 HYSD bars, check for the serviceability limit state of deflection using IS:456 code empirical method.

4. A simply supported reinforced concrete beam of rectangular section 300 mm wide by 450 mm overall depth is used over an effective span of 5 m. The beam is reinforced with 3 bars of 20 mm diameter Fe 415 HYSD grade steel at an effective depth of 400 mm. Two hanger bars of 12 mm diameter are provided. The self weight of the beam together with the dead load is 4 kN/m. Service live load is 10 kN/m. Using M20 grade concrete, estimate the short and long term deflection according to the provisions of the Indian standard code.

5. A reinforced concrete bridge deck slab has an overall depth of 500 m with an effective span of 6.0 m. The slab is reinforced with 25 mm diameter HYSD bars spaced at 150 mm centres at an effective depth of 450 mm. M25 grade concrete with Fe 415 grade HYSD bars are used to cast the slab. The maximum working load moment in the slab is 175 kN·m/m. Check for the limit state of cracking according to the Indian standard code recommendations.

6. A simply supported beam of rectangular section 300 mm wide by 700 mm overall depth has an effective span of 8 m. The beam is reinforced with 4 bars of 25 mm diameter spaced 50 mm apart on tension side at an effective depth of 650 mm. Two nominal hanger bars of 12 mm diameter are provided on the compression side at a cover of 50 mm. The beam is subjected to a service load moment of 140 kN·m at centre of span section. Using M25 grade concrete and Fe 415 HYSD bars, check the beam for the limit states of deflection and cracking using the following methods:
   a. deflection control using empirical method
   b. deflection computations using the theoretical method
   c. maximum width of cracks using the theoretical method

7. A reinforced concrete bridge girder spanning over 16 m and designed to sustain IRC has an overall depth of 1600 mm. The maximum design load working moment in the beam is computed as 2120 kN·m. The width of the beam is 300 mm. The beam is reinforced with 12 bars of 32 mm diameter at an effective depth of 1450 mm. If M25 grade concrete and Fe 415 grade HYSD bars are used, check the beam for the limit state of cracking.

8. A cantilever beam of 5 m span is rectangular in section with width 350 mm and overall depth 750 mm. The beam has been designed to support a working load bending moment of 200 kN·m at support of which 50% is due to dead loads. The loading may be assumed as uniformly distributed over the span. Adopting M25 grade concrete and Fe 500 HYSD bars, compute the following:
   a. the maximum short term elastic deflection
   b. the short term deflection due to only live loads
   c. compliance of deflection to the IS code specifications.

9. A simply supported reinforced concrete slab of effective span 4 m has an overall depth of 200 mm. The service load bending moment at the centre of span is computed as 20 kN·m/m. The characteristic strength of concrete and steel are 25 N/mm² and 415 N/mm² respectively. The slab is reinforced with 0.38% reinforcement at an effective cover of 35 mm. Using IS:456 code specifications, compute the maximum short term deflection under service loads.

10. A simply supported reinforced concrete beam of rectangular section 200 mm wide by 550 mm overall depth is used to resist a total service load of 12 kN/m over an effective span of 4 m. The beam is reinforced with 4 bars of 20 mm diameter at an effective cover of 50 mm on the tension side. Adopting M20 grade concrete and Fe 415 HYSD bars, estimate the maximum short term and long term deflections according to the provisions of the Indian standard code.

REVIEW QUESTIONS

1. Explain the importance of serviceability limit states in limit state design. What happens if serviceability limit states are ignored in the design of reinforced concrete structures?

2. What are the inadequacies in the elastic and ultimate load methods of design which lead to the development of limit state design?

3. How do you check the limit state of deflection using the empirical method specified in the Indian standard code?

4. Bring out the difference between short term and long term deflections of RC structural members? Specify the significance of limiting long term deflections.

5. What are the basic span/depth ratios? Specify the Indian code recommendations of these ratios for different types of structural elements.

6. What are modification factors used for basic span/depth ratios in the estimation of deflections of RC flexural members? Briefly outline their importance in the computation of deflections.

7. Explain the theoretical method of computing the short and long term deflections of RC flexural members.

8. In what way shrinkage and creep affect the deflections of reinforced concrete members? Mention the Indian code recommendations for computing the deflections due to shrinkage and creep.

9. How do you control cracking in reinforced concrete members by using the empirical method suggested in the IS code?

10. Explain briefly the Indian standard code method of estimating the design crack width in reinforced concrete flexural members.

OBJECTIVE QUESTIONS

1. According to the Indian standard code, checking the serviceability limit state in a RC beam includes the computation of
   a. maximum bending moment
   b. ultimate flexural strength
   c. maximum deflection under service loads

2. The span/depth ratio of a reinforced concrete continuous slab reinforced with HYSD bars recommended by the IS code is
   a. 40
   b. 28
   c. 32

3. Maximum permissible final deflection of a reinforced concrete beam including all the effects of loads, creep and shrinkage should not exceed
   a. span/350
   b. span/250
   c. span/480

4. Cracks in reinforced concrete flexural members develop due to
   a. compressive stresses
   b. repeated loads
   c. tensile stresses

5. In a cracked reinforced concrete beam, the short term deflection depends upon
   a. second moment of area of cross-section
   b. effective second moment of area of the section
   c. second moment of area of the cracked section

6. The deflection of a reinforced concrete beam due to shrinkage depends upon
    a. loads on the beam
    b. shrinkage strain
    c. creep coefficient
7. Crack control is possible in reinforced concrete beams by
    a. providing large cover
    b. using large diameter bars
    c. proper detailing of reinforcements
8. The deflection due to creep in a reinforced concrete beam is influenced by
    a. the transient loads on the member
    b. the cross-section of the member
    c. the creep coefficient
9. The maximum permissible crack width in a reinforced concrete member subjected to aggressive environmental conditions should not exceed
    a. 0.2 mm
    b. 0.1 mm
    c. 0.3 mm
10. The computation of crack width in a reinforced concrete beam depends upon
    a. the grade of concrete
    b. minimum cover to the longitudinal bars
    c. modulus of elasticity of steel reinforcement

Limit State Design of Beams

9.1 INTRODUCTION

An economical and satisfactory design of a reinforced concrete beam element requires a comprehensive knowledge of the behaviour of the beam element under different types of loading. According to Bennett[1], design is essentially a creative approach rather than a routine analytical activity in which the structural behaviour is only one of a number of functional, constructional, aesthetic and economic considerations. A successful design should not only satisfy the criterion of safety against total collapse[2,3] of the structures due to various causes, but must ensure that the serviceability[4,5] of the structural element is not impaired while resisting the normal service loads.

Rowe et al[6] have defined the purpose of design as 'the provision of a structure complying with the client's requirements'. The structural engineer is often required to assist the client in defining the requirements more precisely. Once the basic requirements have been defined, the structural engineer becomes a member of an integrated design team. The primary object of structural design is to obtain an optimum structural solution[7] which can result in the greatest overall economy by providing the maximum assistance in satisfying all other requirements of the structure. The general recommendations of the European Concrete Committee[7] are also used in the economical and safe design of reinforced concrete structural elements.

According to Mosley et al[8], the design of a reinforced concrete beam comprises primarily selecting the dimensions of the member which will adequately resist the factored bending moments and shear forces. At the same time, the selected members should satisfy the limit state of serviceability comprising the permissible deflections and cracking at service loads. It is difficult to isolate the two criteria and hence the steps involved are inter-related and broadly condensed into three basic design stages:

1. Preliminary analysis comprising the selection of size of beam element
2. Detailed analysis and design of reinforcements
3. Checking for strength and serviceability requirements.

Limit state design of beams should conform to the specifications of the Indian standard code IS:456-2000[9] and the detailing of reinforcements in the various types of beams should conform to the specifications of the Indian standard special publication SP:34-1987[10].

9.2 DIMENSIONING OF FLEXURAL MEMBERS

The cross-sectional dimensions of reinforced concrete beams are selected based on the following guidelines:

1. The effective and overall depth of the beam is estimated from span/depth ratios to satisfy the limit state of serviceability. Overall depth of width should be in the range of 1.5 to 2.

2. The width of the section should accommodate the required number of bars with sufficient spacing between them with minimum side covers of 20 mm to the links.

3. The depth of the beam should be such that the percentage of steel required is around 75% of that required for balanced section.

4. The minimum number of bars on tension face should be not less than two and not more than six in any one layer.

5. In flanged beams, the depth of the slab is generally taken as 20% of the overall depth.

6. Common widths of beams are 150, 200, 230, 250 and 300 mm. Also the width of the beam should be equal to or less than the dimension of the column supporting the beam. Table 9.1 shows the trial section span/depth ratios to be assumed as a function of span and loading.

Table 9.1: Span/depth ratios for trial section

Sl. No.	Span range	Loading	Span/depth ratio (L/d)
1	3 to 4 m	Light	15 to 20
2	5 to 10 m	Medium to heavy	12 to 15
3	5 to 10 m	Heavy	10 to 12

9.3 DESIGN OF SINGLY REINFORCED BEAMS

1. Design a singly reinforced concrete beam to suit the following data:

   i. *Data:*
      Clear span = 4 m
      Width of supports = 300 mm
      Service load = 5 kN/m
      Materials: M20 grade concrete
                      Fe 415 HYSD bars

   ii. *Stresses:*
      $f_{ck} = 20 \text{ N/mm}^2$
      $f_y = 415 \text{ N/mm}^2$
      Load factor = 1.5 for dead and live loads

iii. *Cross-sectional dimensions*:

Effective depth = (span/20) = (4000/20) = 200 mm

Adopt $d = 200$ mm

$D = 250$ mm

$b = 200$ mm

Effective span = clear span + effective depth

$= (4 + 0.2) = 4.2$ m

Centre of supports = $(4 + 0.3) = 4.3$ m

Hence $L = 4.2$ m

iv. *Loads*:

Self weight g = (0.2 × 0.25 × 25)   = 1.25 kN/m

Live load q = 5.00 kN/m

Total load w = 6.25 kN/m

Design ultimate load w_u = (1.5 × 6.25) = 9.375 kN/m

v. *Ultimate moments and shear forces*:

$$M_u = (0.125\ w_u\ L^2) = (0.125 \times 9.375 \times 4.2^2) = 20.67 \text{ kN·m}$$
$$V_u = (0.5\ w_u\ L) = (0.5 \times 9.375 \times 4.2) = 19.68 \text{ kN}$$

vi. *Tension reinforcements*:

$$M_{u,lim} = 0.138 f_{ck}\ b\ d^2$$
$$= (0.138 \times 20 \times 200 \times 200^2)\ 10^{-6} \text{ kN·m}$$
$$= 22.08 \text{ kN·m}$$

Since $M_u < M_{u,lim}$, section is under-reinforced

$$M_u = (0.87\ f_y\ A_{st}\ d) \left[1 - \frac{A_{st} f_y}{(b d f_{ck})} \right]$$

$$M_u = (0.87 \times 415 A_{st} \times 200) \left[1 - \frac{415 A_{st}}{(200 \times 200 \times 20)} \right]$$

Solving $A_{st} = 350$ mm^2

Provide 3 bars of 16 mm diameter ($A_{st} = 402$ mm^2) and 2 hanger bars of 10 mm diameter.

vii. *Check for shear stress*:

$$\tau_v = (V_u/bd) = (19.68 \times 10^3)/(200 \times 200) = 0.49 \text{ N/mm}^2$$
$$p_t = (100\ A_{st})/(bd) = (100 \times 402)/(200 \times 200) = 1.005$$

Refer to Table 19, IS:456-2000 code and readout the permissible shear stress as

$$\tau_c = 0.63 \text{ N/mm}^2 > \tau_v$$

Provide nominal shear reinforcements using 6 mm diameter two-legged stirrups at a spacing of

$$S_v = \left(\frac{A_{st}\ 0.87\ f_y}{0.4 b} \right) = \left(\frac{2 \times 28 \times 0.87 \times 250}{0.4 \times 200} \right) = 152 \text{ mm}$$

But $s_v > 0.75\,d(> 0.75 \times 200) = 150$ mm

Adopt spacing of stirrups as 150 mm centres.

viii. *Check for deflection control*:

$$p_t = 1.005$$

Refer to Fig. 5.1 and readout the modification factor $K_t = 1.05$ and neglecting the hanger bars, we have

$$(L/d)_{max} = (L/d)_{basic} \times K_t \times K_c \times K_f$$
$$= (20 \times 1.05 \times 1 \times 1)$$
$$= 21$$

$$(L/d)_{actual} = (4200/200) = 21$$

Since

$(L/d)_{actual} = (L/d)_{max},$ deflection control is satisfactory.

ix. *Reinforcement details*:

The details of reinforcement in the beam is shown in Fig. 9.1.

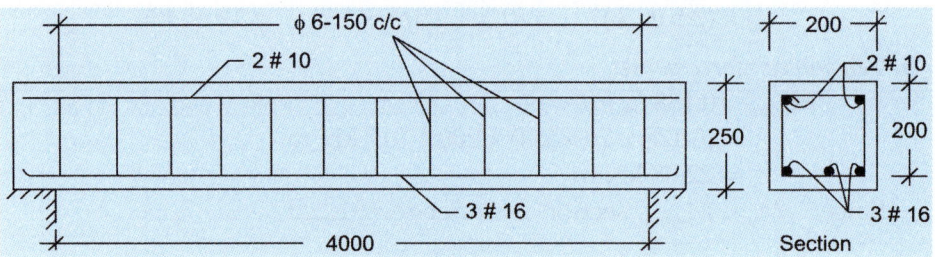

Fig. 9.1: Reinforcement details in singly reinforced beam.

x. *Design using SP-16 design tables*:

$$(M_u/d\,d^2) = (20.67 \times 10^6)/(200 \times 200^2) = 2.58$$

Refer to Table 2 of SP-16 and readout $p_t = 1.005$
$$A_{st} = (p_t bd)/100 = (1.005 \times 200 \times 200)/100 = 350 \text{ mm}^2$$

Hence A_{st} is the same as that computed using theoretical equations.

2. A reinforced concrete beam is to be designed over an effective span of 5 m to support a design service load of 8 kN/m. Adopt M20 grade concrete and Fe 415 HYSD bars and design the beam to satisfy the collapse and serviceability limit states.

i. *Data*:

$L = 5$ m	$f_{ck} = 20 \text{ N/mm}^2$
$q = 8$ kN/m	$f_y = 415 \text{ N/mm}^2$

ii. *Cross-sectional dimensions*:

Refer to Table 6.1 and adopt span/depth ratio of 15 for the given span and range of loading.

$$d = (\text{span}/15) = (5000/15) = 333 \text{ mm}$$

Adopt $d = 350\,mm$
　　　　$D = 400\,mm$
　　　　$b = 200\,mm$

iii. *Loads*:

Self weight of beam　$g = (0.2 \times 0.4 \times 25) = 2.00\,kN/m$
Live load　　　　　q　　　　　　　$= 8.00\,kN/m$
Total working load　w　　　　　　$= 10.00\,kN/m$
Design ultimate load $w_u = (1.5 \times 10)$　$= 10.00\,kN/m$

iv. *Ultimate moments and shear forces*:

$$M_u = (0.125\,w_u L^2) = (0.125 \times 15 \times 5^2) = 46.8\,kN{\cdot}m$$
$$V_u = (0.5\,w_u L)\quad = (0.5 \times 15 \times 5)\quad = 37.5\,kN$$

v. *Reinforcements*:

$$M_{u,lim} = 0.138\,f_{ck}bd^2$$
$$= (0.138 \times 20 \times 200 \times 350^2)\,10^{-6}\,kN{\cdot}m$$
$$= 68\,kN{\cdot}m$$

Since　$M_u < M_{u,lim}$, section is under-reinforced

$$M_u = 0.87\,f_y A_{st}d\left[1 - \left(\frac{A_{st}f_y}{b\,df_{ck}}\right)\right]$$

$$= (46.8 \times 10^6) = 0.87 \times 415 A_{st} \times 350\left[1 - \left(\frac{415 A_{st}}{200 \times 350 \times 20}\right)\right]$$

Solving $A_{st} = 425\,mm^2$

Provide 4 bars of 12 mm diameter ($A_{st} = 452\,mm^2$) as tension reinforcement and 2 bars of 10 mm diameter as hanger bars on compression side.

vi. *Check for shear stress*:

$$V_u = 37.5\,kN$$
$$\tau_v = (V_u/bd) = (37.5 \times 10^3)/(200 \times 350) = 0.53\,N/mm^2$$
$$p_t = (100\,A_{st})/(bd) = (100 \times 452)/(200 \times 350) = 0.645$$

Refer to Table 19, IS:456-2000 code and readout the design shear strength of concrete as

$$\tau_c = 0.51\,N/mm^2$$

Since　$\tau_v > \tau_c$, shear reinforcements are required.

Balanced shear $V_{us} = V_u - (\tau_c bd)$
$$= 37.5 - (0.51 \times 200 \times 350)\,10^{-3}$$
$$= 1.8\,kN$$

As the balance shear is very small, provide nominal stirrups of 6 mm diameter at a spacing given by:

$$s_v > 0.75\,d = (0.75 \times 350) = 262.5\,mm$$

Provide 6 mm diameter stirrups at 250 mm shear supports.

vii. *Check for deflection control*:

$p_t = 0.645$ from Fig. 5.1 readout $K_t = 1.1$

$(L/d)_{max} = (L/d)_{basic} \times K_t \times K_c \times K_f$
$= (20 \times 1.1 \times 1 \times 1) = 22$

$(L/d)_{actual} = (5000/350) = 14.28 < 22$, hence safe.

viii. *Reinforcement details*:

The details of reinforcement in the beam is shown in Fig. 9.2.

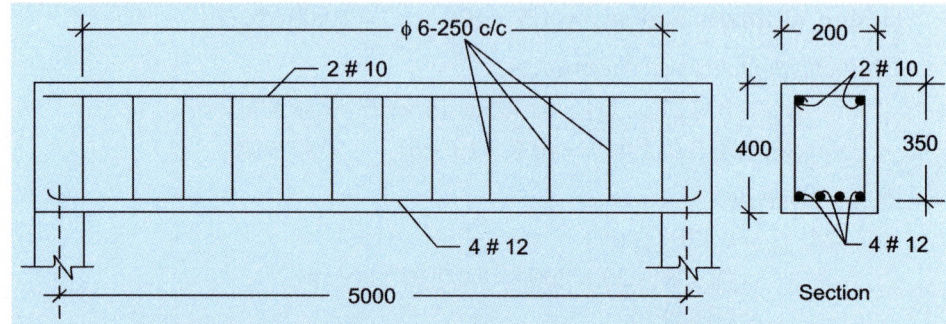

Fig. 9.2: Reinforcement details in singly reinforced beam

ix. *Design of main and shear reinforcements using SP-16 design tables*:

a. Main tension reinforcements:

$$\text{Compute parameter} \left(\frac{M_u}{bd^2}\right) = \left(\frac{46.8 \times 10^6}{200 \times 350^2}\right) = 1.91$$

Refer Table 2, SP-16 and readout the percentage of tension reinforcement at $p_t = 0.606$.

$$A_{st} = \left(\frac{p_t \, b \, d}{100}\right) = \left(\frac{0.606 \times 200 \times 350}{100}\right) = 424 \text{ mm}^2$$

b. Shear reinforcements:

Compute (V_{us}/d) expressed in kN/m

$$\left(\frac{V_{us}}{d}\right) = \left(\frac{1.8}{35}\right) = 0.05$$

As the value of the ratio is very small and outside the purview of values given in Table 62, SP-16, hence provide nominal stirrups using 6 mm diameter bars at a spacing not exceeding $(0.75 \, d) = (0.75 \times 350) = 262.5$ mm.

Adopt a spacing of 250 mm for stirrups.

9.4 DESIGN OF DOUBLY REINFORCED BEAMS

1. Design a reinforced concrete beam to rectangular section using the following data:

Effective span = 5 m

Width of beam = 250 mm

Overall depth   = 500 mm
Service load (DL + LL) = 5 kN/m
Effective cover   = 50 mm

Materials: M20 grade concrete
          Fe 415 HYSD bars

i. *Data*:

    $b = 250$ mm $f_{ck} = 20$ N/mm^2

    $D = 500$ mm $f_y = 415$ N/mm^2

    $d = 450$ mm $E_s = 2 \times 10^5$ N/mm^2

    $d' = 50$ mm

    $L = 5$ m

    $w = 40$ kN/m

ii. *Ultimate moments and shear forces*:

    $M_u = (0.125 \times 1.5 \times 40 \times 5^2) = 187.5$ kN·m

    $V_u = (0.5 \times 1.5 \times 40 \times 5) = 150$ kN

iii. *Main reinforcements*:

      $M_{u,lim} = 0.138 f_{ck} b d^2$

            $= (0.138 \times 20 \times 250 \times 450^2)\, 10^{-6}$ kN·m

            $= 140$ kN·m

Since $M_u > M_{u,lim}$, design a doubly reinforced section.

      $(M_u - M_{u,lim}) = (187.5 - 140) = 47.5$ kN·m

$$f_{sc} = \left[\frac{0.0035(x_{u,max} - d')}{x_{u,max}} \right] E_s$$

$$= \left[\frac{0.0035\{(0.48 \times 450) - 50\}}{(0.48 \times 450)} \right] 2 \times 10^5$$

$$= 538 \text{ N/mm}^2$$

But $f_{sc} \ngtr 0.87 f_y = (0.87 \times 415) = 361$ N/mm^2

$\therefore$ $A_{sc} = \left[\dfrac{(M_u - M_{u,lim})}{f_{sc}(d - d')} \right]$

$$= \left[\frac{47.5 \times 10^6}{361 \times 400} \right] = 329 \text{ mm}^2$$

Provide 2 bars of 16 mm diameter ($A_{sc} = 402$ mm^2)

$$A_{st2} = \left(\frac{A_{sc} f_{sc}}{0.87 f_y} \right) = \left(\frac{329 \times 361}{0.87 \times 415} \right) = 329 \text{ mm}^2$$

$$A_{st1} = \left[\frac{0.36 f_{sc}\, b\, x_{u,lim}}{0.87 f_y} \right]$$

$$A_{st1} = \left[\frac{(0.36 \times 20 \times 250 \times 0.48 \times 450)}{(0.87 \times 415)} \right] = 1077 \text{ mm}^2$$

Total tension reinforcement $A_{st} = (A_{st1} + A_{st2})$
$$= (1077 + 329)$$
$$= 1406 \text{ mm}^2$$

Provide 3 bars of 25 mm diameter ($A_{st} = 1473 \text{ mm}^2$)

iv. *Shear reinforcements*:

$$\tau_v = (V_u/bd) = (150 \times 10^3)/(250 \times 450) = 1.33 \text{ N/mm}^2$$
$$p_t = (100 A_{st})/(bd) = (100 \times 1473)/(250 \times 450) = 1.3$$

Refer to Table 19, IS:456-2000 and readout:
$$\tau_c = 0.68 \text{ N/mm}^2$$

Since $\tau_v > \tau_c$, shear reinforcements are required.

$$V_{us} = [V_u - (\tau_c\, b\, d)]$$
$$= [150 - (0.68 \times 250 \times 450)\,10^{-3}] = 73.5 \text{ kN}$$

Using 8 mm diameter 2 legged stirrups:

$$s_v = \left[\frac{0.87 f_y A_{sv} d}{V_{us}} \right] = \left[\frac{0.87 \times 415 \times 2 \times 50 \times 450)}{73.5 \times 10^3} \right]$$

$$= 221 \text{ mm}$$

$$s_v > 0.75\, d = (0.75 \times 450) = 337.5 \text{ mm}$$

Adopt a spacing of 200 mm near supports gradually increasing to 300 mm towards the centre of span.

v. *Check for deflection control*:

$$(L/d)_{actual} = (5000/450) = 11.1$$
$$(L/d)_{max} = [(L/d)_{basic} \times K_t \times K_c \times K_f]$$
$$p_t = 1.3 \text{ and } p_c = [(100 \times 402)/(250 \times 450)] = 0.35$$

Refer to Fig. 8.1, $K_t = 0.93$,

Fig. 8.2, $K_c = 1.10$, and

Fig. 8.3, $K_f = 1.00$

$$(L/d)_{max} = [(20 \times 0.93 \times 1.10 \times 1.00)] = 20.46$$
$$(L/d)_{actual} < (L/d)_{max}$$

Hence deflection control is satisfied.

vi. *Reinforcement details*:

The reinforcement details in doubly reinforced beam is shown in Fig. 9.3.

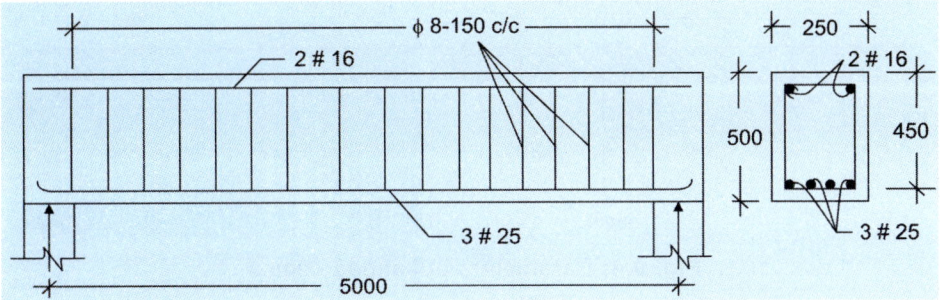

Fig. 9.3: Reinforcement details in doubly reinforced beam

vii. *Design using SP-16 design tables*:

Compute the parameter:

$$\left(\frac{M_u}{b\,d^2}\right) = \left(\frac{187.5 \times 10^6}{250 \times 450^2}\right) = 3.7$$

Refer to Table 50, SP:16 and readout the percentages of tension and compression reinforcements as:

$$p_t = 1.25 \text{ and } p_c = 0.310$$
$$A_{st} = (p_t\,b\,d)/100 = (1.25 \times 250 \times 450)/100 = 1406 \text{ mm}^2$$
$$A_{sc} = (p_t\,b\,d)/100 = (0.310 \times 250 \times 450)/100 = 348 \text{ mm}^2$$

The area of reinforcements are same as that computed using theoretical equations.

9.5 FLANGED BEAMS

9.5.1 Design Parameters of T-Beams

The most common type of reinforced concrete floor and roof system comprises concrete slabs monolithically cast with floor beams in the span range of 5 m to 10 m. In such cases the compressive flange is made up of the width of rib and a portion of the slab length on either side of the rib referred to as the effective width of flange. Figure 9.4 shows the prominent design parameters of flanged T-beams using the notations used in IS:456 code.

a. *Effective width of flange* (b_f): The effective width of flange should, in no case be greater than the breadth of the web plus half the sum of the clear distances to the adjacent beams on either side.

   i. For T-beams, $b_f = (L_0/6) + b_w + 6D_f$

   ii. For L-beams, $b_f = (L_0/12) + b_w + 3D_f$

  iii. For isolated beams, the effective flange width shall be obtained as below but in no case greater than the actual width.

$$\text{T-beam, } b_f = \left[\frac{L_0}{(L_0/b) + 4} + b_w\right]$$

$$\text{L-beam, } b_f = \left[\frac{L_0}{(L_0/b) + 4} + b_w\right]$$

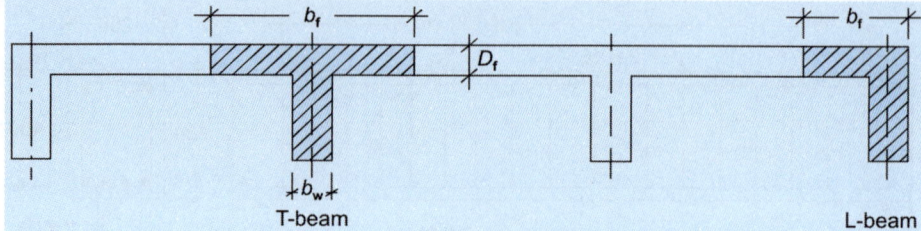

Fig. 9.4: Parameters of flanged beams.

where,

b_f = effective width of the flange

L_0 = distance between points of zero moments in the beam ($L_0 = L$ for simply supported beams and $0.7\ L$ for continuous beams)

L = effective span

b_w = breadth of web

D_f = flange thickness

b = actual width of flange

b. *Effective depth* (d): The basic span/depth ratios outlined in Table 8.1 are applicable for flanged beams modified by using the factor K_f (Refer Fig. 8.3) which may also be termed as reduction factor. For purpose of design, the span/depth ratios of the trial section is generally assumed in the range of 12 to 20 depending upon the span range and degree of loading as given in Table 9.1.

c. *Width of web* (b_w): The web width of tee beam is influenced by the width of the supporting column or the width of the supports like brick concrete or stone masonry walls. The nominal range of width of T-beams varies from 150 mm to 400 mm.

d. *Flange thickness* (D_f): The flange thickness is generally the same as the thickness of the slab between the ribs. The slab thickness depends upon the spacing of ribs, type of loading and is governed by the basic span/depth ratios specified in Table 8.1. Generally the thickness of the slab varies from a minimum of 100 mm to a maximum of 250 mm.

e. *Reinforcement requirements*: The minimum percentage of reinforcement in a flanged beam (Clause 26.5.1.1, IS:456) is based on the width and effective depth. The code specifies the minimum reinforcement as:

$$\left(\frac{A_s}{b_w\, d} \right) = \left(\frac{0.85}{f_y} \right)$$

For Fe 415 HYSD bars, the minimum percentage is about 0.2%. The maximum percentage of tension reinforcement is flanged beams (based on rib width) is limited to 4%.

9.5.2 Design Parameters of L-Beams

The edge beams which are cast monolithic with slabs on one side of the rib only are designated as L-beams. Due to eccentricity of load transferred from the flange as shown in Fig. 9.5. Torsional moments develop in the beams in addition to the bending moments and shear forces.

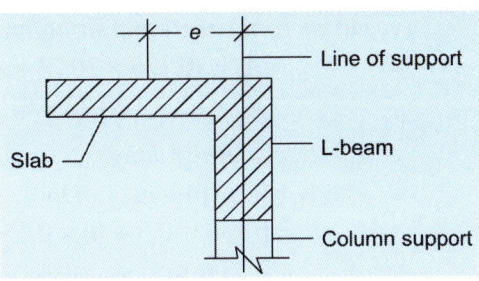

The torsional and hogging bending moments are maximum at the support section.

The support section of the L-beam is the most critical section subjected to combined bending, torsion and shear and this section is designed according to the provisions of the IS:456 code outlined in Section 4.6.

Fig. 9.5: Eccentric load on L-beam

9.5.3 Design Examples

1. A T-beam slab floor of an office comprises a slab 150 mm thick spanning between ribs spaced at 3 m centres. The effective span of the beam is 8 m. Live load on floor is 4 kN/m². Using M20 grade concrete and Fe 415 HYSD bars, design one of the intermediate T-beams.

   i. *Data*:

$L = 8\,\text{m}$ Spacing of T-beams $= 3\,\text{m}$

$D_f = 150\,\text{mm}$ $f_{ck} = 20\,\text{N/mm}^2$

$q = 4\,\text{kN/m}^2$ $f_y = 415\,\text{N/mm}^2$

  ii. *Cross-sectional dimensions*:

From Table 5.1, basic span/depth ratio for simply supported beams is 20. For T-beams, assuming the width of rib = 300 mm and flange width = 3 m, the ratio of web width to flange width is equal to $(300/3000) = 0.1$.

From Fig. 5.3, reduction factor = 0.8.

Hence basic span/depth ratio $= (20 \times 0.8) = 16$

∴ $d = (\text{span}/16) = (8000/16) = 500\,\text{mm}$

Adopt overall depth $D = 550\,\text{mm}$ with cover = 50 mm

Hence the T-beam parameters are:

$d = 500\,\text{mm}$

$D = 550\,\text{mm}$

$b_w = 300\,\text{mm}$

$D_f = 150\,\text{mm}$

iii. *Loads*:

Self weight of slab	$= (0.15 \times 25 \times 3)$	$= 11.25\,\text{kN/m}$
Floor finish	$= (0.6 \times 3)$	$= 1.80$
Shelf weight of rib	$= (0.3 \times 0.4 \times 25)$	$= 3.00$
Plaster finishes		$= 0.45$
Total dead load g		$= 16.50\,\text{kN/m}$
Live load q		$= 4.00\,\text{kN/m}$

Design ultimate load $w_u = 1.5\,(16.50 + 4.0) = 30.75\,\text{kN/m}$

iv. *Ultimate moments and shear forces*:
$$M_u = (0.125 \times 30.75 \times 8^2) = 246 \text{ kN·m}$$
$$V_u = (0.5 \times 30.75 \times 8) \quad = 123 \text{ kN}$$

v. *Effective width of flange*:
   a. $b_f = [(L_0/6) + b_w + 6D_f)$
   $$= [(8/6 + 0.3 + (6 \times 0.15)]$$
   $$= 2.53 \text{ mm}$$
   $$= 2530 \text{ mm}$$
   b. Centre to centre of ribs $= (3 - 0.3) = 2.7 \text{ m}$
   Hence the least of (a) and (b) is $b_f = 2530$ mm.

vi. *Moment capacity of flange*:
$$M_{uf} = 0.36 f_{ck} b_f D_f (d - 0.42 D_f)$$
$$= 0.36 \times 20 \times 2530 \times 150 (500 - 0.42 \times 150)$$
$$= 1194 \times 10^6 \text{ N·mm}$$
$$= 1194 \text{ kN·m}$$
Since $M_u < M_{uf}, x_u < D_f$

Hence the section is considered as rectangular, where $b = b_f$.

vii. *Reinforcements*:

$$M_u = 0.87 f_y A_{st} d \left[1 - \left(\frac{A_{st} f_y}{b d f_{ck}} \right) \right]$$

$$(246 \times 10^6) = (0.87 \times 415 \times A_{st} \times 500) \left[1 - \left(\frac{415 A_{st}}{2530 \times 500 \times 20} \right) \right]$$

Solving $A_{st} = 1417 \text{ mm}^2$

Provide 3 bars of 25 mm diameter ($A_{st} = 1473 \text{ mm}^2$) and two hanger bars of 12 mm diameter on the compression face.

viii. *Shear reinforcements*:
$$\tau_v = (V_u/b_w d) = ((123 \times 10^3)/(300 \times 500)] = 0.82 \text{ N/mm}^2$$
$$p_t = (100 A_{st})/(b_w d) = [(100 \times 1473)/(300 \times 500)] = 0.98$$

Refer to Table 19, IS:456 and readout $\tau_c = 0.60 \text{ N.mm}^2$
Balance shear $V_{us} = [V_u - (\tau_c b_w d)]$
$$= [123 - (0.60 \times 300 \times 500) 10^{-3}]$$
$$= 33 \text{ kN}$$

Using 8 mm diameter two-legged stirrups, the spacing

$$s_v = \left(\frac{0.87 \times 415 \times 2 \times 50 \times 500}{33 \times 10^3} \right) = 547 \text{ mm}$$

But $\quad s_v \not> 0.75 \, d$ or 300 mm, whichever is less.
$$\not> (0.75 \times 500) = 375 \text{ mm}$$

Hence provide 8 mm diameter two-legged stirrups at 300 mm centres throughout the length of the beam.

ix. *Check for deflection control*:

$$p_t = (100A_{st})/(b_w\,d) = [(100 \times 1473)/(300 \times 500)] = 0.98$$
$$(b_w/b_f) = (300/2530) = 0.118$$

Refer to Fig. 8.1, and readout $K_t = 2.00$,
           Fig. 8.2, and readout $K_c = 1.00$, and
           Fig. 8.3, and readout $K_f = 0.80$

$$(L/d)_{max} = [(L/d)_{basic} \times K_t \times K_c \times K_f]$$
$$= [(16 \times 2 \times 1 \times 0.80)] = 25.6$$
$$(L/d)_{provided} = (8000/500) = 16 < 256$$

Hence deflection control is satisfied.

x. *Design using SP-16 design tables*:

$$(M_u/b_f\,d^2) = (246 \times 10^6)/(2530 \times 500^2) = 0.388$$

Refer to Table 2, SP-16 design tables and readout $p_t = 0.111$

$$A_{st} = (p_t\,b_f\,d)/100 = (0.111 \times 2530 \times 500)/100 = 1404 \text{ mm}^2$$

The area of steel is the same as that computed using equations in point vii.

xi. The reinforcement details in the T-beam is shown in Fig. 9.6.

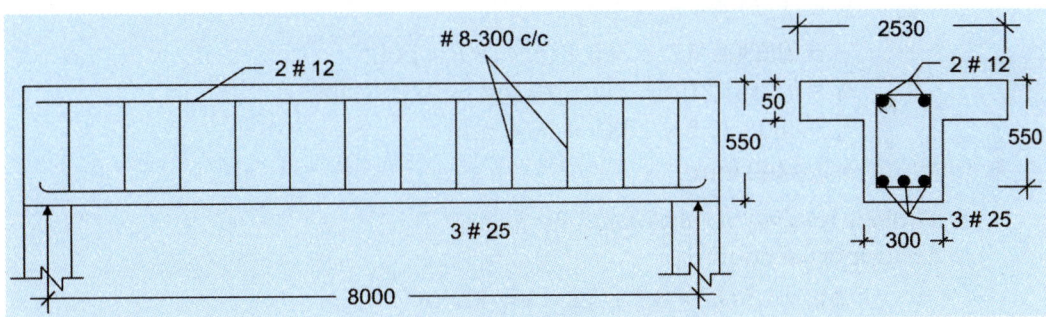

Fig. 9.6: Reinforcement in T-beam

2. Design the L-beam for an office floor to suit the following data (Fig. 9.7).

i. *Data*:

Clear span $L = 8$ m
Thickness of flange $D_f = 150$ mm
Live load $g = 4$ kN/m^2
Spacing of beams $= 3$ m
$f_{ck} = 20$ N/mm^2
$f_y = 415$ N/mm^2
L-beams are monolithic with RC columns
Width of column $= 300$ mm

ii. *Cross-sectional dimensions*:

Since L-beam is subjected to bending, torsion and shear forces, assume a trial section having span/depth ratio of 12.

$$\therefore \quad d = (8000/12) = 666 \text{ mm}$$

Adopt $\quad d = 700 \text{ mm}$

$\quad\quad\quad\quad D = 750 \text{ mm}$

$\quad\quad\quad\quad b_w = 300 \text{ mm}$

iii. *Effective span*:

Effective span is the least of:

(i) centre to centre of supports = $(9 + 0.3) = 9.3$ m

(ii) clear span + effective depth = $(8 + 0.7) = 8.7$ mm

$\quad\quad \therefore \ L = 8.3$ mm

iv. *Loads*:

Dead load of slab	$= (0.15 \times 25 \times 0.5 \times 3)$	$= 5.60$ kN/m
Floor finish	$= (0.6 \times 0.5 \times 3)$	$= 0.90$
Shelf weight of rib	$= (0.3 \times 0.6 \times 25)$	$= 4.50$
Live load	$= (4 \times 0.5 \times 3)$	$= 6.00$
Total working load w		$= 18.00$ kN/m

v. *Effective flange width*:

Effective flange width (b_f) is the least of the following values:

a. $\quad b_f = (L_0/12) + b_w + 3D_f)$

$\quad\quad\quad = (0.8300/12) + 300 + (3 \times 140) = 1442$

b. $\quad b_f = b_w + 0.5$ times the spacing between ribs

$\quad\quad b_f = 300 + (0.5 \times 2700) = 1650$ mm

$\quad \therefore \ b_f = 1442$ mm

vi. *Ultimate bending moment and shear forces*:

At support section:

$$M_u = 1.5 \, (17 \times 8.3^2)/12 = 147 \text{ kN·m}$$
$$V_u = 1.5 \, (0.5 \times 17 \times 8.3) = 106 \text{ kN}$$

At centre of span section:

$$M_u = 1.5 \, (17 \times 8.3^2)/24 = 73 \text{ kN·m}$$

vii. *Torsional moment at support section*:

Torsional moment is produced due to dead load of slab and live load on it, i.e.

(working load/m – self-weight of the rib = $(17 - 4.50) = 12.50$ kN/m

$\therefore$ Total ultimate load on slab = $1.5(12.50 \times 8.3) = 156$ kN

Total ultimate shear force = $(0.5 \times 156) = 78$ kN

Distance of centroid of shear force from the centre line of the beam (Refer Fig. 9.7) = $0.5 \times 1442 - 150 = 571$ mm

Ultimate torsional moment $T_u = (78 \times 0.571) = 44.5$ kN·m

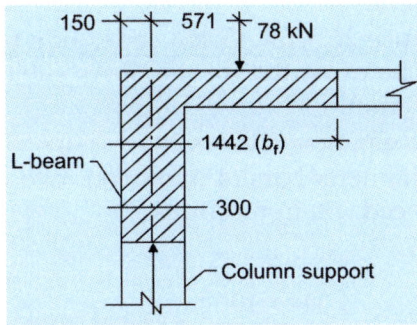

Fig. 9.7: Eccentric load on L-beam.

viii. *Equivalent bending moment and shear forces*:

According to IS:456-2000, Clause 41.4.2, at the support section, the equivalent bending moment is computed as

$$M_{el} = (M_u + M_t)$$

where, $M_t = T_u \left[\dfrac{1 + (D/b)}{1.7} \right] = 44.5 \left[\dfrac{1 + (750/300)}{1.7} \right] = 92 \text{ kN·m}$

∴ $M_{el} = (147 + 92) = 239$ kN·m

$V_e = V_u + 1.6 \, (T_u/b)$

    $= 106 + 1.6 \, (44.5/0.3)$

    $= 344$ kN

ix. *Main longitudinal reinforcement*:

Support section is designed as rectangular section to resist the hogging equivalent bending moment $M_{el} = 239$ kN·m

$M_{u,lim} = 0.138 f_{ck} \, b \, d^2$

    $= (0.138 \times 20 \times 300 \times 700^2) \times 10^{-6}$

    $= 405.7$ kN·m

Since $M_u < M_{u,lim}$, the section is under-reinforced:

$$M_{el} = 0.87 f_y A_{st} d \left[1 - \left(\frac{A_{st} f_y}{b d f_{ck}} \right) \right]$$

$$(239 \times 10^6) = (0.87 \times 415 A_{st} \times 700) \left[1 - \left(\frac{415 A_{st}}{300 \times 700 \times 20} \right) \right]$$

Solving, $A_{st} = 1056.3$ mm^2

Provide 3 bars of 22 mm diameter on the tension side ($A_{st} = 1140$ mm^2)

Area of steel required at centre of span to resist a moment of $M_u = 73$ kN·m will be less than the minimum given by

$$A_{st(min)} = \left(\frac{0.85 b_w \, d}{f_y} \right) = \left(\frac{0.85 \times 300 \times 700}{415} \right) = 430 \text{ mm}^2$$

Provide 2 bars of 20 mm diameter ($A_{st} = 628$ mm^2)

x. *Side face reinforcement*:

According to Clause 26.5.1.7, IS:456 code, side face reinforcement of 0.1% of web area is to be provided for members subjected to torsion, when the depth exceeds 450 mm.

∴ Area of reinforcement = $(0.001 \times 300 \times 750) = 225 \text{ mm}^2$

Provide 10 mm diameter bars (4 numbers), two on each face as horizontal reinforcement spaced 200 mm centres.

xi. *Shear reinforcements*:

$$\tau_{ve} = \left(\frac{V_e}{b_w d}\right) = \left(\frac{344 \times 10^3}{300 \times 700}\right) = 1.63 \text{ N/mm}^2$$

$$p_t = \left(\frac{100 A_{st}}{b_w d}\right) = \left(\frac{100 \times 1140}{300 \times 700}\right) = 0.542\%$$

From Table 19 (IS:456), readout:

$$\tau_c = 0.49 \text{ N/mm}^2 < \tau_{ve}$$

Hence, shear reinforcements are required.

Using 10 mm diameter two-legged stirrups with side covers of 25 mm and top and bottom covers of 50 mm, we have $b_1 = 250$ mm, $d_1 = 650$ mm, $A_{sv} = (2 \times 78.5) = 157 \text{ mm}^2$.

The spacing s_v is computed using the equations specified in Clause 41.4.3, IS:456-2000 code.

$$s_v = \left[\frac{(0.87 f_y A_{sv} d_1)}{\left(\frac{T_u}{b_1}\right) + \left(\frac{V_u}{2.5}\right)}\right]$$

$$= \left[\frac{0.87 \times 415 \times 157 \times 650}{\left(\frac{44.5 \times 10^6}{250}\right) + \left(\frac{106 \times 10^3}{2.5}\right)}\right] = 167 \text{ mm}$$

or

$$s_v = \left[\frac{A_{sv} \, 0.87 f_y}{(\tau_{ve} - \tau_c)n}\right]$$

$$= \left[\frac{(157 \times 0.87 \times 415)}{(1.63 - 0.49)300}\right] = 165 \text{ mm}$$

Provide 10 mm diameter two-legged stirrups at a minimum spacing given by (Clause 26.5.1.7, IS:456).

Adopt the minimum spacing based on shear and torsion consideration computed as $s_v = 160$ mm.

xii. *Check for deflection control*:

$$p_t = \left(\frac{100 \times 1140}{300 \times 700}\right) = 0.54\%$$

$$p_c = \left(\frac{100 \times 628}{300 \times 700}\right) = 0.299$$

$(b_w/b_f) = (300/1442) = 0.208$

Refer to Fig. 8.1, $K_t = 1.2$,
          Fig. 8.2, $K_c = 1.1$, and
          Fig. 8.3, $K_f = 0.8$

$$(L/d)_{max} = [(L/d)_{basic} \times K_t \times K_c \times K_f]$$
$$= [(9.20 \times 1.2 \times 1.1 \times 0.8)]$$
$$= 21.12$$

$(L/d)_{actual} = (8300/700) = 11.85 < 21.12$

Hence deflection control is satisfied.

xiii. *Reinforcement details*:

Figure 9.8 shows the details of reinforcement in the L-beam.

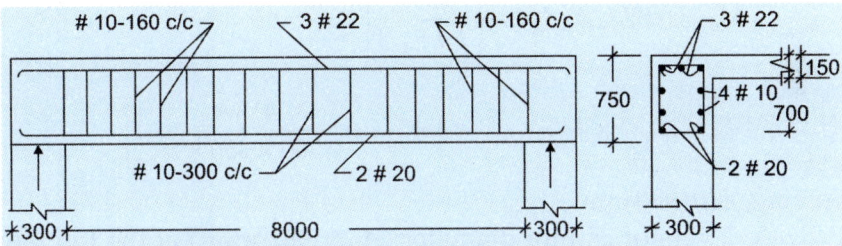

Fig. 9.8: Reinforcement details in L-beam

9.6 CANTILEVER BEAMS

Cantilever beams are generally used to support the canopy or chazza slabs of larger span at the entrance of buildings. Cantilever beams are designed for the maximum moments and shear forces developed at the support section which is normally a reinforced concrete column.

Example: Design a cantilever beam using the following data:

  i. *Data*:

Clear span = 3 m
Working live load = 15 kN/m
Cantilever beam is monolithic with reinforced concrete column 300 mm wide × 600 mm deep at the junction of the column and beam.
$f_{ck} = 20 \, \text{N/mm}^2$
$f_y = 415 \, \text{N/mm}^2$

 ii. *Cross-sectional dimensions*:

According to IS:456-2000 code, the trial section depth is determined by the span/depth ratio of 7.

        depth = (span/7) = (3000/7) = 428 mm

Adopt effective depth $d = 450$ mm
Overall depth $D = 500$ mm
Width $b = 300$ mm

iii. *Loads*:

Self weight of beam = $(0.3 \times 0.5 \times 25)$ = 3.75 kN/m
Live load = 15.00 kN/m
Finishes etc. = 1.25 kN/m
Total working load w = 20.00 kN/m

iv. *Ultimate moment and shear force*:

$M_u = 1.5 (0.5\, wL^2) = (1.5 \times 0.5 \times 20 \times 3^2) = 135$ kN·m
$V_u = 1.5 (wL) = (1.5 \times 30 \times 3) = 90$ kN

v. *Main reinforcements*:

$M_{u,lim} = 0.138 f_{ck}\, b\, d^2$
$= (0.138 \times 20 \times 300 \times 450^2)\, 10^{-6}$
$= 167$ kN·m

Since $M_u < M_{u,lim}$, section is under-reinforced:

$$M_u = 0.87 f_y A_{st} d \left[1 - \left(\frac{A_{st} f_y}{b d f_{ck}} \right) \right]$$

$$(135 \times 10^6) = (0.87 \times 415 \times A_{st} \times 450) \left[1 - \left(\frac{415 A_{st}}{300 \times 450 \times 20} \right) \right]$$

Solving, $A_{st} = 980$ mm^2
Provide 2 bars of 25 mm diameter ($A_{st} = 982$ mm^2) at the top tension face and 2 hanger bars of 12 mm diameter at the compression face.

vi. *Shear reinforcements*:

$$\tau_v = \left(\frac{V_u}{b_w d} \right) = \left(\frac{90 \times 10^3}{300 \times 450} \right) = 0.66 \text{ N/mm}^2$$

$$p_t = \left(\frac{100 A_{st}}{b d} \right) = \left(\frac{100 \times 982}{300 \times 450} \right) = 0.72$$

From Table 19 (IS:456), readout:
$\tau_c = 0.55$ N/mm$^2 < \tau_{ve}$
Hence shear reinforcements are required.
Refer to Table 19, IS:456 and readout $\tau_c = 0.60$ N/mm^2
Balance shear $V_{us} = [V_u - (\tau_c b d)]$
$= [90 - (0.55 \times 300 \times 450) \times 10^{-3}]$
$= 15.75$ kN
Using 8 mm diameter two-legged stirrups, the spacing

$$s_v = \left(\frac{0.87 f_y A_{sv} d}{V_{us}} \right) = \left(\frac{0.87 \times 415 \times 2 \times 50 \times 450}{15.75 \times 10^3} \right) = 1031 \text{ mm}$$

But $s_v \not> 0.75\, d = (0.75 \times 450) = 337.5$ mm but not greater than 300 mm.
Adopt 8 mm diameter two-legged stirrups at 300 mm centres.

vii. *Anchorage length at supports*:

Anchorage length required is computed as

$$L_d = \left(\frac{0.87 f_y \phi}{4\tau_{bd}}\right) = \left(\frac{0.87 \times 415 \times 25}{4 \times 1.2 \times 1.6}\right) = 1175 \text{ mm}$$

The main tension reinforcement is extended into the column to a length of 500 mm and bent down at 90° and extended up to 500 mm as shown in Fig. 9.9, so that the total anchorage length is 1200 mm with the 90° bend being equivalent to $8\phi = (8 \times 25) = 200$ mm.

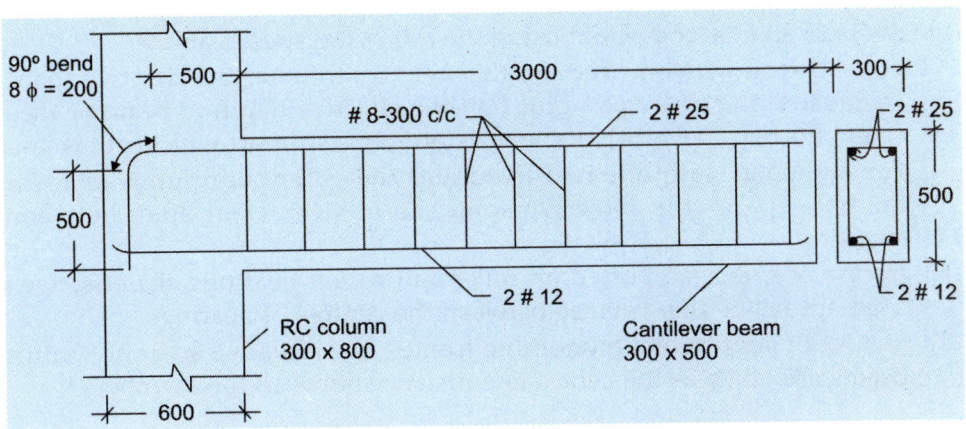

Fig. 9.9: Reinforcement details in cantilever beam

viii. *Check for deflection control*:

$$p_t = \left(\frac{100 A_{st}}{bd}\right) = \left(\frac{100 \times 982}{300 \times 450}\right) = 0.727$$

Refer to Fig. 8.1, $K_t = 1.08$,
  Fig. 8.2, $K_c = 1.00$, and
  Fig. 8.3, $K_f = 1.00$

$$(L/D)_{max} = [(L/d)_{basic} \times K_t \times K_c \times K_f]$$
$$= [(7 \times 1.08 \times 1 \times 1)]$$
$$= 7.56$$

$$(L/D)_{actual} = (3000/450) = 6.66 < 7.56$$

Hence deflection control is satisfied.

ix. *Reinforcement details*:

Figure 9.9 shows the reinforcement details in the cantilever beam.

x. *Design using SP-16 design tables*:

$$(M_u/b\,d^2) = (136 \times 10^6)/(300 \times 450^2) = 2.22$$

Refer to Table 2, SP-16 and readout the value of $p_t = 0.725$

$$A_{st} = (p_t\,b\,d)/100 = (0.725 \times 300 \times 450)/100 = 979 \text{ mm}^2$$

The area of steel is same as that computed using theoretical equations.

9.7 CONTINUOUS BEAMS

9.7.1 Effective Span

Continuous beams are commonly used in multistorey buildings of several bays in perpendicular directions. IS:456, Clause 22.2 specifies the effective span which depends upon the width of supports.

    a. If the support width is less than 1/12 span, then the effective span shall be as per freely supported beams, i.e. clear span plus the effective depth or centre to centre of supports whichever is less.

    b. If the supports are wider than 1/12, the clear span or 600 mm whichever is less, the effective span is computed using the following specifications:

        i. For end span with one free and the other continuous, the effective span shall be equal to the clear span plus half the effective depth of beam or the clear span plus half the width of the discontinuous support whichever is less.

        ii. For end span with one end fixed and the other continuous or for intermediate spans, the effective span shall be the clear span between the supports.

        iii. In case of spans supported on roller and rocker bearings, the effective span shall always be the distance between the centres of bearings.

        iv. In case of continuous monolithic frames, the effective span of continuous beams are taken as the centre line distance between the members.

9.7.2 Span/Depth ratio

In case of continuous beams, the IS:456-2000 code recommends a basic span/depth ratio of 26 which can be modified by the factors K_t, K_c and K_f detailed in Section 8.2.2 for conforming to the serviceability limit state of deflection limits of span/250 specified in Clause 23.2.

Normally continuous beams carry heavy dead and live loads and consequently the span/depth ratio recommended in practical designs are normally between 15 and 20.

The use of upper limit of span/depth ratio of 26 results in shallower depths and requires high percentage of reinforcements tending toward over-reinforced sections.

9.7.3 Bending Moment and Shear Force Coefficients

Analysis of continuous beams for computation of design bending moments and shear forces preceeds the design process. Rigorous analysis of moments and shear forces using moment distribution, Kani's rotation contribution or stiffness or flexibility matrix methods, involve lengthy computations and are time consuming procedures. However, the Indian standard code, IS:456-2000 permits the use of moment and shear force coefficients given in Tables 9.2 and 9.3 (Tables 12 and 13, IS:456-2000) for computing the design bending moments and shear forces in continuous beams supporting uniformly distributed loads over three or more spans which do not differ by more than 15% of the longest span. If redistribution of moments is used in the design process or unsymmetrical loading is used, the bending moment and shear force coefficients cannot be used and a rigorous analysis is required.

Table 9.2: Bending moment coefficients (Table 12, IS:456-2000)

Type of load	Span moments		Support moments	
	Near middle of end span	At middle of interior span	At support next to end support	At other interior supports
Dead load and imposed load (fixed)	$+\dfrac{1}{12}$	$+\dfrac{1}{16}$	$-\dfrac{1}{10}$	$-\dfrac{1}{12}$
Imposed load (not fixed)	$+\dfrac{1}{10}$	$+\dfrac{1}{12}$	$-\dfrac{1}{9}$	$-\dfrac{1}{9}$

Note: For obtaining the bending moment, the coefficient shall be multiplied by the total design load and effective span.

Table 9.3: Shear force coefficients (Table 13, IS:456-2000)

Type of load	At end support	At support next to the end support		At all other interior supports
		Outer side	Inner side	
Dead load and imposed load (fixed)	0.4	0.6	0.55	0.5
Imposed load (not fixed)	0.45	0.6	0.6	0.6

Note: For obtaining the shear force, the coefficient shall be multiplied by the total design load.

9.7.4 Design Example

Design a continuous reinforced concrete beam of rectangular section to support a dead load of 10 kN/m and a service live load of 15 kN/m over three simply supported spans of 8 m each. Adopt M20 grade concrete and Fe 415 HYSD bars.

   i. *Data*:

     $L = 8\,\text{mm}$ $f_{ck} = 20\,\text{N/mm}^2$

     $g = 10\,\text{kN/m}$ $f_y = 415\,\text{N/mm}^2$

     $q = 15\,\text{kN/m}$

  ii. *Cross-sectional dimensions*:

     Depth = (span/12) = (8000/20) = 666 mm

     Adopt $d = 650\,\text{mm}$

               $D = 700\,\text{mm}$

               $b = 300\,\text{mm}$

 iii. *Loads*:

     Self-weight of the beam $= (0.3 \times 0.7 \times 25) = 5.25\,\text{kN/m}$

     Dead load $= 10.00\,\text{kN/m}$

     Finishes $= 0.75\,\text{kN/m}$

     Total dead load g $= 16.00\,\text{kN/m}$

     Live load q $= 15\,\text{kN/m}$

 iv. *Bending moments and shear forces*:

     Referring to the bending moment and shear force coefficients (Tables 9.2 and 9.3).

Negative moment at interior support

$$M_u(-ve) = 1.5 \left[\frac{gL^2}{10} + \frac{gL^2}{9} \right]$$

$$= 1.5 \left[\frac{16 \times 8^2}{10} + \frac{15 \times 8^2}{9} \right] = 314 \text{ kN} \cdot \text{m}$$

Positive BM at centre of span:

$$M_u(+ve) = 1.5 \left[\frac{gL^2}{12} + \frac{gL^2}{10} \right]$$

$$= 1.5 \left[\frac{16 \times 8^2}{12} + \frac{15 \times 8^2}{10} \right] = 182 \text{ kN} \cdot \text{m}$$

Maximum shear force at the support section

$$V_u = 1.5 \times 0.6 \, (g + q)$$
$$= 1.5 \times 0.6 \times 8 \, (16 + 15) = 223.2 \text{ kN}$$

v. *Limiting moment of resistance*:

$$M_{u,lim} = 0.138 f_{ck} \, b \, d^2$$
$$= (0.138 \times 20 \times 300 \times 650^2) \times 10^{-6}$$
$$= 350 \text{ kN·m}$$

Since $M_u < M_{u,lim}$, section is under-reinforced:

vi. *Main reinforcements*:

$$M_u = 0.87 f_y A_{st} d \left[1 - \left(\frac{A_{st} f_y}{bdf_{ck}} \right) \right]$$

$$(314 \times 10^6) = (0.87 \times 415 A_{st} \times 650) \left[1 - \frac{415 A_{st}}{(300 \times 650 \times 20)} \right]$$

Solving $A_{st} = 980 \text{ mm}^2$
(for −ve BM)

Provide 4 bars of 25 mm diameter ($A_{st} = 1964 \text{ mm}^2$) at the top tension face near supports. The area of steel required for the positive moment of 182 kN·m is 850 mm².

Provide 2 bars of 25 mm diameter at the bottom tension face at centre of span sections.

vii. *Shear reinforcements*:

$$\tau_v = \left(\frac{V_u}{bd} \right) = \left(\frac{223.2 \times 10^3}{300 \times 650} \right) = 1.14 \text{ N/mm}^2$$

$$p_t = \left(\frac{100 A_{st}}{bd} \right) = \left(\frac{100 \times 1964}{300 \times 650} \right) = 1.007$$

Refer to Table 19 (IS:456), readout:
$$\tau_c = 0.62 \text{ N/mm}^2 < \tau_v$$
Hence shear reinforcements are required.
Balance shear = $[223.2 - (0.62 \times 300 \times 650) \, 10^{-3}$
$$= 102.3 \text{ kN}$$
Using 8 mm diameter two-legged stirrups, the spacing

$$s_v = \left(\frac{0.87 f_y A_{sv} d}{V_{us}}\right) = \left(\frac{0.87 \times 415 \times 2 \times 50 \times 650}{102.3 \times 10^3}\right) = 229 \text{ mm}$$

Adopt 8 mm diameter two-legged stirrups at 200 mm centres near supports gradually increasing to 300 mm toward the centre of span

viii. *Check for deflection control*:
At centre of span:

$$p_t = \left(\frac{100 A_{st}}{b d}\right) = \left(\frac{100 \times 982}{300 \times 650}\right) = 0.50\%$$

Refer to Fig. 8.1 readout the modification factor $K_t = 1.2$.
Neglecting bars in compression side, $K_c = 1.0$ and $K_f = 1.0$.
$$(L/d)_{max} = [(L/d)_{basic} \times K_t \times K_c \times K_f]$$
$$= [(26 \times 1.2 \times 1.0 \times 1.0)] = 31.2$$
$$(L/d)_{actual} = (8000/650) = 12.3 < 31.2$$
Hence, deflection control is satisfied.

ix. *Reinforcement details*:
The details of reinforcement in the continuous beam is shown in Fig. 9.10.

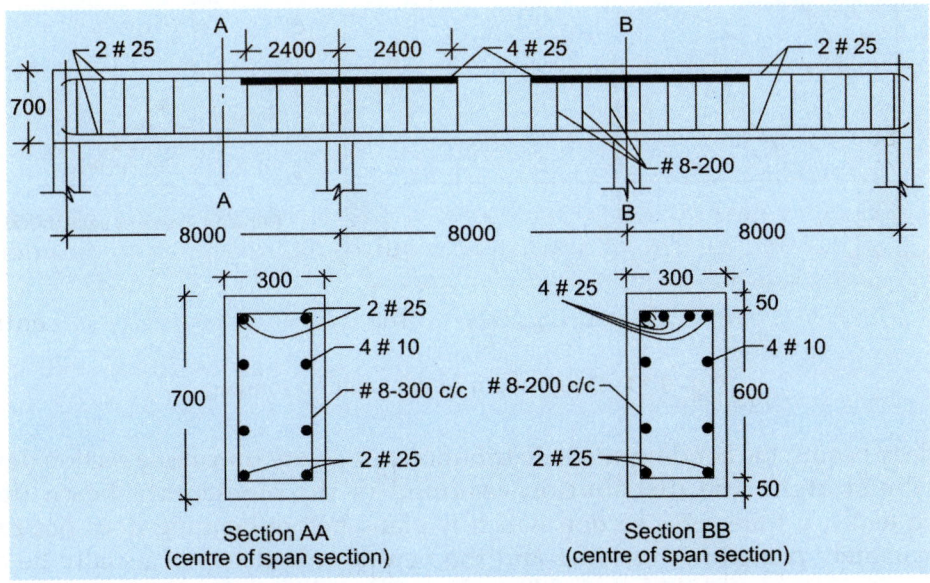

Fig. 9.10: Reinforcement details in continuous beam.

x. *Design of shear reinforcements using SP-16 design tables*:

Compute $V_{us} = 102.3$ kN and $d = 65$ cm

Compute parameter (V_{us}/d) kN/cm

$\qquad (V_{us}/d) = (102.3/65) = 1.57$

Refer to Table 62 of SP-16 and readout the spacing of 8 mm diameter two-legged stirrups as $s_v = 23$ cm = 230 mm.

The spacing of stirrups is the same as that obtained by theoretical calculation.

9.8 DEEP BEAMS

9.8.1 Design Aspects of Deep Beams

According to the Clause 29.1, IS:456-2000, a beam is classified as deep when the ratio of effective span to overall depth (l/D) is less than 2 for simply supported members and 2.5 for continuous members. Deep beams are structural elements loaded as traditional simple beams in which a significant amount of the load is carried to the supports by a compression force combining the load and the reaction. A beam is considered as deep, if the depth is large in relation to the span of the beam. A typical deep beam under uniform loading is shown in Fig. 9.11.

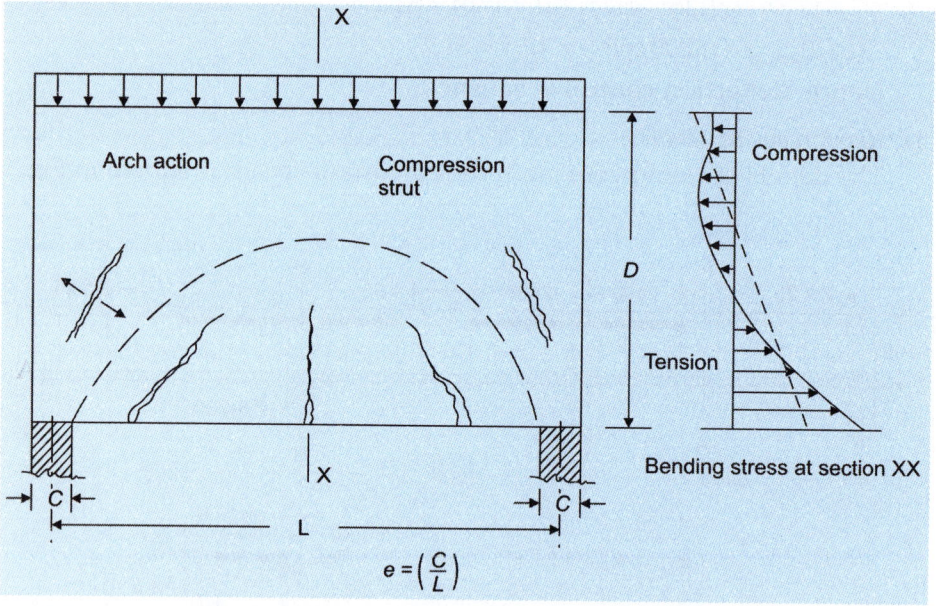

Fig. 9.11: Deep beam under uniform loading

In deep beams, the bending stress distribution across any transverse section deviates from the straight line distribution assumed in the elementary beam theory. Consequently, a transverse section which is plane before bending does not remain approximately plane after bending and the neutral axis does not usually lie at the mid depth. In case of deep beams, shear flexure[11] and shear modes[12] dominated by tensile cleavage failure are common. The ultimate failure due to shear[13] is generally brittle in nature in contrast to the ductile behaviour and progressive flexural failure with large number of cracks prevalent in normal beams.

In many cases of industrial and multistoreyed structures, deep beams are invariably used due to structural requirements. Floor slabs under horizontal load, short span beams carrying heavy loads, and transfer girders are examples of deep beams. The side walls of coal bunkers also behave as deep beams. The method of design of deep beams is different from that of the simple traditional beams having large span depth ratios since several influencing parameters have to be considered in the design of deep beams.

The behaviour of deep beams is significantly different in comparison with the simple beams and many assumptions made for the design of simple beams are not valid for deep beams since the structural behaviour of deep beams under transverse loads is complex. The major differences are listed below:

1. The stress and strain distribution is not linear since plane sections do not remain plane after bending.
2. The structural behaviour is influenced by both support to depth and span to depth ratios.
3. Shear deformation cannot be neglected as in simple beams.
4. At ultimate limit stage, the stress block in concrete compression zone cannot be assumed to be parabolic in shape as in simple beams.
5. Conventional shear investigations are not applicable for deep beams.

A simple elementary design procedure is presented here based on the investigations of Kong[14] as reported by Subramanian[15]. The design steps are as follows:

1. Calculate the bending moment as in the case of an ordinary beam:
   a. Simply supported beams with uniformly distributed load (w_u)

   $$M_{max} = \left[\frac{w_u L^2}{8} \right]$$

   b. Continuous beams with uniformly distributed load (w_u) (As per ACI:318-19)

   (i) Mid span: $M_{max} = \left[\dfrac{w_u L^2}{24} \right] (1 - e^2)$

   (ii) Face of support: $M_{max} = \left[\dfrac{w_u L^2}{24} \right] (2 - 3e + e^2)$

   where, e = ratio of the width of the support to the effective span of the beam.

2. Area of reinforcement

   $$A_{st} = \left[\frac{M_u}{0.87 f_y z} \right]$$

where, M_u = factored applied moment

f_y = characteristic yield strength of steel

z = lever arm which to be taken as (Clause 29.2, IS:456)

$z = [0.2L + 0.4D]$, if $(1 \leq L/D \leq 2)$ and $0.6L$, if $(L/D < 1)$ for single span beams

$z = [0.2L + 0.3D]$, if $(1 \leq L/D \leq 2.5)$ and $0.5L$, if $(L/D < 1)$ for continuous beams

The positive and negative reinforcements have to be arranged according to the specifications prescribed in Clause 29.3, IS:456 code.

9.8.2 Design Example

Design a single span deep beam to suit the following data:

Effective span = 6 m

Overall depth = 6 m

Width of support = 0.6 m

Total factored load on beam including self-weight = 600 kN/m

Materials: M20 grade concrete

Fe 415 HYSD bars

i. *Design parameters*:

$L = 6\,m$

$w_u = 600\ kN/m$

$e = (0.6/6) = 0.1$

$D = 6\,m$

$M_u = [0.125 \times 600 \times 6^2] = 2700\ kN{\cdot}m$

ii. Ratio of $(L/D) = (6/6) = 1 \leq 1.5$, hence compute the lever arm as:

$z = [0.2L + 0.4D] = [(0.2 \times 6) + (0.4 \times 6)] = 3.6$

Area of reinforcement $A_{st} = \left(\dfrac{M_u}{0.87 f_y z} \right) = \left[\dfrac{2700 \times 10^6}{0.87 \times 415 \times 3600} \right] = 2078\ mm^2$

Minimum horizontal reinforcement = $(0.002 \times 600 \times 6000) = 7200\ mm^2$

Zone of depth = $(0.25D - 0.05L) = (0.25 \times 6 - 0.05 \times 6) = 1.2\ m$

The tension reinforcements of 7200 mm² is arranged within a depth of 1.2 m from the tension face.

Adopt 24 bars of 20 mm diameter in 6 rows of 4 bars each.

According to Clause 32.5, IS:456 code, minimum vertical and horizontal reinforcements have to be provided in deep beams.

Vertical reinforcement area = $(0.0012 \times 600 \times 1000) = 720\ mm^2$.

Horizontal reinforcements = $(0.002 \times 600 \times 6000) = 7200\ mm^2$.

Provide 8 mm diameter four-legged stirrups at 250 mm centres and 10 mm diameter bars horizontally at 300 mm centres.

The reinforcement details are shown in Fig. 9.12.

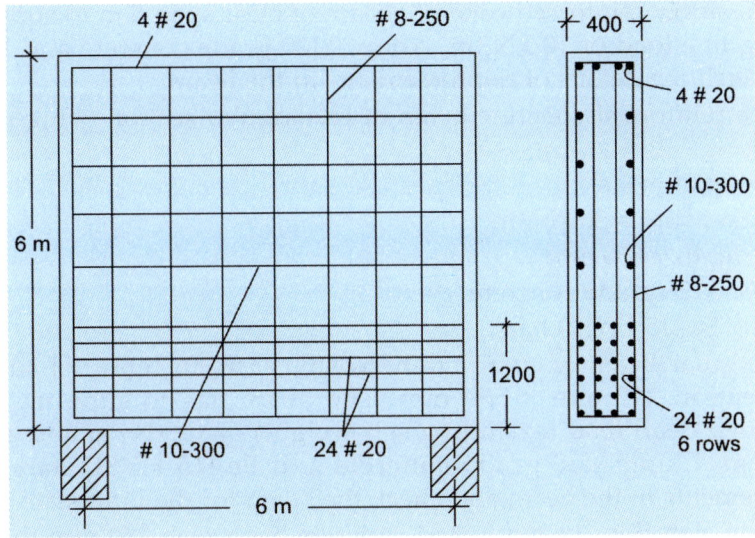

Fig. 9.12: Reinforcement details in single span deep beam.

REFERENCES

1. Bennett EW, *Structural Concrete Elements*, Chapman & Hall, London, 1973, pp 149–170.
2. Bate SCC, Why limit state design? *Concrete*, March 1968, London, pp 103–108.
3. Krishna Raju N, Limit state design for structural concrete, *Proc Institution Engineers* (India), Vol 51, January 1971, pp 138–43.
4. ACI:318M-11, Building code requirements for structural concrete (ACI Standard), *American Concrete Institute*, Farmington Hills, Michigan, USA, 2011.
5. Unnikrishna Pillai S and Devdas Menon, Reinforced Concrete Design, 3 edn, Tata McGraw-Hill, New Delhi, 2009, pp 87–96.
6. Rowe RE, Cranston WB and Best BC, New concepts in the design of concrete, *Structural Engineer*, Vol 43, 1965, pp 339–403.
7. Arora JS, *Introduction to Optimum Design*, McGraw-Hill International, New York, 1989.
8. Mosley WH, Bungey JH and Hulse R, *Reinforced Concrete Design to Euro Code-2*, 7 edn, Palgrave Macmillan, London, New York, 2012, pp 176–216.
9. IS:456-2000, Indian Standard Code of Practice for Plain and Reinforced Concrete (4th Revision), BIS, New Delhi, 2000, p 37.
10. SP:34-1987, Handbook of Concrete Reinforcement and Detailing, BIS, New Delhi, 1987.

ASSIGNMENT

1. Design a singly reinforced concrete beam to support a working live load of 6 kN/m over a clear span of 3 m. Adopt M20 grade concrete and Fe 415 HYSD bars. Adopt width of supports = 200 mm.

2. Design a singly reinforced concrete beam of clear span 5 m to support a design working live load of 10 kN/m. Adopt M20 grade concrete and Fe 415 HYSD bars. Sketch the details of reinforcements in the beam.

3. Design a reinforced concrete beam of rectangular section using the following data:
   Effective span = 8 m
   Serving live load = 30 kN/m
   Overall depth = 650 mm
   Materials: M25 grade concrete
           Fe 415 HYSD bars.

4. A rectangular beam of span 7 m between centre to centre of supports has a cross-section 250 mm × 550 mm. The beam has to support a uniformly distributed dead load (excluding self-weight) of 15 kN/m and a live load of 20 kN/m. Using M20 grade concrete and Fe 415 HYSD bars, design the reinforcements in the beam and check the beam for the limit state of deflection.

5. A T-beam slab floor of reinforced concrete has a slab 150 mm thick spanning between the beams which are spaced 3 m apart. The beams have a clear span of 10 m and the end bearings are 450 mm thick walls. The live load on the floor is 4 kN/m². Adopting M20 grade concrete and Fe 415 HYSD bars, design one of the intermediate T-beams and sketch the details of reinforcements in the beam.

6. Design a L-beam for an office floor using the following data:
   Clear span = 6 m, distance between centre to centre of supports = 6.3 m. The L-beams are cast monolithically with RCC columns, spacing of beams = 2.75 m, loading on office floor = 4 kN/m². Thickness of floor slab = 100 mm, width of column = 300 mm. Adopt M20 grade concrete and Fe 415 HYSD bars.

7. Design a cantilever beam of span 4 m to support a working live load of 15 kN/m. The cantilever beam is monolithic with reinforced concrete columns 300 mm wide × 750 mm deep at the junction of the column and beam. Adopt M25 grade concrete and Fe 415 HYSD bars. Sketch the details of reinforcements in the cantilever beam.

8. Design a continuous concrete beam of rectangular section to support a dead load of 8 kN/m and a working live load of 12 kN/m over three simply supported spans of 10 m each. Adopt M25 grade concrete and Fe 415 HYSD bars.

9. A transfer girder has to be designed as a deep beam. Using the following data, design the deep beam and sketch the details of reinforcements.
   Effective span = 9 m
   Overall depth of beam = 5 m
   Width of beam = 500 mm
   Uniformly distributed live load = 200 kN/m
   Grade of concrete = M25
   Type of reinforcements = Fe 415 HYSD bars

10. Design a deep continuous beam using the following data:
    Number of spans = 2
    Effective length of each span = 10 m
    Overall depth of beam = 4.5 m

Width of beam = 500 mm
Uniformly distributed load = 200 kN/m
Concentrated load at centre of spans = 200 kN
Grade of concrete = M25
Type of reinforcements = Fe 415 HYSD bars

REVIEW QUESTIONS

1. What are the main controlling parameters in the selection of cross-sectional dimensions of reinforced concrete beams?
2. List the various factors influencing the ultimate moment of resistance of a reinforced concrete beam.
3. Specify the reasons for the Indian standard code imposing maximum limits for the spacing and percentage area of reinforcements.
4. What are the partial safety factors used for concrete and steel reinforcements? Specify the reason for a higher safety factor for concrete than steel.
5. Why is it necessary to impose minimum and maximum limits on flexural reinforcements used in beams? Specify the values recommended in the IS code.
6. Under what situations you would resort to the use of doubly reinforced beams? Specify the advantages of doubly reinforced beams in comparison with singly reinforced beams.
7. What are the design parameters to be considered in the design of T- and L-beams. What section in an L-beam is the most critical section and list the various force and moments acting on this section.
8. Explain the terms: (a) Equivalent bending moment, and (b) Equivalent shear force with reference to the design of beams subjected to combined bending, torsion and shear.
9. What are deep beams and how do you classify them? What special design considerations are used in the design of reinforcements in deep beams?
10. How do you check for deflection control in flexural members? What are the factors to be considered in the computation of deflections in reinforced concrete beams.

OBJECTIVE QUESTIONS

1. The cross-section of a reinforced concrete beam should preferably be designed as
   a. over-reinforced
   b. balanced
   c. under-reinforced
2. The span/depth ratio recommended in the Indian standard code for the design of a cantilever beam is
   a. 20
   b. 10
   c. 7

3. When a designed beam does not conform to the limit state of deflection, the alternative procedure is to
    a. increase the tension reinforcement
    b. increase the width of the member
    c. increase the depth of the member

4. The maximum permissible long term deflection including the effects of all loads, creep and shrinkage in a reinforced concrete beam should be limited to
    a. Span/300
    b. Span/150
    c. Span/250

5. Limit state design stipulates that the designed beam should satisfy the limit states of
    a. strength
    b. serviceability
    c. strength and serviceability

6. In case of cellar floors, where there is limited headroom, the ideal type of beam to be used is
    a. flanged beam
    b. singly reinforced beam
    c. doubly reinforced beam

7. According to the Indian standard code, the maximum amount of tension and compression reinforcements in a reinforced concrete beam, expressed as a percentage of the cross-sectional area is limited to a value of
    a. 2%
    b. 0.15%
    c. 4%

8. L-beam should be designed for
    a. flexure
    b. torsion
    c. flexure, shear and torsion

9. In case of cantilever beams, the main reinforcements to resist tension are generally provided at the
    a. top of the member
    b. soffit of the member
    c. middle of the member

10. In case of continuous deep beams, the ratio of effective span to overall depth is
    a. greater than 2
    b. less than 2.5
    c. equal to 3

Limit State Design of Slabs

10.1 GENERAL ASPECTS

The most common type of structural element used to cover floors and roofs are slabs in which the thickness is considerably smaller than their lateral dimensions[1]. They are frequently used in buildings, top and bottom of tanks, decks in bridge structures, raft foundations and staircase slabs. Slabs support and transmit loads to the walls or beams supporting them. In multistoreyed buildings, the floor slabs act as deep horizontal girders and their action as rigid diaphragms of great stiffness is important in restricting the lateral deformation of multistoreyed frames by resisting lateral wind and earthquake loads[2]. The maximum volume of concrete that goes into a structure is in the form of floor, roof slabs and footings. Due to this, the slightest reduction in the design depth of slabs will lead to considerable economy.

Slabs may have different shapes and support conditions. They can be solid, ribbed and waffle types. They can be supported on all the sides, only on opposite sides or on columns. They are classified depending upon the distribution of loads to the edges. A slab is generally designed as a flexural element considering strip of one metre width. As slabs are thin compared to the beams, the serviceability limit state is normally critical in slabs rather than the ultimate strength or collapse limit states of flexure and shear. Slabs have a large surface area compared to their volume and hence they are more affected by shrinkage and temperature stresses. These secondary stresses can be resisted by the use of distribution reinforcement. Normally shear stresses are not critical in slabs and hence shear reinforcements are not generally necessary.

10.2 TYPES OF SLABS

Slabs are classified into different types depending upon their support conditions as simply supported, continuous or cantilevered. Slabs may also be in different shapes such as square, rectangular, circular, trapezoidal and triangular, while the rectangular slabs are more frequently used for floors and roofs.

One-way slabs[3] are those supported on two opposite sides so that the loads are carried in one direction only. A common example of one-way slab is the verandah slab spanning in the shorter direction with main reinforcements. Two-way slabs[4] are supported on all the four sides with dimensions such that the loads are carried to the supports along both directions. Two-way slabs are common in the floors of multistorey buildings. Cantilevered slabs are generally used for chajjas over window and in balconies projecting from the buildings. In T-beam-slab floors, the slab is continuous over T-beams and designed as a continuous slab with positive moments at mid-span and negative moments over supports.

Flat slabs[5] are generally multi-span slabs, which are directly supported on columns at regular intervals without beams. In case of basements, where headroom available is limited, flat slabs can be conveniently used. Flat slabs are commonly used for garages where limited headroom is available. Grid or coffered or waffle slabs[6] consist of small beams spaced at short intervals in perpendicular directions integrally cast with a thin slab. This type of floor is generally used for large conference halls and commercial buildings requiring column free space. The grid floor is generally supported at the edges on solid walls or beams supported on columns at regular intervals. Figure 10.1 shows the various types of slabs normally used in the construction industry.

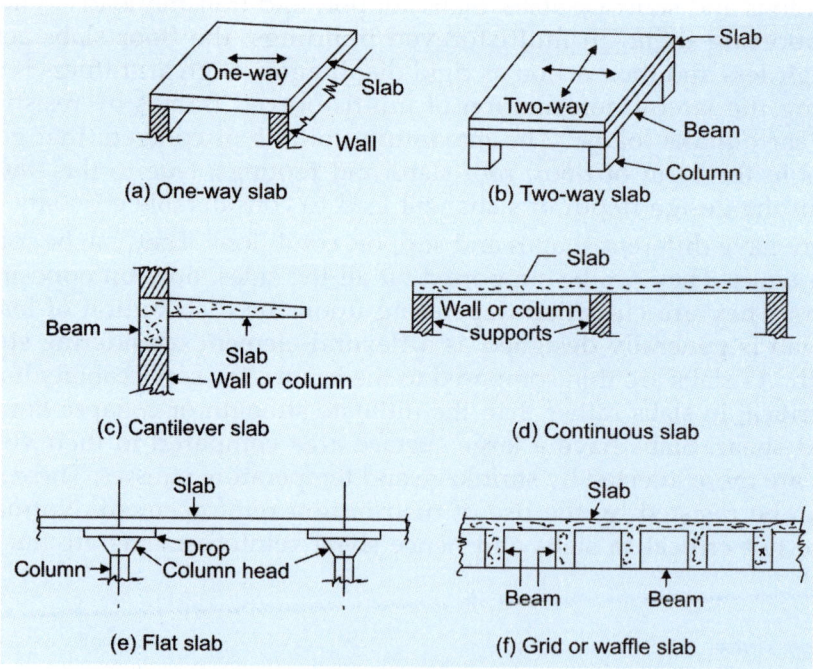

Fig. 10.1. Various types of RC slabs

10.3 DESIGN OF ONE-WAY SLABS

10.3.1 Design Parameters

Reinforced concrete slabs supported on two opposite sides or on all four sides with the ratio of long to short span exceeding 2 are referred to as *one-way slabs*. The slabs are designed as beams of unit width for a given type of loading and support conditions. The span/depth ratios specified in IS:456-2000 code for beam is also applicable for slabs.

The percentage of reinforcement in slabs is generally low in the range of 0.3% to 0.5%. Hence the use of modification factor (K_t) for tension reinforcement results in the span/depth ratio in the range of 25 to 30 for one-way slabs. The thickness of the slab should be such that the shear force at supports is resisted by concrete alone without recourse to shear reinforcements. The permissible shear stress is increased by the use of shear enhancement factor (k) specified in Clause 40.2.1.1, IS:456-2000 code.

Due to practical considerations, the depth of slab selected is generally greater than the minimum depth required for balanced section and hence the slab is under-reinforced. Consequently the reinforcements in the slab can be computed by using equations specified in Annexure-G of the code or by using Charts or Tables of SP:16. The slab designed for flexure is checked for shear stresses and limit state of deflection.

10.3.2 Design Example

Design a one-way slab with a clear span of 3.5 m, simply supported on 200 mm thick concrete masonry walls to support a live load of 4 kN/m². Adopt M20 grade concrete and Fe 415 HYSD bars.

i. *Data*:

Clear span = 3.5 m $f_{ck} = 20$ N/mm²

Width of supports = 200 mm $f_y = 415$ N/mm²

Live load = 4 kN/m²

Floor finish = 1 kN/m²

ii. *Depth of slab*:

Assume depth d = (span/25) = (3500/25) = 140

Assuming a clear cover of 20 mm and using 10 mm diameter bars, we have:

Effective depth d = 140 mm

Overall depth D = 165 mm

iii. *Effective span*:

The least value of

a. Clear span + effective depth = (3.5 + 0.14) = 3.64 m

b. Centre to centre of supports = (3.5 + 0.20) = 3.70 m

Hence L = 3.64 m.

iv. *Loads*:

Self-weight of slab = (0.165 × 25) = 4.125 kN/m

Floor finish   = 1.000

Live load   = 4.000

Total service load w = 9.125 kN/m

Ultimate load w_u = (1.5 × 9.125) = 13.69 kN/m

v. *Ultimate moments and shear forces*:

$$M_u = (0.125 \, w_u \, L^2) = (0.125 \times 13.69 \times 3.64^2) = 22.67 \text{ kN·m}$$
$$V_u = (0.5 \, w_u \, L) \quad = (0.5 \times 13.69 \times 3.64) \quad = 24.92 \text{ kN}$$

vi. *Limiting moment of resistance*:

$$M_{u,lim} = 0.138 \, f_{ck} bd^2$$
$$= (0.138 \times 20 \times 10^3 \times 140^2) \times 10^{-6}$$
$$= 54 \text{ kN·m}$$

Since $M_u < M_{u,lim}$, section is under-reinforced.

vii. *Main reinforcements*:

$$M_u = (0.87 \, f_y A_{st} d)\left[1 - \left(\frac{A_{st} f_y}{bd f_{ck}}\right)\right]$$

$$(22.67 \times 10^6) = (0.87 \times 415 A_{st} \times 140)\left[1 - \left(\frac{415 A_{st}}{10^3 \times 140 \times 20}\right)\right]$$

Solving, $A_{st} = 524 \text{ mm}^2$

Using 10 mm diameter bars, the spacing

$$s_v = \left(\frac{1000 \, a_{st}}{A_{st}}\right) = \left(\frac{1000 \times 78.5}{524}\right) = 150 \text{ mm}$$

Adopt a spacing of 150 mm and alternate bars are bent up at supports.

viii. *Distribution reinforcement*:

$$A_{st} = 0.12\% \text{ of gross cross-sectional area}$$
$$= (0.0012 \times 10^3 \times 165) = 198 \text{ mm}^2$$

Provide 8 mm diameter bars at 250 mm centres ($A_{st} = 201 \text{ mm}^2$)

ix. *Check for shear stress*:

$$\tau_v = (V_u/bd) = (24.92 \times 10^3)/(10^3 \times 140) = 0.178 \text{ N/mm}^2$$
$$p_t = (100 \, A_{st}/bd) = (100 \times 0.5 \times 524)/(10^3 \times 140) = 0.187$$

Permissible shear stress in slab (Table 19, IS:456)

$$k\tau_c = (1.27 \times 0.28) = 0.30 \text{ N/mm}^2 > \tau_v$$

Hence, the shear stresses are within safe permissible limits.

x. *Check for deflection control*:

$$(L/d)_{max} = (L/d)_{basic} \times K_t \times K_c \times K_f$$
$$p_t = (100 \times 524)/(10^3 \times 140) = 0.37$$

Refer to Fig. 8.1, $K_t = 1.40$,

Fig. 8.2, $K_c = 1.00$, and

Fig. 8.3, $K_f = 1.00$

$$\therefore (L/d)_{max} = [(20 \times 1.40 \times 1.00 \times 1.00)] = 28$$
$$(L/d)_{actual} = (3600/140) = 26 < 28$$

Hence, the limit state of deflection is satisfied.

xi. *Design of reinforcement using SP-16 design tables:*

$$\left(\frac{M_u}{bd^2}\right) = \left(\frac{22.67 \times 10^6}{10^3 \times 140^2}\right) = 115$$

Refer to Table 2, SP-16 and readout the percentage of steel corresponding to Fe 415 N/mm² and f_{ck} = 20 N/mm².

$$p_t = 0.343\%$$

Hence $A_{st} = (p_t bd)/100$
$$= (0.343 \times 10^3 \times 140)/(100)$$
$$= 480 \, mm^2$$

xii. *Reinforcement details:*

The reinforcement details in the slab is shown in Fig. 10.2.

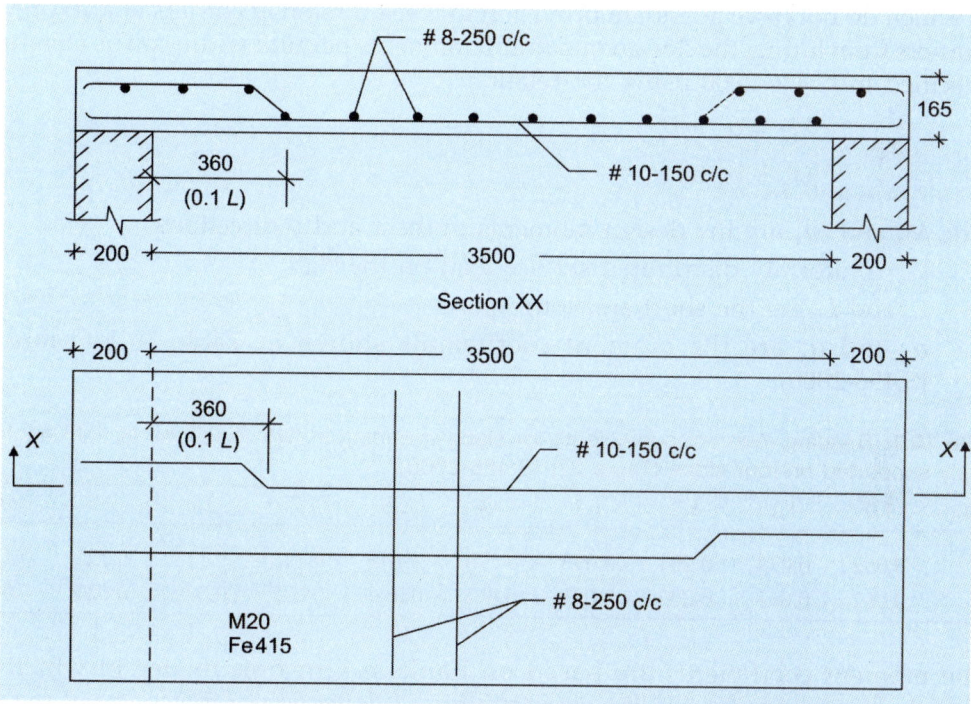

Fig. 10.2: Details of reinforcements in one-way slab

10.4 DESIGN OF TWO-WAY SLABS

10.4.1 Design Parameters

In case of multistorey buildings with column and beam construction, the floor and roof slabs are supported on all the four sides. Two-way slabs should have their longer span not exceeding two times the shorter span so that significant flexural bending develops in mutually perpendicular directions. The flexural moments are maximum at the centre of the slab with the larger magnitude of moment developing along the shorter span.

The moments developed in the slab are influenced by the following factors:

a. Short and long span length (L_x and L_y).

b. Type of supporting edges such as free, fixed, continuous etc.

c. Magnitude and type of load on the slab such as concentrated, uniformly distributed etc.

The magnitude of reinforcement in the direction of short and long spans are designed to resist the maximum design developed in the mutually perpendicular directions.

10.4.2 Simply Supported Slabs

In case of simply supported slabs with free edges, loading on the slab results in sagging of the slab toward the centre of span and lifting of the slab at corners due to nonuniform variation of load transmitted to the supports. Hence simply supported slabs which do not have adequate provision to resist torsion at corners and to prevent the corners from lifting, the design maximum moments per unit width can be computed as specified in IS:456-2000 using the relation:

$$M_x = \alpha_y \, w L_x^2$$

$$M_y = \alpha_y \, w L_x^2$$

where, M_x and M_y are the design moments in the x and y directions

w = uniformly distributed service load on the slab

L_x and L_y are the short and long spans

α_x and α_y are the moment coefficients shown in Table 10.1 (Table 27, IS:456-2000).

Table 10.1: Bending moment coefficients for slabs spanning in two directions at right angles, simply supported on four sides (Table 27, IS:456-2000)

L_y/L_x	1.0	1.1	1.2	1.3	1.4	1.5	1.75	2.0	2.5	3.0
α_x	0.062	0.074	0.084	0.093	0.099	0.104	0.113	0.118	0.122	0.124
α_y	0.062	0.061	0.059	0.055	0.051	0.046	0.037	0.029	0.020	0.014

The moment coefficients are based on Rankine–Groshoff theory in which the moments are evaluated using the compatibility of deflection at the centre of the slab strips in the orthogonal directions. According to IS:456 code, at least 50% of the tension reinforcement provided at mid-span should extend to the supports. The remaining 50% should extend to or within 0.1 L_x or 0.1 L_y of the support.

10.4.3 Two-Way Restrained Slabs with Corners Held Down

Slabs which are supported in such a way that the corners are prevented from lifting are referred to as *restrained slabs*. The slabs may be continuous or discontinuous over the edges. The edges of the two-way slab are assumed to be supported rigidly against vertical translation. The design moments in restrained slabs are computed by using the moment coefficients shown in Table 10.2 (Table 26, IS:456-2000 code).

Table 10.2: Bending moment coefficients for rectangular panels supported on four sides with provision for torsion at corners (Table 26, IS:456-2000)

Case no.	Type of panel and moments considered	Short span coefficients α_x (Values of L_y/L_x)								Long span coefficients α_y for all values of L_y/L_x
		1.0	1.1	1.2	1.3	1.4	1.5	1.75	2.0	
1.	Interior panels									
	Negative moment at continuous edge	0.032	0.037	0.043	0.047	0.051	0.053	0.060	0.065	0.032
	Positive moment at mid-span	0.024	0.028	0.032	0.036	0.039	0.041	0.045	0.049	0.024
2.	One short edge continuous									
	Negative moment at continuous edge	0.037	0.043	0.048	0.051	0.055	0.057	0.064	0.068	0.037
3.	One long edge discontinuous									
	Negative moment at continuous edge	0.037	0.044	0.052	0.057	0.063	0.067	0.077	0.085	0.037
	Negative moment at mid-span	0.028	0.033	0.039	0.044	0.047	0.051	0.059	0.065	0.028
4.	Two adjacent edges discontinuous									
	Negative moment at continuous edge	0.047	0.053	0.060	0.065	0.071	0.075	0.084	0.091	0.047
	Negative moment at mid-span	0.035	0.040	0.045	0.049	0.053	0.056	0.069	0.035	–
5.	Two short edges discontinuous									
	Negative moment at continuous edge	0.045	0.049	0.052	0.056	0.059	0.060	0.065	0.069	–
	Negative moment at mid-span	0.035	0.037	0.040	0.043	0.044	0.045	0.049	0.052	0.035
6.	Two long edges discontinuous									
	Negative moment at continuous edge	–	–	–	–	–	–	–	–	0.045
	Negative moment at mid-span	0.035	0.043	0.051	0.057	0.063	0.068	0.080	0.088	0.035
7.	Three edges discontinuous (One long edge continuous)									
	Negative moment at continuous edge	0.057	0.064	0.071	0.076	0.080	0.084	0.091	0.097	–
	Negative moment at mid-span	0.043	0.048	0.053	0.057	0.060	0.064	0.069	0.073	0.043
8.	Three edges discontinuous (One short edge continuous)									
	Negative moment at continuous edge	–	–	–	–	–	–	–	–	0.057
	Negative moment at mid-span	0.043	0.051	0.059	0.065	0.071	0.076	0.087	0.096	0.043
9.	Four edges discontinuous									
	Negative moment at mid-span	0.056	0.064	0.072	0.079	0.085	0.089	0.100	0.107	0.056

The following assumptions are made in the division of slab into strips and provision of reinforcement:

1. The positive moment of reinforcement is uniformly distributed over the middle strip extending over 75% of the span as shown in Fig. 10.3.
2. Edge strips comprise a width equal to $(L_x/8)$ and $(L_y/8)$.

3. Minimum reinforcement as specified in IS code for slabs should be provided in edge strips.

4. At the corners of simply supported slabs, torsion reinforcement comprising 75% of the area required for maximum mid-span moment in the slab is provided in each of the four layers in the form of a mesh extending to a length of one-fifth of the shorter span.

10.4.4 Span/Depth Ratio in Two-Way Slabs

In case of two-way slabs, due to flexural bending in mutually perpendicular directions, the magnitude of moments will be smaller than in one-way slabs resulting in smaller percentages of reinforcement. Due to this modification factor K_t is higher resulting in the increase of span/depth ratio based on practical knowledge, the following span/overall depth ratios have been recommended in the IS:456 code, Clause 24.1 for two-way slabs with shorter span up to 3.5 m using Fe 415 HYSD bars and for loading class up to 3 kN/m².

1. Simply supported slabs = 28
2. Continuous slabs = 32

10.4.5 Deflection and Crack Control

In two-way slabs, the deflections will be smaller than in one-way slabs due to flexural rigidity of the slab in mutually perpendicular directions. The deflection in two-way slabs is influenced by span/depth ratio and for computations of modification factors, the shorter span and the percentage of steel in that direction should be considered.

If empirical rules regarding the detailing of reinforcements specified in the IS:456 code, Clause 26 is followed, crack control will be achieved and only in special structures where abnormal loads are applied, actual computation of crack width will be necessary as detailed in Annexure F of the IS:456 code.

10.4.6 Design Example

Design a two-way slab for an office floor of size 3.5 m × 4.5 m with discontinuous and simply supported edges on all the sides with corner prevented from lifting and supporting a service live load 4 kN/m². Adopt M20 grade concrete and Fe 415 HYSD bars.

    i. *Data*:

$L_x = 3.5$ $(L_y/L_x) = (4.5/3.5)$

$L_y = 4.5$ $= 1.28 < 2$

$f_{ck} = 20 \text{ N/mm}^2$

$f_y = 415 \text{ N/mm}^2$

Since the ratio of long to short span is less than 2, the slab should be designed as two-way slab with provision for torsion at corners.

    ii. *Depth of slab*:

As the loading class exceeds the value of 3 kN/m², adopt a span/depth ratio of 25.

Depth (span/25)= (3500/25) = 140 mm

Adopt effective depth $d = 140$ mm
Overall depth $D = 165$ mm

iii. *Effective span*:
Effective span = (clear span + effective depth)
$$= (3.5 + 0.14) = 3.64 \text{ m}$$

iv. *Loads*:
Self-weight of slab = (0.165×25) = 4.125 kN/m^2
Live load   = 4.000
Floor finish   = 0.600
Service load w = 8.725 kN/m^2
$\therefore$ Design ultimate load $w_u = (1.5 \times 8.725) = 13.08$ kN/m^2

v. *Ultimate design moments and shear forces*:
Refer to Table 10.2 (Table 26, IS:456 code) and readout the moment
coefficients for $(L_y/L_x) = 1.28$
$$\alpha_x = 0.077 \text{ and } \alpha_y = 0.056$$
$\therefore$
$$M_{ux} = (\alpha_x w_u L_x^2) = (0.077 \times 13.08 \times 3.64^2) = 13.34 \text{ kN} \cdot \text{m}$$
$$M_{uy} = (\alpha_y w_u L_x^2) = (0.056 \times 13.08 \times 3.64^2) = 9.70 \text{ kN} \cdot \text{m}$$
$$V_{ux} = (0.5 w_u L_x) = (0.5 \times 13.08 \times 3.64) = 23.8 \text{ kN}$$

vi. *Check for depth*:
$$M_{max} = 0.138 f_{ck} b d^2$$

$$d = \sqrt{\frac{13.34 \times 10^6}{0.138 \times 20 \times 10^3}} = 69.52 \text{ mm} < 140 \text{ mm}$$

Hence the effective depth selected is sufficient to resist the design ultimate
moment.

vii. *Reinforcements (short and long span)*:

$$M_u = (0.87 f_y A_{st} d \left[1 - \left(\frac{A_{st} f_y}{b d f_{ck}} \right) \right]$$

$$(13.34 \times 10^6) = (0.87 \times 415 A_{st} \times 140) \left[1 - \left(\frac{415 A_{st}}{10^3 \times 140 \times 20} \right) \right]$$

Solving, $A_{st} = 980$ mm^2
Adopt 10 mm diameter bars at 250 mm centres ($A_{st} = 315$ mm^2) in short span
direction. Using 10 mm diameter bars in the long span direction,
effective depth = $(140 - 10) = 130$ mm

$$(9.7 \times 10^6) = (0.87 \times 415 A_{st} \times 130) \left[1 - \left(\frac{415 A_{st}}{10^3 \times 130 \times 20} \right) \right]$$

Solving, $A_{st} = 213$ mm^2
Provide 10 mm diameter bars at 300 mm centres ($A_{st} = 262$ mm^2) in the long
span direction.

viii. *Check for shear stress*:

Considering the short span and unit width of slab

$$\tau_v = (V_u/b\,d) = (23.8 \times 10^3)/(10^3 \times 140) = 0.17\,\text{N/mm}^2$$
$$p_t = (100\,A_{st})/(b\,d) = (100 \times 315)/(10^3 \times 140) = 0.225$$

Refer to Table 19, IS:456 and readout

$$k\,\tau_c = (1.27 \times 0.31) = 0.39\,\text{N/mm}^2 > \tau_v$$

Hence the shear stresses are within safe permissible limits.

ix. *Check for deflection control*:

Considering unit width of slab in the short span direction L_x

$$(L/d)_{max} = 20$$

For $\quad p_t = 0.225$, from Fig. 8.1, $K_t = 1.6$
$$(L/d)_{max} = (20 \times 1.6) = 32$$
$$(L/d)_{actual} = (3640/140) = 26 < 32$$

Hence deflection control is satisfied.

x. *Check for crack control*:

(i) Reinforcement provided is more than the minimum percentage of $0.12\% = (0.012 \times 165 \times 1000) = 198\,\text{mm}^2$

(ii) Spacing of main reinforcement $\ngtr 3\,d$, i.e. $(3 \times 140) = 420\,\text{mm}$

(iii) Diameter of reinforcement $< (D/8) < (165/8) < 20.6\,\text{mm}$. Hence cracks will be within safe permissible limits.

xi. *Torsion reinforcement at corners*:

Area of torsion steel at each of the corners in 4 layers is computed as $(0.75 \times 315) = 236\,\text{mm}^2$.

Length over which torsion steel is provided $= (1/5)$ short span $= (1/5)(3500) = 700\,\text{mm}$.

Provide 6 mm diameter bars at 120 mm centres for a length of 700 mm at all four corners in 4 layers.

xii. *Reinforcement in edge strips*:

$$A_{st} = 0.12\% = (0.0012 \times 10^3 \times 165) = 198\,\text{mm}^2/\text{m}$$

Provide 10 mm diameter bars at 300 mm centres ($A_{st} = 262\,\text{mm}^2$).

xiii. *Design using SP-16 design tables*:

Refer to Table 40, SP-16 and readout the reinforcement spacing for 10 mm diameter bars (nearest depth = 150 mm).

10 mm diameter bars at 250 mm centres for short span.

10 mm diameter bars at 300 mm centres for long span.

xiv. *Reinforcement details*:

The details of reinforcement in the two-way slab is shown in Fig. 10.3.

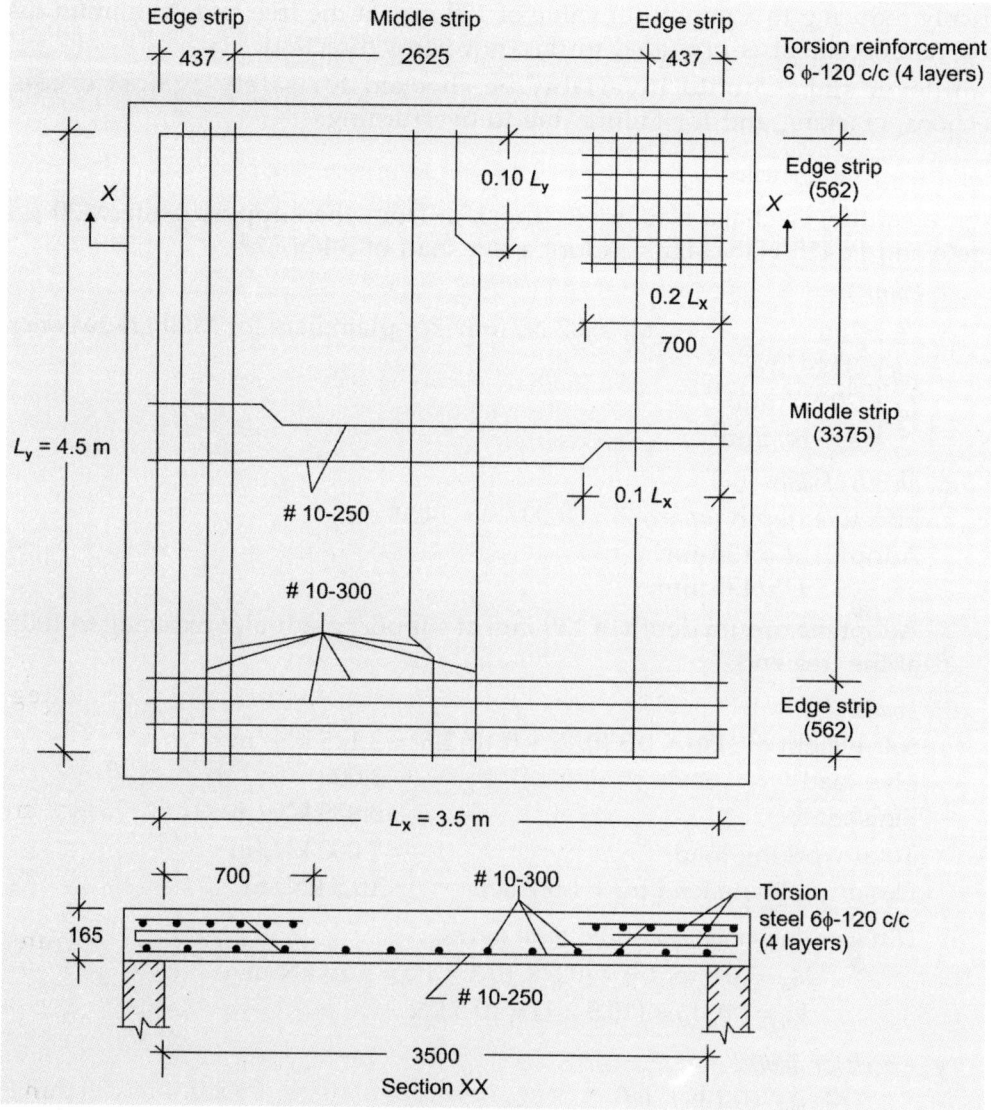

Fig. 10.3: Reinforcement details in two-way slabs (with provision for torsion at corners)

10.5 DESIGN OF CANTILEVER SLABS

10.5.1 General Features

Reinforced concrete slabs projecting from fixed end and free at the other end are referred to as cantilever slabs. Chajjas and balconies projecting from lintel beams or floor slabs are designed as one-way slabs considering them as a cantilever fixed or continuous at the supports. In general, the depth of cantilever slab is based on span/depth ratio of 7 specified in IS:456-2000. It is important to provide the required anchorage length near supports to the main reinforcements to prevent failure due to anchorage.

In cantilever slabs, maximum thickness is generally provided at the fixed end and gradually reducing to a minimum value of 100 mm at the free end. Minimum distribution reinforcement is provided in the transverse direction.

Cantilever slabs should invariably be checked for safety against excessive deflections, cracking and for failure due to overturning.

10.5.2 Design Example

Design a cantilever chajja slab projecting 1 m from the support using M20 grade concrete and Fe 415 HYSD bars. Adopt a live load of 3 kN/m².

  i. *Data*:

$L = 1$ m $\qquad\qquad$ $\tau_{bd} = 1.2$ N/mm² for plain bars for M20 grade concrete

$q = 3$ kN/m²

$f_{ck} = 20$ N/mm²

$f_y = 415$ N/mm²

 ii. *Depth of slab*:

Effective depth (span/7) = (1000/7) = 142.8 mm

Adopt $\quad d = 150$ mm

$\qquad\qquad D = 175$ mm

Adopt maximum depth of 150 mm at support gradually reducing to 100 mm at the free end.

iii. *Loads*:

Self-weight of slab = 0.5 (0.15 + 0.10) 2.5 = 3.125 kN/m

Live load $\qquad\qquad\qquad\qquad\qquad\qquad$ = 3.000

Finishes $\qquad\qquad\qquad\qquad\qquad\qquad\quad$ = 0.875 kN/m

Total working load $\qquad\qquad\qquad\qquad$ = 7.000 kN/m

Design ultimate load $w_u = 1.5(7.00)$ $\quad$ = 10.5 kN/m

 iv. *Ultimate design moments and shear forces*:

$$M_u = (0.5\, w_u\, L^2) = (0.5 \times 10.5 \times 1^2) = 5.25 \text{ kN·m}$$
$$V_u = (w_u\, L) = (10.5 \times 1) = 10.5 \text{ kN}$$

  v. *Check for depth*:

$$M_{u,lim} = 0.138\, f_{ck} b d^2$$
$$= (0.138 \times 20 \times 10^3 \times 150^2)\, 10^{-6}$$
$$= 62.1 \text{ kN·m}$$

Since $\quad M_u < M_{u,lim}$, section is under-reinforced.

 vi. *Reinforcements*:

$$M_u = 0.87\, f_y A_{st} d \left[1 - \left(\frac{A_{st} f_y}{b d f_{ck}}\right)\right]$$

$$(5.25 \times 10^6) = (0.87 \times 415 A_{st} \times 150)\left[1 - \left(\frac{415 A_{st}}{10^3 \times 150 \times 20}\right)\right]$$

Solving, $A_{st} = 105$ mm² $< A_{st,min}$.

Hence provide 10 mm diameter bars at 300 mm centres ($A_{st} = 262$ mm²) in the span direction and the same as distribution reinforcement.

vii. *Anchorage length*:

$$L_d = \left(\frac{0.87\,f_y\,\phi}{4\tau_{bd}}\right) = \left(\frac{0.87 \times 415 \times 10}{4 \times 1.2 \times 1.6}\right) = 470 \text{ mm}$$

viii. *Check for deflection control*:

$$(L/d)_{max} = [(L/d)_{basic} \times K_t] \text{ and } K_c = K_f = 1.00$$

$$p_t = \left(\frac{100 A_{st}}{b\,d}\right) = \left(\frac{100 \times 262}{10^3 \times 150}\right) = 0.174 \text{ mm}$$

Refer to Fig. 8.1, readout $K_t = 1.8$
Hence $(L/d)_{max} = (9.7 \times 1.8) = 12.6$
   $(L/d)_{actual} = (1000/150) = 6.66 < 12.6$

Hence the slab satisfies the deflection criteria.

ix. *Reinforcement details*:
   The reinforcement details in the cantilever slab is shown in Fig. 10.4.

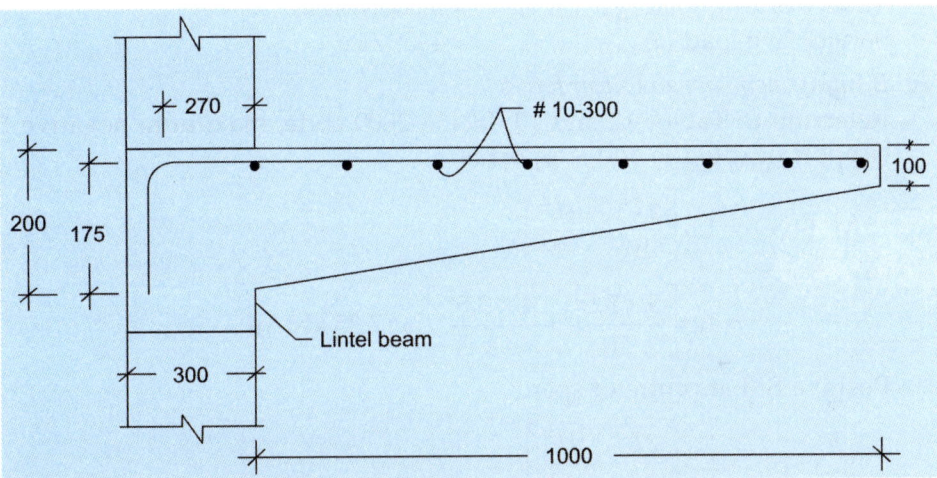

Fig. 10.4: Reinforcement details in cantilever slab

10.6 DESIGN OF CONTINUOUS SLABS

10.6.1 General Features

In multistorey buildings comprising T-beam and slab floors, the slabs are continuous over the beams which are spaced at regular intervals. A simplified approach to design of continuous slabs is presented in IS:456-2000 code in which moment and shear force coefficients are recommended for computation of moments and shear forces which are the same as that for continuous beams.

The coefficients are applicable only for substantially uniformly distributed loads over three or more spans which do not differ by more than 15% of the longest span. When coefficients specified in Tables 12 and 13 of the code are used for computation of bending moments, redistribution of moments is not permitted. If the spans are significantly different, a rigorous analysis using traditional methods such as moment distribution is made to compute the design maximum moments and shear forces.

10.6.2 Design Examples

Design a one-way slab for an office floor which is continuous over T-beams spaced at 3.5 m intervals. Assume a live load 4 kN/m² and adopt M20 grade concrete and Fe 415 HYSD bars.

i. *Data*:

$L = 3.5$ m $f_{ck} = 20$ N/mm²
$q = 4$ kN/m² $f_y = 415$ N/mm²

ii. *Depth of slab*:

Assuming a span/depth ratio of 26 (Clause 23.2.1 of IS:456)
Effective depth $d =$ (span/26)= (3500/26) = 135 mm.
Adopt $d = 140$ mm $D = 165$ mm

iii. *Loads*:

Self-weight of slab = (0.165 × 25) = 4.125 kN/m²
Finishes   = 0.875 kN/m²
Total working load (g)   = 5.000 kN/m²
Service live load (q)   = 4 kN/m²

iv. *Bending moments and shear forces*:

Referring to Tables 12 and 13, IS:456-2000 code, maximum negative BM at support next to the end support is:

$$M_u\,(-ve) = 1.5\left[\frac{gL^2}{10} + \frac{qL^2}{9}\right]$$

$$= 1.5\left[\frac{5 \times 3.5^2}{10} + \frac{4 \times 3.5^2}{9}\right] = 17.35 \text{ kN} \cdot \text{m}$$

Positive BM at centre of span

$$M_u\,(+ve) = 1.5\left[\frac{gL^2}{12} + \frac{qL^2}{10}\right]$$

$$= 1.5\left[\frac{5 \times 3.5^2}{12} + \frac{4 \times 3.5^2}{10}\right] = 15 \text{ kN} \cdot \text{m}$$

Maximum shear force at the support
$$V_u = 1.5 \times 0.6\,(g + q)\,L$$
$$= (1.5 \times 0.6)\,(5 + 4)\,3.5 = 28.35 \text{ kN}$$

v. *Check for depth of slab*:
$$M_{u,lim} = 0.138\,f_{ck}bd^2$$
$$= (0.138 \times 20 \times 10^3 \times 140^2)\,10^{-6}$$
$$= 54.1 \text{ kN·m}$$
Since $M_u < M_{u,lim}$, section is under-reinforced.

vi. *Reinforcements (short and long span)*:

$$M_u = (0.87\,f_y\,A_{st}\,d\left[1 - \left(\frac{A_{st}f_y}{bd\,f_{ck}}\right)\right]$$

$$(17.35 \times 10^6) = (0.87 \times 415 A_{st} \times 140) \left[1 - \left(\frac{415 A_{st}}{10^3 \times 140 \times 20} \right) \right]$$

Solving, $A_{st} = 360$ mm^2.

Provide 10 mm diameter bars at 150 mm centres ($A_{st} = 524$ mm^2). The same reinforcement is provided for positive BM at mid-span.

Distribution steel $= (0.0012 \times 10^3 \times 165) = 198$ mm^2

Provide 10 mm diameter bars at 300 mm centres ($A_{st} = 262$ mm^2).

vii. *Check for shear stress*:
$$\tau_v = (V_u/bd) = (28.35 \times 10^3)/(10^3 \times 140) = 0.20 \text{ N/mm}^2$$
$$p_t = (100 \, A_{st})/(bd) = (100 \times 262)/(10^3 \times 140) = 0.187$$

Refer to Table 19, IS:456 and readout:
$$k\tau_c = (1.27 \times 0.30) = 0.38 \text{ N/mm}^2$$

Since $\tau_v < \tau_c$, the sab is safe against shear stresses.

viii. *Check for deflection control*:

Considering the end and interior spans
$$(L/d)_{max} = [(L/d)_{basic} \times K_t \times K_c \times K_f]. \text{ Also } K_c = K_f = 1.00$$

$$p_t = \left(\frac{100 \times 393}{10^3 \times 140} \right) = 0.28$$

From Fig. 8.1, read out $K_t = 1.5$

$$\therefore \qquad (L/d)_{max} = \left(\frac{20 + 26}{2} \right) 1.5 = 34.5$$

$$(L/d)_{actual} = (3500/140) = 25 < 34.5$$

Hence the slab is safe against deflection control.

ix. *Reinforcement details*:

The reinforcement details in the continuous slab is shown in Fig. 10.5.

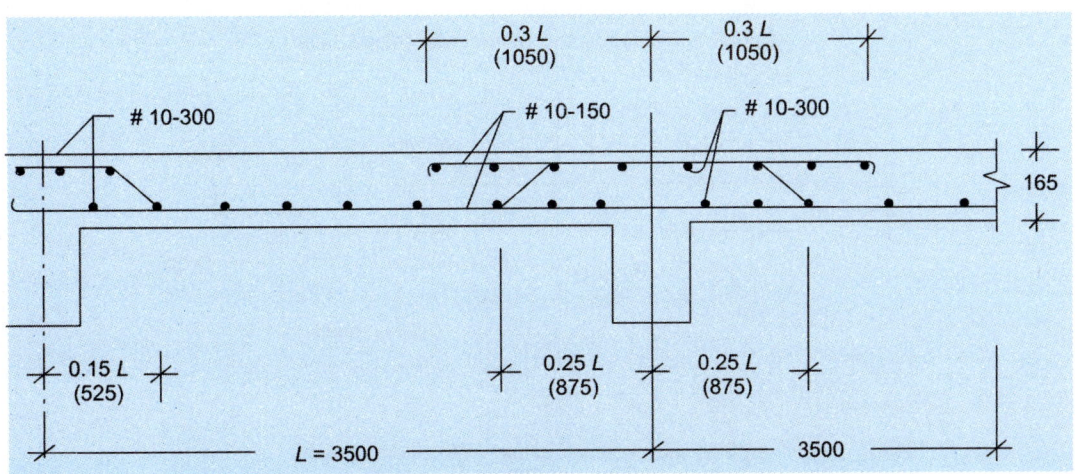

Fig. 10.5: Reinforcement details in one-way continuous slab

10.7 DESIGN OF FLAT SLABS

10.7.1 Types of Flat Slabs

A flat slab is a reinforced concrete slab with or without drops, supported generally without beams by column with or without flared column heads.

The different types of slabs shown in Fig. 10.6 are referred to as:

a. Slabs without drops and column heads.

b. Slabs without drops and column with column head.

c. Slabs with drops and column with column head.

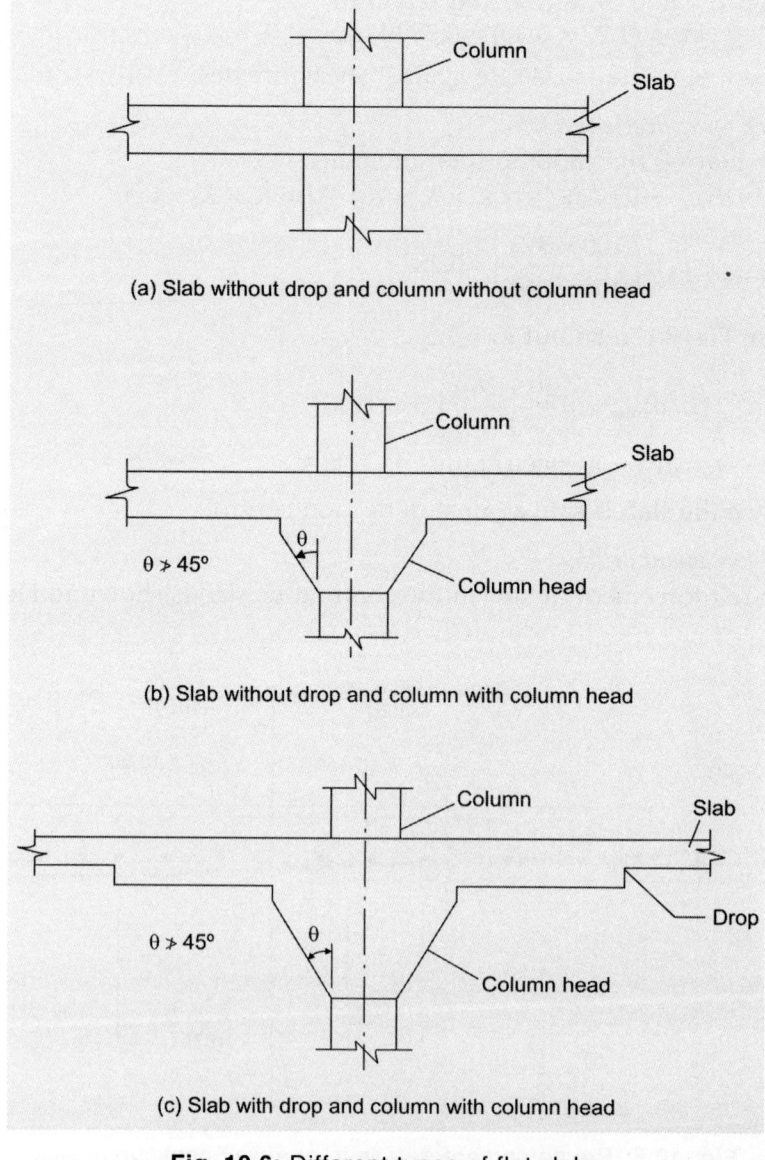

(a) Slab without drop and column without column head

(b) Slab without drop and column with column head

(c) Slab with drop and column with column head

Fig. 10.6: Different types of flat slabs

10.7.2 Panel Divisions

The flat slab panel is divided into column strip and middle strip in Fig. 10.7.

a. *Column strip*: Column strip is a design strip having a width of $0.25 L_2$ but not greater than $0.25 L_1$ on each side of the column centre line where L_1 is the span in the direction, moments are being determined, measured centre to centre of supports and L_2 is the span transverse to L_1 measured centre to centre of support.

b. *Middle strip*: Middle strip is a design strip bounded on each of its opposite sides by the column strip.

c. *Panel*: Panel is that part of the slab bounded on each of its four sides by the centre line of columns or centre lines of adjacent spans.

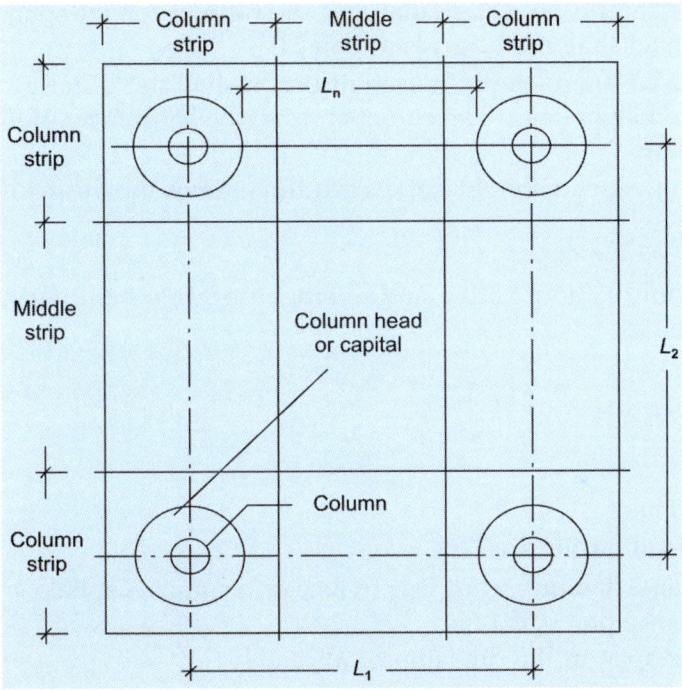

Fig. 10.7: Division of flat slab into column and middle strips

10.7.3 Thickness of Flat Slabs

The minimum thickness of a flat slab is 125 mm.

For slabs with drops (span/depth) = 40

For slabs without drops (span/depth) = 36

The IS code 456-2000 specifies that longer span should be considered for calculation of the thickness of the slab.

10.7.4 Drops and Column Head

a. *Drops*: The drops rectangular in plan should have a length in each direction not less than one-third of the panel length in that direction. For exterior panels, the width of drops at right angles to the non-continuous edge and measured from the centre line of the columns shall be equal to one-half the width of drop for interior panels.

b. *Column heads*: Where column heads are provided, that portion of a column head which lies within the largest right circular cone or pyramid that has a vertex angle of 90° and can be included entirely within the outlines of the column and the column head shall be considered for design purposes.

10.7.5 Direct Design Method

The following limitations apply to the slab systems designed by the direct design method:

a. There must be at least three continuous spans in each direction.

b. The panels should be rectangular and the ratio of longer span to shorter span within a panel shall not be greater than 2.0.

c. The successive span length in each direction shall not differ by more than one-third of the longer span. The end spans may be shorter but not longer than the interior spans.

d. The design live load should not exceed three times the design dead load.

Design moments for a bar

The absolute sum of the positive and average negative bending moment in each direction is given as

$$M_0 = \left(\frac{W \times L_n}{8} \right)$$

where,

M_0 = total moment

W = design load on the area ($L_2 \times L_n$)

L_n = clear span extending from face to face of columns, capitals, brackets or walls but not less than $0.65 L_1$

L_1 = length of span in the direction of M_0 and

L_2 = length of span transverse to L_1.

The recommendations of IS:456-1978 result in the following percentage of moments in the column and middle strips for an interior panel.

Moments in interior panel

Type of moments =	Bending moment distribution % M_0	
	Column strip	Middle strip
Negative moments =	$(0.65 \times 0.75 \times M_0)$ 49%	15%
Positive moments =	$(0.35 \times 0.60 \times M_0)$ 21%	15%

Moments in exterior panel

The moments in the exterior panel are influenced by the flexural stiffness of columns and slab.

The total design moment M_0 is distributed in the following properties:

$$\text{Interior negative design moment} = 0.75 - \left(\frac{0.10}{1 + \dfrac{1}{\alpha_c}} \right)$$

$$\text{Exterior negative design moment} = \left(\frac{0.65}{1 + \dfrac{1}{\alpha_c}} \right)$$

$$\text{Positive design moment} = 0.63 - \left(\frac{0.28}{1 + \dfrac{1}{\alpha_c}} \right)$$

where, α_c = ratio of flexural stiffness of exterior columns to the flexural stiffness of the slab at a joint taken in the direction, moments are being determined and is given by:

$$\alpha_c = \frac{\Sigma K_c}{K_s}$$

where,

ΣK_c = sum of the flexural stiffness of the columns meeting at the joint, and

K_s = flexural stiffness of the slab, expressed as moment per unit rotation.

At an exterior support, the column strip shall be designed to resist the total negative moment in the panel at that support.

10.7.6 Equivalent Frame Method

The structure is analysed as a continuous frame with the following assumptions:

a. The structure is considered to be made up of equivalent frames longitudinally and transversely consisting of row of columns and strip of slab with a width equal to the distance between the centre lines of the panel on each side of the row of columns.

b. Each frame is analysed by any established method like moment distribution or any other suitable method. Each strip of floor and roof may be analysed as a separate frame with the columns above and below assumed fixed at their extremities.

c. The relative stiffness is computed by assuming gross cross-section of the concrete alone in the calculation of the moment of inertia.

d. Any variation of moment of inertia along the axis of the slab on account of provision of drops should be considered. In case of recessed or coffered slab which is made solid in the region of the columns, the stiffening effect may be ignored provided the solid part of the slab does not extend more than $0.15L_{ef}$ into the span measured from the centre line of the columns. The stiffening effect of flared column heads may be ignored.

10.7.7 Shear in Flat Slab

The critical section for shear is at a distance $(d/2)$ from the periphery of the column/capital/drop panel, perpendicular to the plane of the slab, where d is the effective depth of the section. The shape in plan is geometrically similar to the support immediately below the slab.

The nominal shear stress in flat slabs is computed as $(V/b_0 d)$, where V is the shear force due to design load and b_0 is the periphery of the critical section and d is the effective depth.

When shear reinforcement is not provided, the calculated shear stress at the critical section shall not exceed $k_s \cdot \tau_c$

where,

$k_s = (0.5 + \beta_c)$ but not greater than 1

β_c = ratio of short side to long side of the column/capital and

$\tau_c = 0.25 \sqrt{f_{ck}}$ in limit state method of design and

$\quad = 0.16 \sqrt{f_{ck}}$ in working stress method of design.

When the shear stress exceeds this value, suitable shear reinforcement according to the provisions of the code should be provided.

10.7.8 Design Example

Design the interior panel of a flat slab with drops for an office floor to suit the following data

    i. *Data*:

        Size of office floor = 25 m × 25 m

        Size of panels = 5 m × 5 m

        Loading class = 4 kN/m^2

        Materials: M20 grade concrete

                    Fe 415 HYSD bars

   ii. *Slab thickness*:

        According to IS:456 code, for two-way continuous slabs:

        Thickness of slab (span/40) = (5000/40) = 125 mm

        Adopt thickness of slab in middle strip = 150 mm

        Thickness of slab at drops = (150 + 50) = 200 mm

        Column head diameter is computed as

                $D \not> 0.25L = (0.25 \times 5) = 1.25$ m

        Length of drop $\not< (L/3)$ in either direction

                $\not< (5/3) = 1.66$ mm

        Adopt drop width = 2.5 m

        ∴ Column strip = drop width = 2.5 m

          Middle strip width = 2.5 m

      Span of flat slab = $L_1 = L_2 = 5$ m

iii. *Loads*:

Self-weight of slab in middle strip = (0.15×25) = 3.75 kN/m²

Dead load due to extra thickness of slab at drops = (0.05×25) = 1.25 kN/m²

Live load = 4.00

Finishes = 1.00

Working load w = 10.00 kN/m²

Design ultimate load $w_u = (1.5 \times 10) = 15$ kN/m²

iv. *Ultimate bending moment*:

$$M_0 = \left(\frac{W L_n}{8}\right)$$

where $L_n = (5 - 1.25) = 3.75$ m $> 0.65\, L_1 > (0.65 \times 5) = 3.25$

and $L_1 = L_2 = 5$ m

∴ $W = (w_u L_2 L_n) = (15 \times 5 \times 3.75) = 281.25$ kN

$$M_0 = \left(\frac{281.25 \times 3.75}{8}\right) = 132 \text{ kN} \cdot \text{m}$$

For interior panels with drops:

Column strip moments

Negative BM = 49% of M_0

$\qquad = (0.49 \times 132) = 65$ kN·m

Positive BM = 21% of M_0

$\qquad = (0.21 \times 132) = 28$ kN·m

Middle strip moments

Negative BM = 15% of $M_0 = (0.15 \times 132) = 20$ kN·m

Positive BM = 15% of $M_0 = (0.15 \times 132) = 20$ kN·m

v. *Check for thickness of slab*:

a. Thickness of slab required at drops

$$d = \sqrt{\frac{M_u}{0.138 f_{ck} b}} \quad \text{where, } b = 2500 \text{ mm}$$

$$= \sqrt{\frac{65 \times 10^6}{0.138 \times 20 \times 2500}} = 97 \text{ mm}$$

Effective depth provided $d = 170$ mm

Overall depth $D = 200$ mm

b. Thickness of slab required in middle strips

$$d = \sqrt{\frac{20 \times 10^6}{0.138 \times 20 \times 2500}} = 53.8 \text{ mm}$$

Effective depth provided $d = 120$ mm

Overall depth $D = 150$ mm

vi. *Check for shear stress*:

Shear stress is checked near the column head at section $(D + d)$.

Total load on the circular area with $(D + d)$ as diameter is

$$W_1 = (\pi/4)(D + d)^2 w_u$$
$$= (\pi/4)(1.25 + 0.17)^2\ 15$$
$$= 23.74\ \text{kN}$$

Shear force $= [(\text{Total load}) - (\text{load on circular area})]$
$$= [(15 \times 5 \times 5) - (23.74)]$$
$$= 351.26\ \text{kN}$$

Shear force/metre width of perimeter:

$$V_u = \left(\frac{351.26}{1.25 + 0.17}\right) = 78.8\ \text{kN/m}$$

$\therefore$ Shear stress $\tau_v = (V_u/b\,d) = [(78.8 \times 10^3)/(10^3 \times 170)]$
$$= 0.46\ \text{N/mm}^2$$

According to Clause 31.6.3.1, IS:456-2000

Permissible shear stress $= (k_s\ \tau_c)$

where, $k_s = (0.5 + \beta_c)$ and $\beta_c = (L_1/L_2) = (5/5) = 1$

$\therefore$ $k_s = (0.5 + 1) = 1.5 \ngtr 1.0$

$\therefore$ $k_s = 1.0$

$$\tau_c = 0.25\sqrt{f_{ck}} = 0.25\sqrt{20} = 1.12\ \text{N/mm}^2$$
$$k_s\ \tau_c = (1 \times 1.12) = 1.12\ \text{N/mm}^2$$

Hence $\tau_v < k_s\ \tau_c$

Hence the slab is safe against shear failure.

vii. *Reinforcements in column and middle strips*:

a. Column strip:

$$M_u = (0.87\ f_y\ A_{st}\ d)\left[1 - \left(\frac{A_{st} f_y}{b\,d\,f_{ck}}\right)\right]$$

$$(65 \times 10^6) = (0.87 \times 415 A_{st} \times 170)\left[1 - \left(\frac{415 A_{st}}{2500 \times 170 \times 20}\right)\right]$$

Solving, $A_{st} = 1122\ \text{mm}^2$ for negative bending moment

$\therefore A_{st}/\text{metre} = (1122/2.5) = 449\ \text{mm}^2$

Adopt 12 mm diameter bars at 250 mm centres ($A_{st} = 452\ \text{mm}^2$)

A_{st} for positive moment

$$(28 \times 10^6) = (0.87 \times 415 A_{st} \times 170)\left[1 - \left(\frac{415 A_{st}}{2500 \times 170 \times 20)}\right)\right]$$

Solving, $A_{st} = 672\ \text{mm}^2$

$\therefore A_{st}/\text{metre} = (672/2.5) = 269\ \text{mm}^2$

Provide 10 mm diameter bars at 270 mm centres ($A_{st} = 285\ \text{mm}^2$).

b. Column strip:

A_{st} for positive and negative moments is computed as:

$$(20 \times 10^6) = (0.87 \times 415 A_{st} \times 120)\left[1 - \left(\frac{415 A_{st}}{2500 \times 120 \times 20)}\right)\right]$$

Solving, $A_{st} = 474 \text{ mm}^2$

$\therefore A_{st}/\text{metre} = (474/2.5) = 190 \text{ mm}^2$

Provide 10 mm diameter bars at 300 mm centres ($A_{st} = 262 \text{ mm}^2$).

viii. *Reinforcement details*:

The details of reinforcement in the flat slab is shown in Fig. 10.8.

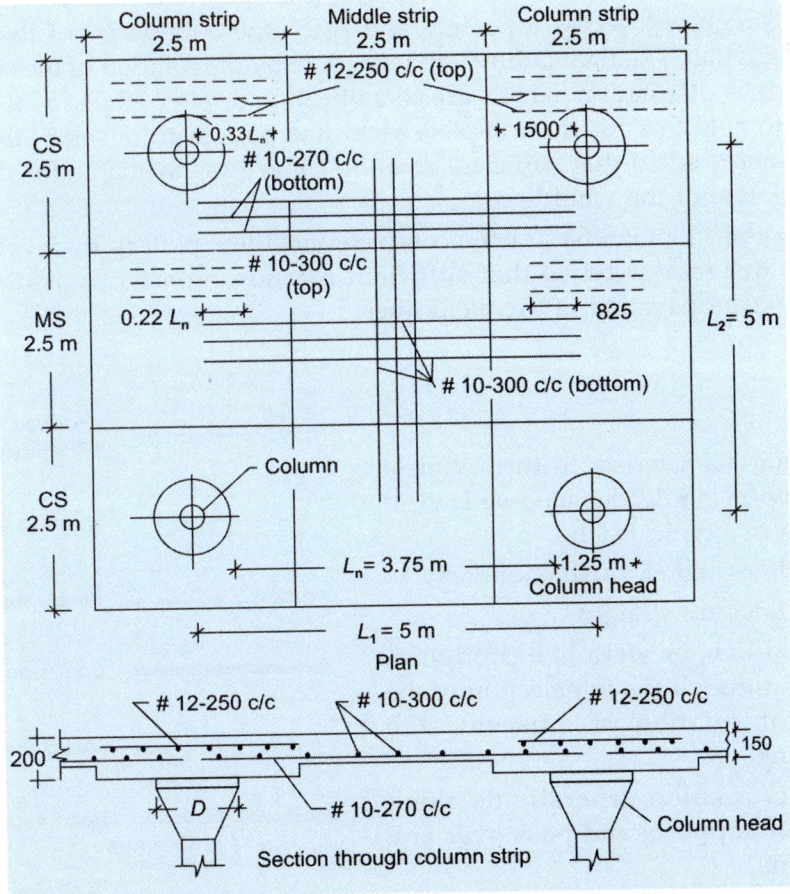

Fig. 10.8: Reinforcement details in flat slab

10.8 YIELD LINE ANALYSIS OF SLABS

10.8.1 General Features

Reinforced concrete slabs when loaded to failure develop a characteristic pattern of cracks depending upon the shape of the slab, support conditions and distribution of reinforcements in the slab. An ingenious method of estimating the ultimate load capacity of slabs was developed by the Danish engineer Ingerslav[7] in 1923. The method was further developed and improved by engineers like Johanssen[8], Wood[9] and Jones[10]. The typical crack pattern developed in an isotropically reinforced concrete slab is shown in Fig. 10.9.

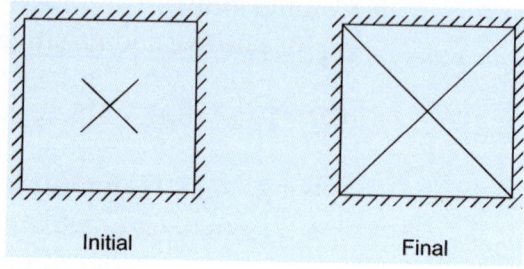

Initial   Final

Fig. 10.9: Yield line pattern in a simply supported square slab

As the load is increased on the slab, the region of highest moment will yield first and the yield lines developed at the centre will progress towards the edges and reach the boundaries of the slab before collapse of the slab. The final failure will take place by the rotation of the slab elements about the axes of rotation which are usually the support edges of the slab. It is important to note that for the complete yield line pattern to develop, the slab must be under-reinforced so that sufficient rotation capacity is available for the initiation and propagation of the yield lines.

It is important to note that for the complete yield line pattern to develop, the slab must be under reinforced so that sufficient rotation capacity is available for the initiation and propagation of the yield lines.

10.8.2 Characteristic Features of Yield Lines

The following characteristic features of yield lines are helpful in selecting a possible yield line pattern in a typical slab.

1. Yield lines end at a slab boundary.
2. Yield lines are straight.
3. A yield line, or yield line produced, passes through the intersection of the axes of rotation of adjacent slab elements.
4. Axes of rotation generally lie along lines of supports and pass over any columns.

Fig. 10.10 shows the notations used to represent the yield line and supports.

The yield line patterns developed in slabs of different shapes and with different edge conditions are shown in Fig. 10.11. Negative

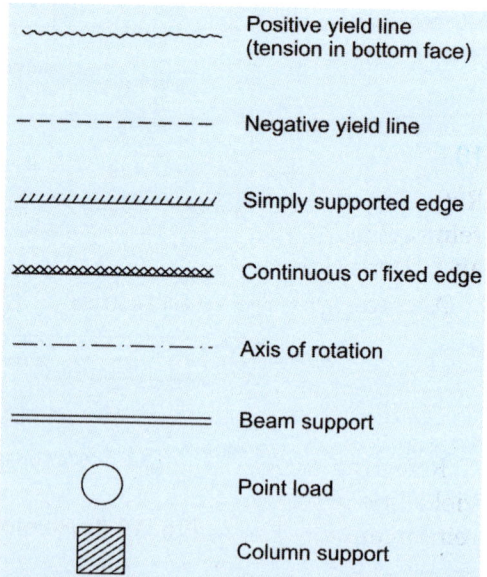

Positive yield line (tension in bottom face)

Negative yield line

Simply supported edge

Continuous or fixed edge

Axis of rotation

Beam support

Point load

Column support

Fig. 10.10: Notations for yield lines and supports

yield lines from near the supports in case of slabs are fixed or continuous at the edges.

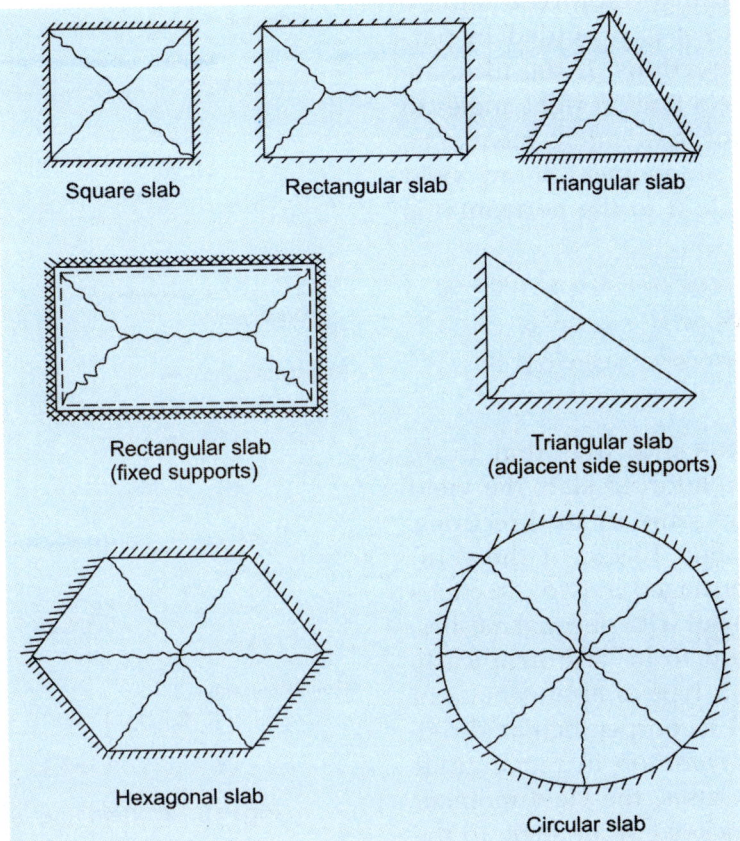

Fig. 10.11: Typical yield line patterns in reinforced concrete slabs

10.8.3 Yield Moments

Referring to Fig. 10.12a, when the yield line is at right angle to the direction of the reinforcement, the yield line ultimate moment is given by the governing equation for an under-reinforced section of a flexural member.

According to the condition of IS:456-1978

$$m = M_u = 0.87 f_y A_{st} d \left[1 - \frac{A_{st} f_y}{b d f_{ck}} \right]$$

Referring to Fig. 10.12b, if a yield line *ab* has an ultimate *m* per unit length and the yield line *ab* makes an angle α with the yield line *cd* which is at right angle to the reinforcement. The yield moment *m* is calculated as follows:

$$m_\alpha \cdot ab = m \cos \alpha \times cd$$

$$m = m \cos \alpha \times \left(\frac{cd}{ab} \right) = m \cos^2 \alpha$$

If there is more than one mesh of reinforcement

$$m_x = \Sigma m \cos^2 \alpha$$

In isotropically reinforced square slabs equal steel is provided in perpendicular directions. If the ultimate moment of yield lines at right angles to the direction of the reinforcement is m, then the ultimate moment of any yield line at an angle α to the horizontal is given by:

$$m_\alpha = m \cos^2\alpha + m \cos^2 (90°-\alpha)$$
$$= m \cos^2\alpha + m \sin^2\alpha$$
$$= m (\cos^2\alpha + \sin^2\alpha)$$
$$= m$$

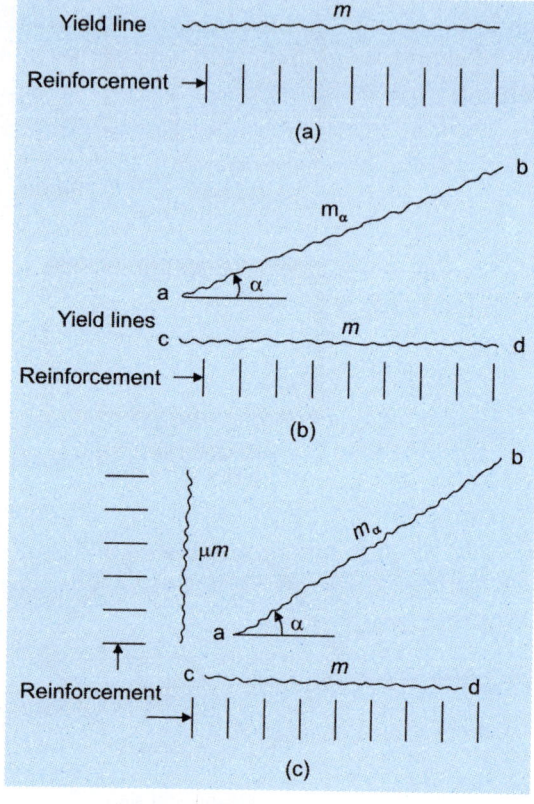

Fig. 10.12: Yield line moments

The criterion indicates that in an isotropically reinforced slab, the yield moment is the same in all directions. Referring to Fig. 10.12c, if the reinforcement is arranged in two directions at right angles but with unequal meshes, the slab is said to be orthotropically reinforced. This type of arrangement of different steel in perpendicular directions is very common in rectangular slabs. In such cases, the yield moment along a line inclined at an angle to the horizontal is given by:

$$m = m \cos^2\alpha + \mu m \cos^2 (90° - \alpha)$$
$$= m \cos^2\alpha + \mu m \sin^2\alpha$$

10.8.4 Ultimate Loads on Slabs

There are two methods of determining the ultimate load capacity of slabs. They are based on the principles of (a) virtual work (b) equilibrium.

The virtual work method is based on the principle that the applied loads causing a small virtual displacement is equal to the internal work done or energy dissipated in rotation along the yield lines. It is generally assumed that the elastic deformations in the slab are negligible and all the plastic deformation takes place at the yield lines.

In the equilibrium method, the equilibrium of the individual segments of slab formed by the yield lines under the action of the applied loads and moments and forces acting on the edges of the segments are considered.

Both the virtual work and equilibrium methods give an upper bound to the collapse load on the slab. Hence it is essential that all possible yield line patterns have to be investigated to find the lowest value of the ultimate load.

If a correct yield line pattern has been assumed, the lower bound solution will coincide with the upper bound solution but lower bound solutions are not available except for a few simple cases of slabs. Test results have shown that the actual failure loads of slabs is greater than the predicted values by yield line analysis because the membrane action. Hence the upper bound solutions resulting from yield line analysis can be used with a reasonable degree of safety.

10.8.5 Yield Line Analysis by Virtual Work Method

1. *Isotropically reinforced square slab simply supported and supporting uniformly distributed load*: The principle of virtual work method is to equate the internal work done due to rotation of yield lines to the external work done due to the loads having a virtual displacement.

$$\text{External work done} = \Sigma(W\delta)$$

where, W = loads

δ = virtual displacements

$$\text{Internal work done} = \Sigma(M\theta) = \Sigma(mL\theta)$$

where, m = ultimate moment per unit length of yield line

L = length of yield line

Refer to Fig. 10.13.

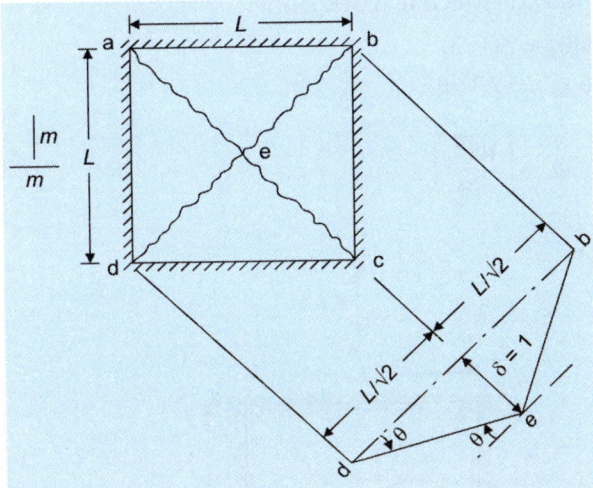

Fig. 10.13: Yield line pattern in a square slab (simply supported)

The square slab is isotropically reinforced, the ultimate moment along the yield line is also m, thus the total work done in yield line ac is given by:

$$\Sigma(M\theta)_{ac} = \Sigma m{\cdot}L{\cdot}\theta = m\sqrt{2}\,L\frac{\sqrt{2}}{L} = 4\,m$$

The work done in yield line bd is the same as in ac.

Total internal work done = $\Sigma(M\theta)$ = 8 m.

For a virtual displacement, δ = 1 at e, the centre of gravity of each of the triangular elements deflects by 1/3

$\therefore$ $\qquad$ $\Sigma(W\delta) = 1/3\,wL^2$

where, w = uniformly distributed load on slab. Equating, we get

$$\Sigma(M\theta) = \Sigma(W\delta)$$

We have:

$$m = \left(\frac{wL^2}{24}\right)$$

2. *Isotropically reinforced square slab fixed on all edges and subjected to a uniformly distributed load:*

Referring to Fig. 10.14, since the edges are fixed, negative yield lines will form along the edges.

Internal work done along the positive yield lines ac and db is given by:

$$\Sigma(M\theta) = 8 \, m \text{ (Refer to previous problem)}$$

Internal work done along the negative yield lines ab, bc, cd and de is given by:

$$\Sigma(M\theta) = 4[mL \, (2/L)] = 8 \, m$$

$\therefore$ Total internal work done = $\Sigma(M\theta) = 16 \, m$

External work done = $\Sigma(W\delta) = 1/3 \, wL^2$

Equating internal to external work done:

$$\Sigma(M\theta) = \Sigma(W\delta)$$
$$16 \, m = 1/3 \, wL^2$$

$$m = \left(\frac{wL^2}{48}\right)$$

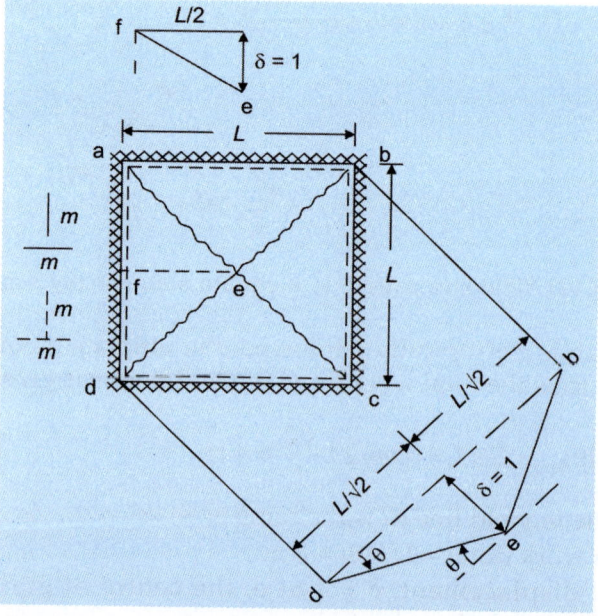

Fig. 10.14: Yield line pattern in a square slab (fixed)

3. *Triangular slab simply supported on adjacent edges and subjected to uniformly distributed load and isotropically reinforced:* Referring to Fig. 10.15, the triangular slab acb is simply supported at ac and cb. The yield line formed is cd. Unit displacement is given for point d. Since slab is isotropically reinforced $m_x = m_y = m$.

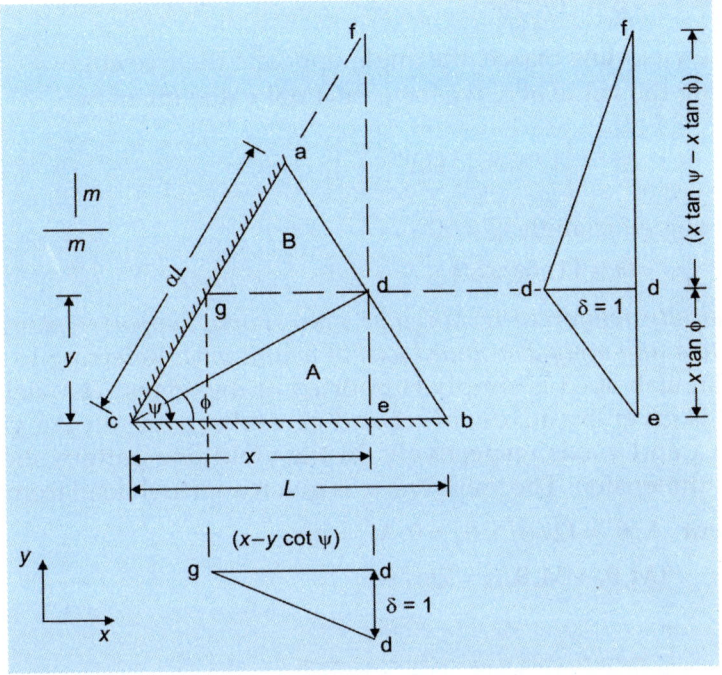

Fig. 10.15: Triangular slab simply supported on adjacent edges

For element A, $\theta_{Ax} = 1/de = 1/x\phi$ $\tan \theta_{Ay} = 0$.

$\therefore (M_x\theta_x + M_y\theta_y)_A = x \cdot m \cdot \theta_x = m \cot \phi$

For element B, $\theta_{Bx} = 1/df = 1(x \tan \psi - x \tan \phi)$

and $\theta_{By} = 1/gd = 1/(x - y \cot \psi)$

$\therefore (M_xQ_x + M_yQ_y)_B = m\left[\dfrac{1}{\tan \psi - \tan \phi} + \dfrac{1}{\cot \phi - \cot \psi}\right]$

$$= m\left[\dfrac{1 + \tan \psi \cdot \tan \phi}{\tan \psi - \tan \phi}\right] = m \cot(\psi - \phi)$$

Thus $\Sigma(M\theta) = m[\cot(\psi - \phi) + \cot \phi]$

and $\Sigma(W\delta) = 1/6\, w\alpha L^2 \sin \psi$

Equating, we get $\Sigma(M \cdot \theta) = \Sigma(W \cdot \delta)$

We have:

$$m = \dfrac{w\alpha L^2 \cdot \sin \psi}{6[\cot(\psi - \phi) + \cot \phi]}$$

$$m = 1/6 w\alpha L^2 \sin\phi \cdot \sin(\psi - \phi)$$

For a maximum value of m, $\dfrac{dm}{d\phi} = 0$

$\therefore \cos\phi \cdot \sin(\psi - \phi) = \sin\phi \cdot \cos(\psi - \phi)$

$\tan\phi = \tan(\psi - \phi)$

$\phi = (1/2)\psi$

Hence, the yield line bisects the angle opposite the free edge. Substituting the value of ϕ, we have the final value given by:

$$M = \left\{ 1/6\, w\alpha L^2 \sin^2\left(\frac{\psi}{2}\right) \right\}$$

In a rignt angled triangle $\psi = 90°$.

Then $\qquad m = (1/6\, w\alpha L^2)$

4. *Orthotropically reinforced rectangular slab, simply supported along its edges and subjected to a uniformly distributed load of w/unit area:* Referring to Fig. 10.16, the rectangular slab abcd is simply supported at the edges. The yield line pattern assumed is given by ae, de, bf, cf and ef, and m and μm are yield moments along the x and y axis respectively. In the yield line pattern shown 'βL' is an unknown dimension. The yield line *ef* is given a virtual displacement of unity.

For element A, $\theta_x = (2/\alpha L)$, $\theta_y = 0$, $M_x = mL$

$\qquad (M_x\theta_x + M_y\theta_y)_A = 2m/\alpha$

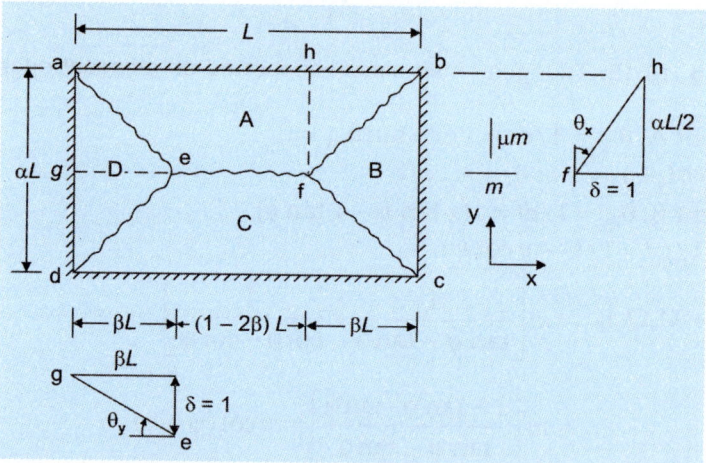

Fig. 10.16. Rectangular slab simply supported at the edges

For element D, $\theta_x = 0$, $\theta_y = (1/\beta L)$, $M_y = (\alpha L\mu m)$

$\therefore \qquad (M_x\theta_x + M_y\theta_y)_D = (\alpha\mu m/\beta)$

Since element A and C, B and D are similar

$$\Sigma(M\cdot\theta) = 2\left(\frac{2m}{\alpha} + \frac{\alpha\mu m}{\beta}\right)$$

The external work done is given by

$$\Sigma(W\delta) = wL^2 \left[\frac{2\beta\alpha}{3} + \frac{\alpha(1 - 2\beta)}{2} \right]$$

Equating $\Sigma(M\theta) = \Sigma(W\delta)$, we get

$$m = \frac{1}{12}\alpha^2 L^2 \left(\frac{3\beta - 2\beta^2}{2\beta + \mu\alpha^2} \right)$$

If the work equation is of the form

$$m = w \left[\frac{f_1(x_1 x_2)}{f_2(x_1 x_2)} \right]$$

For a maximum value of m

$$\left(\frac{\delta m}{\delta x} \right) = 0$$

This is obtained for the condition

$$\frac{f_1(x_1 x_2)}{f_2(x_1 x_2)} = \frac{\dfrac{\delta}{\delta x}[f_1(x_1 x_2)]}{\dfrac{\delta}{\delta x}[f_2(x_1 x_2)]}$$

Using this criterion for a maximum value of m,

$$\frac{\delta m}{\delta \beta} = 0$$

Hence, we have

$$\left(\frac{3\beta - 2\beta^2}{2\beta + \mu\alpha^2} \right) = \left(\frac{3 - 4\beta}{2} \right)$$

Cross multiplying, we get the quadratic as

$$4\beta^2 + 4\mu\alpha^2 b - 3\mu\alpha^2 = 0$$

The positive root of this quadratic

$$\beta = 1/2 \left[\sqrt{(3\mu\alpha^2 + \mu^2\alpha^4)} - \mu\alpha^2 \right]$$

Substituting the value of β in the above above expression for m, we have

$$m = \frac{w\alpha^2 L^2}{24} \left[\sqrt{(3 + \mu \cdot \alpha^2)} - a\sqrt{\mu} \right]^2$$

5. *Isotropically reinforced circular slab, simply supported and uniformly loaded all round:* Referring to Fig. 10.17, a circular slab of radius 'r' is simply supported at the edges and supports a uniformly distributed load of w/unit area. In circular slabs, the failure will take place by the formation of an infinite number of positive yield lines running radially from the centre to the circumference, resulting in the formation of a flat cone at collapse.

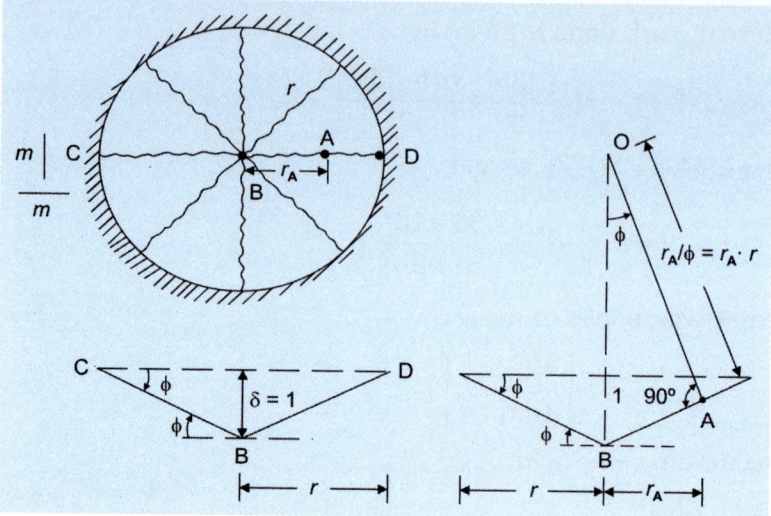

Fig. 10.17: Circular slab simply supported at the edges

For unit displacement at the centre of slab,
external work done = $\Sigma(W\delta) = (\pi r^2 \cdot w/3)$
For a central displacement of unity

$$\angle AOB = \phi = 1/r$$

$\therefore$ Length $OA = (r_A/\phi) = r_A r$

Change of slope in the tangential direction at A, per unit length of arc is equal to the angle between the two normal unit lengths of arc apart at A is given by $(1/r_A \cdot r)$. Total change of slope in one complete revolution is given by

$$\Sigma\theta = (2\pi r_A \times 1/r_A r) = 2\pi/r$$

Internal work done in rotation at yield lines = $\Sigma(mL\theta)$, since all the yield lines are of equal length, work done is given by

$$mL\Sigma\theta = mr(2\pi/r) = 2\pi m$$

Equating internal work to external work done, we have:

$$(1/3)\pi r^2 w = 2\pi m$$

$$m = \left(\frac{wr^2}{6}\right)$$

10.8.6 Yield Line Analysis by Equilibrium Method

1. *Square slab, isotropically reinforced and subjected to a uniformly distributed load:* The assume yield line pattern is shown in Fig. 10.18. Considering the equilibrium of the triangular element C, by taking, moments about the edge ab we have

$$mL = (1/2)L \cdot (L/2)(w)(L/6)$$

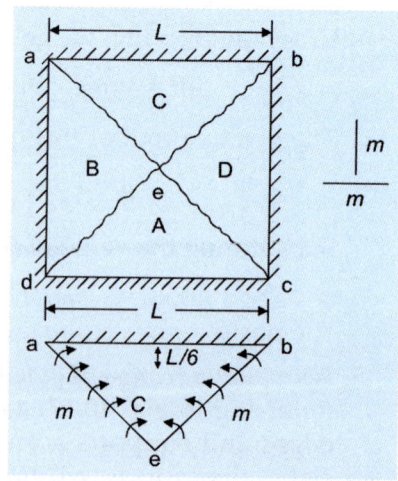

Fig. 10.18: Equilibrium of element C

$$\therefore \qquad m = \left(\frac{wL^2}{24}\right)$$

2. *Rectangular slab orthotropically reinforced and subjected to a uniformly distributed load:* The assumed yield line pattern is shown in Fig. 10.19. Considering the equilibrium of the trapezoidal element A.

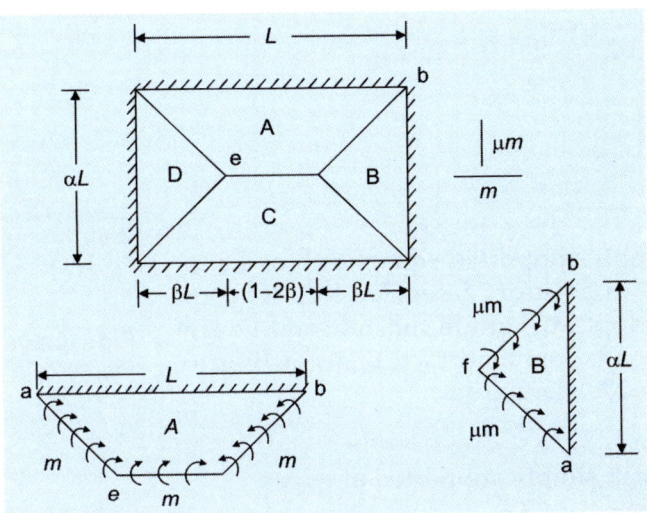

Fig. 10.19: Equilibrium of elements in a rectangular slab

$$mL = w\left[(1 - 2\beta)L \cdot \frac{\alpha L}{2} \cdot \frac{\alpha L}{4} + 2\frac{\beta L}{2} \cdot \frac{\alpha L}{2} \cdot \frac{\alpha L}{6}\right]$$

$$m = \left[\frac{w\alpha^2 L^2}{24} \cdot (3 - 4\beta)\right]$$

Taking moments about bc for element B:

$$\mu m \cdot L = 1/6 w\alpha\beta^2 L^3$$

$$m = \left(\frac{w\beta^2 L^2}{6\mu}\right)$$

Equating the two equilibrium expressions, we have

$$\left[\frac{\alpha^2(3 - 4\beta)}{24}\right] = \left(\frac{\beta^2}{6\mu}\right)$$

or $4\beta^2 + 4\mu\alpha^2\beta - 3\mu\alpha^2 = 0$

The positive root of this quadratic in β is

$$\beta = \frac{1}{2}\left[\sqrt{3\mu\alpha^2 + \mu^2\alpha^4)} - \mu\alpha^2\right]$$

Substituting the value of β in the equilibrium expression, we have:

$$m = \frac{w\alpha^2 L^2}{24}\left[\sqrt{(3 + \mu\alpha^2)} - \alpha\sqrt{\mu}\right]^2$$

3. *Hexagonal slab isotropically reinforced and subjected to uniformly distributed load:* The isotropically reinforced hexagonal slab is shown in Fig. 10.20. Considering the equilibrium of element A, we have:

$$mL = \left(\frac{1}{2}L \cdot \frac{\sqrt{3}L}{2} \cdot \frac{1}{3} \cdot \frac{\sqrt{3}L}{2}\right) \cdot w$$

$$\therefore \quad m = \left(\frac{wL^2}{8}\right)$$

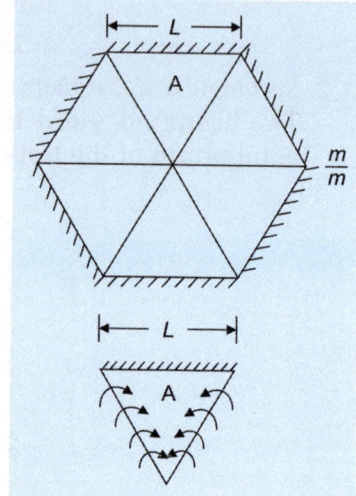

Fig. 10.20: Equilibrium of elements in a hexagonal slab

10.8.7 Design Examples

1. Design a simply supported square slab of 4.5 m side length to support a service live load of 4 kN/m². Adopt M20 grade concrete and Fe 415 HYSD bars. Assume load factors according to IS:456-2000 code standards.

   i. *Data:*
   Square slab simply supported at edges
   Side length $L = 4.5$ m
   Live load $q = 4$ kN/m²
   $f_{ck} = 20$ N/mm²
   $f_y = 415$ N/mm²

   ii. *Depth of slab:*
   For simply supported slab using Fe 415 HYSD bars, according to Clause 24.1 IS:456-2000 code
   (Span/overall depth) = $(35 \times 0.8) = 28$
   ∴ Overall depth $D = (\text{span}/28) = (4500/28) = 160$ mm
   Adopt overall depth $D = 160$ mm and effective depth $d = 135$ mm

   iii. *Loads:*
   Self-weight of slab (0.16×25) = 4.00 kN/m²
   Live load   = 4.00 kN/m²
   Floor finish   = 1.00 kN/m²
   Total service load w = 9.00 kN/m²
   ∴ Design ultimate load $w_u = (1.5 \times 9) = 13.5$ kN/m

   iv. *Ultimate moments and shear forces:*
   The yield or ultimate moment capacity of a simply supported square slab is given by the relation
   $$m = M_u = (w_u L^2/24) = (13.5 \times 4.5^2)/24$$
   $$= 11 \cdot 39 \text{ kN·m}$$
   Ultimate shear $V_u = (0.5 w_u L) = (0.5 \times 13.5 \times 4.5)$
   $$= 30.375 \text{ kN/m}$$

v. *Limiting moment capacity of the slab*:

$$M_{u,lim} = 0.138 f_{ck} bd^2$$
$$= (0.138 \times 20 \times 10^3 \times 135^2) 10^{-6}$$
$$= 50.3 \text{ kN·m}$$

Since $M_u < M_{u,lim}$, the section is under-reinforced.

vi. *Reinforcement*:

$$M_u = 0.87 f_y A_{st} d \left[1 - \left(\frac{A_{st} f_y}{bd f_{ck}} \right) \right]$$

$$(11.39 \times 10^6) = (0.87 \times 415 A_{st} \times 135) \left[1 - \left(\frac{415 A_{st}}{10^3 \times 135 \times 20} \right) \right]$$

Solving $A_{st} = 241 \text{ mm}^2$
Adopt 10 mm diameter bars at 300 mm centres ($A_{st} = 262 \text{ mm}^2$).

viii. *Check for shear stress*:

$$\tau_v = \left(\frac{V_u}{bd} \right) = \left(\frac{30.375 \times 10^3}{1000 \times 135} \right) = 0.225 \text{ N/mm}^2$$

$$p_t = \left(\frac{100 A_{st}}{b_w d} \right) = \left(\frac{100 \times 262}{1000 \times 135} \right) = 0.194\%$$

Refer to Table 19 to IS:456-2000 and readout the permissible shear stress as
$$k_s \tau_c = (1.28 \times 0.31) = 0.39 \text{ N/mm}^2$$

Since $k_s \tau_c > \tau_v$, the shear stresses are within safe permissible limits.

2. Design a rectangular slab 5 m × 4 m in size and simply supported at the edges to support a service load (live) of 4 kN/m². Assume coefficient of orthotropy (μ) as 0.7. Adopt M20 grade concrete and Fe 415 HYSD bars.

i. *Data*:

$L = 5$ m	$\mu = 0.7$
$\alpha L = 4$ m	$f_{ck} = 20 \text{ N/mm}^2$
$\alpha = 0.8$	$f_y = 415 \text{ N/mm}^2$

ii. *Depth of slab*:
Overall depth = (span/28) = (4000/28) = 143 mm
Adopt overall depth $D = 150$ mm
Effective depth $d = 125$ mm

iii. *Loads*:

Self-weight of slab = (0.15 × 25)	= 3.75 kN/m²
Live load	= 4.00
Floor finish	= 1.25
Total service load w	= 9.00 kN/m²

Design ultimate load $w_u = (1.5 \times 9) = 13.5 \text{ kN/m}^2$

iv. *Ultimate moments and shear forces*:

$$M_u = m = \left(\frac{w_u \, \alpha^2 L^2}{24}\right)\left[\sqrt{(3+\alpha^2)} - \alpha\sqrt{\mu}\right]^2$$

$$= \left(\frac{13.5 \times 16}{24}\right)\left[\sqrt{(3+0.7 \times 0.64)} - 0.8\sqrt{0.7}\right]^2$$

$$= 12.68 \text{ kN·m/m}$$

$$V_u = (0.5 w_u \, L) = (0.5 \times 13.5 \times 4) = 27 \text{ kN/m}$$

v. *Limiting moment capacity of the slab*:

$$M_{u, \text{lim}} = 0.138 f_{ck} \, bd^2$$

$$= (0.138 \times 20 \times 10^3 \times 135^2)$$

$$= 50.3 \text{ kN·m}$$

Since $M_u < M_{u, \text{lim}}$, the section is under-reinforced.

vi. *Reinforcement*:

$$M_u \text{ (short span)} = (0.87 f_y A_{st} d)\left[1 - \left(\frac{A_{st} f_y}{bd f_{ck}}\right)\right]$$

$$(12.68 \times 10^6) = (0.87 \times 415 A_{st} \times 125)\left[1 - \left(\frac{415 A_{st}}{10^3 \times 125 \times 20}\right)\right]$$

Solving, $A_{st} = 295 \text{ mm}^2$

Adopt 10 mm diameter bars at 250 mm centres ($A_{st} = 262 \text{ mm}^2$) in the short span direction.

$$A_{st} \text{ (long span)} = \mu A_{st}$$

$$= (0.7 \times 250) = 175 \text{ mm}^2$$

$$A_{st} \text{ (minimum)} = (0.0012 \times 10^3 \times 150) = 180 \text{ mm}^2$$

Provide 10 mm bars at 300 mm centres ($A_{st} = 262 \text{ mm}^2$)

vii. *Check for shear stress*:

$$\tau_v = \left(\frac{V_u}{bd}\right) = \left(\frac{27 \times 10^3}{10^3 \times 125}\right) = 0.216$$

$$p_t = \left(\frac{100 A_{st}}{b_w \, d}\right) = \left(\frac{100 \times 315}{10^3 \times 125}\right) = 0.252$$

Permissible shear stress from Table 19, IS:456

$$(k_s \cdot \tau_c) = (1.30 \times 0.36) = 0.468 \text{ N/mm}^2$$

Since $(k_s \cdot \tau_c) > \tau_v$, the shear stresses are within safe permissible limits.

3. A right angled triangular slab is simply supported at the adjacent edges AB and BC. The side $AB = 4$ m, $BC = 3$ m and $CA = 5$ m. The slab is isotropically

reinforced with 10 mm diameter bars at 100 mm centres both ways at an average effective depth of 120 mm. The overall depth of the slab is 150 mm. If $f_{ck} = 20\,\text{N/mm}^2$ and $f_y = 415\,\text{N/mm}^2$, estimate the safe permissible service live load on the slab.

i. *Data*:

$$AB = 4\,\text{m} \qquad\qquad f_{ck} = 20\,\text{N/mm}^2$$
$$BC = 3\,\text{m} \qquad\qquad f_y = 415\,\text{N/mm}^2$$
$$CA = 5\,\text{m} \qquad\qquad d = 120\,\text{mm}$$
$$L = 4\,\text{m} \qquad\qquad D = 150\,\text{mm}$$
$$\alpha L = 3\,\text{m}$$

Reinforcements provided (10 mm diameter) at 100 mm centres both ways.

$$A_{st} = \left(\frac{1000 \times 78.5}{100}\right) = 785\,\text{mm}^2/\text{m}$$

ii. *Yield or ultimate moment*:

$$m = M_u = 0.87\,f_y\,A_{st}d\left[1 - \left(\frac{A_{st}f_y}{bdf_{ck}}\right)\right]$$

$$= (0.87 \times 415 \times 785 \times 120)\left[1 - \left(\frac{785 \times 415}{1000 \times 120 \times 20}\right)\right]$$

$$= 29 \times 10^6\,\text{N·mm}$$

$$= 29\,\text{kN·m}$$

iii. *Ultimate load on slab*:

$$w_u = \left(\frac{6\,M_u}{\alpha L^2}\right) = \left(\frac{6 \times 29}{0.75 \times 16}\right) = 14.5\,\text{kN/m}^2$$

iv. *Service live load*

Total service load = $(14.5/1.5) = 9.66\,\text{kN/m}^2$
Dead load of slab = $(0.15 \times 25) = 3.75\,\text{kN/m}^2$
Service live load = $(9.66 - 3.75) = 5.91\,\text{kN/m}^2$

4. A hexagonal slab of side length 4 m is simply supported at the edges and it is isotropically reinforced with 12 mm diameter bars at 150 mm centres both ways, at an average effective depth of 120 mm. The overall depth of the slab is 150 mm. Calculate the ultimate load capacity of the slab and also the safe permissible live load if M20 grade concrete and Fe 415 HYSD bars are used.

i. *Data*:

Hexagonal slab, simply supported at edges having side length $L = 4$ m
12 mm diameter bars provided at 150 mm centres.

$$\therefore \qquad A_{st} = \left(\frac{1000 \times 113}{150}\right) = 753\,\text{mm}^2/\text{m}$$

$$D = 150\,\text{mm} \qquad\qquad d = 120\,\text{mm}$$
$$f_{ck} = 20\,\text{N/mm}^2 \qquad\qquad f_y = 415\,\text{N/mm}^2$$

ii. *Yield or ultimate moment*:

$$m = M_u = 0.87 f_y A_{st} d \left[1 - \left(\frac{A_{st} f_y}{b d f_{ck}} \right) \right]$$

$$= (0.87 \times 415 \times 753 \times 120) \left[1 - \left(\frac{785 \times 415}{10^3 \times 120 \times 20} \right) \right]$$

$$= 28.38 \times 10^6 \text{ N·mm}$$

$$= 28.38 \text{ kN·m}$$

iii. *Ultimate load on slab*:

$$w_u = \left(\frac{8m}{L^2} \right) = \left(\frac{8 \times 28.38}{4^2} \right) = 14.19 \text{ kN/m}^2$$

iv. *Service live load*:
Total service load = $(14.19/1.5) = 9.46 \text{ kN/m}^2$
Self-weight of slab = $(0.15 \times 25) = 3.75 \text{ kN/m}^2$
$\therefore$ Safe permissible live load is given by:
$q = (9.46 - 3.75) = 5.71 \text{ kN/m}^2$

5. Design a circular slab of 4 m diameter which is simply supported at the edges. Adopt service load as 4 kN/m² and M20 grade concrete with Fe 415 HYSD bars. Assume load factors according to IS:456-2000 code.

i. *Data*:
Circular slab simply supported at edges
Diameter of slab = 4 m
Radius $r = 2$ m
Live load $q = 4 \text{ kN/m}^2$
$f_y = 415 \text{ N/mm}^2$

ii. *Depth of slab*:
Overall depth of slab $D = (\text{span}/28) = (4000/28) = 142.8$ mm
Adopt overall depth $D = 150$ mm
Effective depth $d = 120$ mm

iii. *Ultimate loads*:
Self-weight of slab (0.15×25) = 3.75 kN/m²
Live load   = 4.00 kN/m²
Floor finish   = 1.25 kN/m²
Total service load w = 9.00 kN/m²
Ultimate design load $w_u = (1.5 \times 9) = 13.5$ kN/m²

iv. *Ultimate moments and shear forces*:
The yield or ultimate moment capacity of a circular slab simply supported at edges is given by the relation

$$M_u = \left(\frac{w_u \cdot r^2}{6} \right) = \left(\frac{13.5 \times 2^2}{6} \right) = 9 \text{ kN·m/m}$$

v. *Limiting moment capacity of the slab*:

$$M_{u, \text{lim}} = 0.138 f_{ck} bd^2$$
$$= (0.138 \times 20 \times 10^3 \times 120^2) \, 10^{-6}$$
$$= 39.74 \text{ kN·m}$$

Since $M_u < M_{u, \text{lim}}$, the section is under-reinforced.

vi. *Reinforcement*:

$$M_u = (0.87 f_y A_{st} d \left[1 - \left(\frac{A_{st} f_y}{bd f_{ck}} \right) \right]$$

$$(9 \times 10^6) = (0.87 \times 415 A_{st} \times 120) \left[1 - \left(\frac{415 A_{st}}{10^3 \times 120 \times 20} \right) \right]$$

Solving $A_{st} = 215 \text{ mm}^2$

Adopt 10 mm diameter bars at 300 mm centres ($A_{st} = 262 \text{ mm}^2$).

vii. *Check for shear stress*:

$$V_u = (0.5 \, w_u \, L) = (0.5 \times 13.5 \times 4) = 27 \text{ kN}$$

$$\tau_v = \left(\frac{V_u}{bd} \right) = \left(\frac{27 \times 10^3}{10^3 \times 120} \right) = 0.225 \text{ N/mm}^2$$

$$= \left(\frac{100 A_{st}}{bd} \right) = \left(\frac{100 \times 262}{10^3 \times 120} \right) = 0.218$$

Refer to Table 19, IS:456-2000 and readout the permissible shear stress

$$(k_s \tau_c) = (1.30 \times 0.33) = 0.429 \text{ N/mm}^2$$

Since $(k_s \tau_c) > \tau_v$, the shear stresses are within safe permissible limits.

6. A rectangular slab 4.5 m × 6.5 m is simply supported at the edges. The coefficient of orthotropy $\mu = 0.75$. If the ultimate design load is 12 kN/m², estimate the ultimate moment capacity of the slab in the short span direction by deriving the expression:

$$w = (24 m / L_y^2) (\mu / \tan^2 \phi)$$

From first principle, where

w = ultimate design load

m = ultimate moment capacity of th slab in the short span direction

μ = coefficient of orthotropy

L_y = short span length

ϕ = angle made by the shorter yield line with the side L_y

i. *Data*:

Short span length $L_y = 4.5 \text{ m}$

Long span length $\alpha L_y = L_x = 6.5 \text{ m}$

Ultimate design load $w = 12 \text{ kN/m}^2$

Coefficient of orthotropy $\mu = 0.75$

Ultimate moment capacity of slab in the short span direction = m kN·m/m
Ultimate moment capacity of slab in the long span direction = μm kN·m/m

ii. *Derivation on relation between ultimate load and ultimate moment*:
Referring to Fig. 10.21.

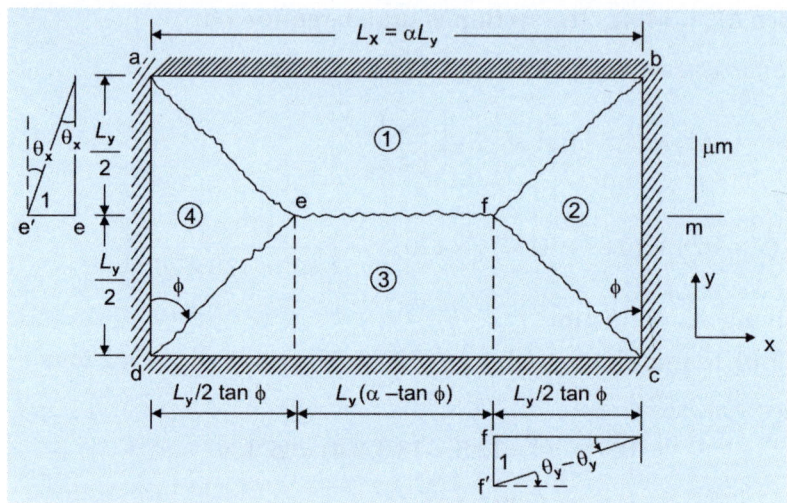

Fig. 10.21: Rectangular slab with simply supported edges

External work done = $\Sigma(W\theta)$

Yield line ef is given in unit direction $\delta = 1$

Work done by the elements 1, 2, 3 and 4 are computed as detailed below:

For element 1:

$$\text{Work done} = \left[L_y(\alpha - \tan\phi)L_y/2 \times \frac{1}{2}\right] + \left[\left(2 \times \frac{1}{2}\right) \times (L_y/2\tan\phi)\right]$$
$$\times (L_y/2 \times 1/3)]$$
$$= 1/4\, L_y^2(\alpha - \tan\phi) + (1/12)L_y^2\tan\phi$$

For element 2:

$$\text{Work done} = (1/2)L_y(L_y/2)\tan\phi \cdot 1/3)]$$
$$= (1/12)L_y^2\tan\phi$$

∴ Total external work done in elements 1, 2, 3 and 4 is obtained as:

$\Sigma(W\delta) = 2$ (work done in element 1 and work done in element 2)

$$\Sigma(W\delta) = (1/2)wL_y^2(\alpha - 1/3\tan\phi) \tag{1}$$

Internal work done on yield lines *ae, bf, cf* and *de* by rotation of the elements 1, 2, 3 and 4 is obtained as follows:

For element 1:

$$\theta_x = (2/L_y),\ \theta_x = 0,\ M_x = mL_x = m\alpha L_y$$

∴ $M_x\theta_x + M_y\theta_y = 2\,m\alpha$

For element 2:

$$\theta_y = (2/L_y\tan\phi),\ \theta_x = 0,\ \text{and } M_y = \mu \cdot m \cdot L_y$$

∴ $(M_x\theta_x + M_y\theta_y) = 2\,\mu m/\tan\phi$

∴ Total internal work done in all the yield lines is:

$\Sigma(M\theta) = 2[2\,m\alpha + (2\,\mu m/\tan\phi)]$

$= 4\,m[\alpha + (\mu/\tan\phi)]$ (2)

For equilibrium, we must equate:

$\Sigma(M\delta) = \Sigma(M\cdot\theta)$

$(1/2)wL_y^2\,(\alpha - 1/3\tan\phi) = 4\,m(\alpha + \mu/\tan\phi)$

$(wL_y^2/24m) = (\alpha + \mu/\tan\phi)/(3\alpha - \tan\phi)$ (3)

For a maximum value of 'm', we have

$dm/d\,(\tan\phi) = 0$

Differentiating RHS of Eq. (3), we have:

$(\alpha + \mu/\tan\phi)/(3\alpha - \tan\phi) = (-\mu/\tan^2\phi)/(-1) = (\mu/\tan^2\phi)$

$\alpha\cdot\tan^2\phi + 2\mu\cdot\tan\phi - 3\alpha\cdot\mu = 0$

The positive root of the quadratic is

$\tan\phi = \sqrt{3\mu + \mu^2/\alpha^2} - \mu/\alpha$

which gives the values $\tan\phi$ for minimum collapse load.
Substituting this in Eq. (3), the collapse load is expressed as

$w = (24\cdot m/L_y^2)\,(\mu/\tan^2\phi)$ (4)

iii. *Example*:

In this given example using numerical values

$w = 12\ \text{kN/m}^2$

$\mu = 0.75$

$L_x = 6.5\ \text{m}$

$L_y = 4.5\ \text{m}$

∴ $\alpha = (6.5/4.5) = 1.44$

$\tan\phi = \sqrt{(3\mu + \mu^2/\alpha^2)} - (\mu/\alpha)$

$= \sqrt{(3\times0.75) + (0.75^2/1.44^2)} - (0.75/1.44)$

$= 1.06$

∴ $m = (w\,L_y^2/24)\,(\tan^2\phi/\mu)$

$= (12\times4.5^2)/(24)\,(1.06^2/0.75)$

$= 15.16\ \text{kN·m/m}$

If the slab is fixed on all the four sides and m' is the moment capacity of the negative reinforcement, then the relation between the ultimate moment and ultimate load on the slab is expressed by the relation:

$w = [24(m + m')/L_y^2]\,[\mu/\tan^2\phi]$

If $m = m'$, then

$w = (48\,m/L_y^2)\,(\mu/\tan^2\phi)$

7. A two-way RCC slab is rectangular having a size 4 m × 5 m with two longer edges fixed in position and the two shorter edges are simply supported. Derive

the relation between moment capacity of slab and ultimate load by first principles and hence design the slab for a working live load of 3 kN/m² by yield line theory. Assume μ = 0.8. Adopt M15 grade concrete and HYSD bars.

i. *Data*:

Short span length L_y = 4 m

Long span length L_y = 5 m

Coefficient of orthotropy = μ = 0.8

Working live load q = 3 kN/m²

Longer edges are fixed and shorter edges are simply supported

Materials : M15 grade concrete
: Fe 415 HYSD bars

ii. *Stresses*

f_{ck} = 15 N/mm²

f_y = 415 N/mm²

iii. *Derivation of relation*:

Referring to Fig. 10.22.

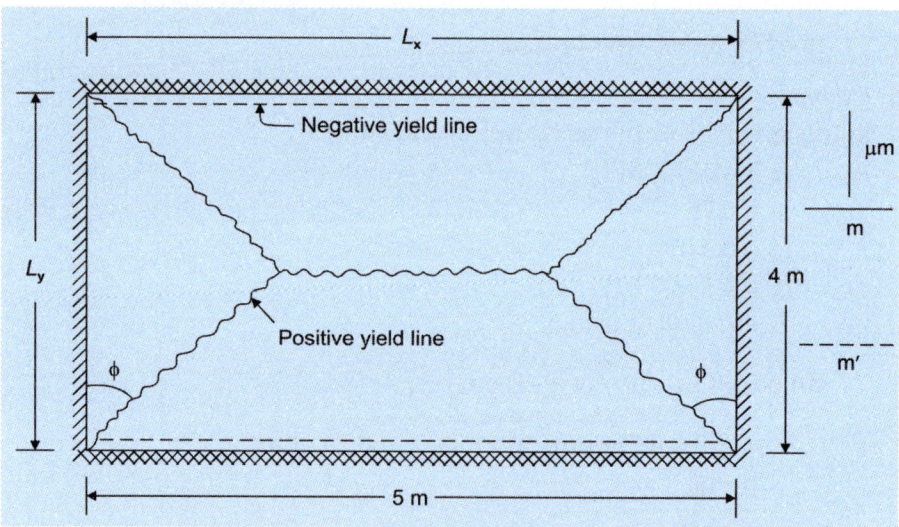

Fig. 10.22: Rectangular slab with fixed and simply supported edges

External work done is obtained as

$$\Sigma(W\delta) = w \cdot (1/2)L_y^2 \big)(\alpha - 1/3\tan\phi)$$

Internal work done by rotation of:

a. Positive yield lines = $4\,m(\alpha + \mu/\tan\phi)$

b. Negative yield lines = $4\alpha \cdot m'$

∴ Total internal work done ($M\theta$) is expressed as:

$$\Sigma(M\theta) = 4\alpha(m + m') + (4m\mu/\tan\phi)$$

Equating $\Sigma(W\delta) = \Sigma(M \cdot \theta)$

$$w \cdot (1/2)L_y^2(\alpha - 1/3\tan\phi) = 4\alpha(m + m') + (4m\mu/\tan\phi)$$

Assuming $m = m'$

$$w(1/2)L_y^2(\alpha - 1/3\tan\phi) = 4m(2\alpha + \mu/\tan\phi)$$

or $\qquad (wL_y^2)(24\,m) = (2\alpha + \mu/\tan\phi)(3\alpha - \tan\phi)$

For a maximum value 'm' differentiating the RHS of the Eq. (3), we have:

$$2\alpha + (\mu/\tan\phi)/(3\alpha - \tan\phi) = (-\mu/\tan^2\phi)/(-1) = (\mu/\tan^2\phi)$$

or $\qquad 2\alpha\tan^2\phi + 2\mu\tan\phi - 3\alpha\mu = 0$

The positive root of this quadratic

$$\tan\phi = \sqrt{(1.5\mu + \mu^2/4\alpha^2)} - (\mu/2\alpha)$$

and $\qquad m = (wL_y^2/24)(\tan^2\phi/\mu)$

iv. *Example*:

$$\alpha = (L_x/L_y) = (5/4) = 1.25$$
$$L_x = 5\,\text{m}, L_y = 4\,\text{m}, \mu = 0.8$$

v. *Thickness of slab*:

For a two-way slab (IS:456):

Effective depth $d = (\text{span}/35) = (4000/35) = 114$ mm

Adopt $d = 120$ mm and overall depth $D = 150$ mm

vi. *Loads*:

Self-weight of slab = (0.15×24) = 3.6 kN/m²

Finishes $\qquad\qquad\qquad$ = 0.6 kN/m²

Live load $\qquad\qquad\qquad$ = 3.0 kN/m²

Total working load $\qquad$ = 7.2 kN/m²

∴ Ultimate load $w_u = (1.5 \times 7.2) = 10.8$ kN/m²

vi. *Moment of resistance*:

$$\tan\phi = \sqrt{1.5\mu + \mu^2/4\alpha^2)} - \mu/2\alpha)$$

$$= \sqrt{(1.5 \times 0.8) + (0.8^2/4 \times 1.25^2)} - (0.8/2 \times 1.25)$$

$$= 0.82$$

∴ $\qquad m = (wL_y^2/24)(\tan^2\phi/\mu)$

$$= (10.8 \times 4^2/24)(0.82^2/0.8)$$

$$= 6.05 \text{ kN·m/m}$$

For RCC slab with $d = 120$ mm

$$M_{u,\,\text{lim}} = 0.148 f_{ck}bd^2$$

$$= (0.148 \times 15 \times 1000 \times 120^2)/10^{-6}$$

$$= 31.968 \text{ kN·m}$$

Since $m_u < M_{u,\,\text{lim}}$, the section is under-reinforced.

∴ $(6.05 \times 10^6) = 0.87 f_y A_{st} d\,[1 - A_{st}f_y/bdf_{ck})]$

$$= 0.87 \times 415 \times A_{st} \times 120[1 - (A_{st} \times 415/1000 \times 120 \times 15)]$$

Solving, $A_{st} = 150$ mm²

But minimum quantity of steel = 0.12%

$$= (0.12 \times 1000 \times 150/100) = 180 \text{ mm}^2/\text{m}$$

Adopt 6 mm diameter bars at 150 mm centres both ways and also over the fixed edges as negative reinforcement.

8. A uniformly loaded isotropically reinforced concrete square slab is simply supported on three sides and unsupported on the fourth. If w = load per unit area at collapse of slab and m = positive plastic moment per unit width, show that for the yield line pattern shown in Fig. 10.22, the minimum upper bound solution is given by the relation:

$$w = \left(\frac{14.2\,m}{L^2}\right), \text{when } \tan\phi = 1.3$$

where, ϕ = angle made by the inclined yield line with the edge.

   i. *Data*:

   Side length of slab = $L_x = L_y = L$

   Moment capacity of slab = m

   Ultimate load = w

   Slab is isotropically reinforced.

   Three adjacent edges are simply supported and the remaining edge is unsupported.

   ii. *Derivation of relation*:

   Referring to Fig. 10.23, the external work done is computed for elements 1, 2 and 3.

   For element 1, we have:

   $$(W\delta) = [(0.5 \times 0.5\,L \tan\phi \times (1/3)]w = \left[\frac{w \cdot L^2 \cdot \tan\phi}{12}\right]$$

   For element 2, we have:

   $$(W\delta) = [(0.5\,w \times 0.5\,L(L - 0.5\,L \tan\phi] + [(0.5\,w$$
   $$+ 0.5\,L \tan\phi \times 0.5\,L(1.3)]$$

   $$= \left(\frac{wL^2}{4}\right)\left(1 - \frac{\tan\phi}{3}\right)$$

   ∴ Total external work done for the elements 1, 2 and 3 is given by:

   $$\Sigma(W\delta) = \left(\frac{wL^2 \tan\phi}{12}\right) + [2 \cdot (w \cdot L^2/4)\,(1 - \tan\phi/3)]$$

   $$= \left(\frac{wL^2}{2}\right) + [1 - \tan\phi/6)]$$

   Internal work done by rotation of yield lines is computed for the elements 1, 2 and 3.

   For element 1, we have:

   $$\theta_y = (2/L \cdot \tan\phi), \ \theta_y = 0 \ \text{and} \ M_x = m \cdot L$$
   $$(M_x\theta_x + M_y\theta_y) = (2/L \cdot \tan\phi)\,(mL) = (2m/\tan\phi)$$

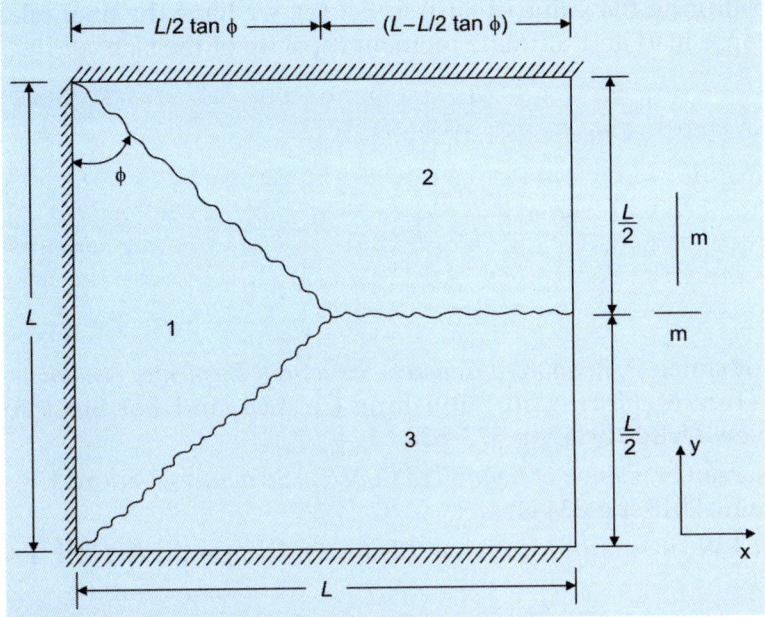

Fig. 10.23: RC slab with three edges simply supported and one edge free

For elements 2 and 3, we have:
$$\theta_y = 2/L, \ \theta_x = 0 \ \text{and} \ M_y = mL$$
$$(M_x\theta_x + M_y\theta_y) = 2(2/L)(mL) = 4\,m$$

∴ Total internal work done is given by the equation:
$$\Sigma(M\theta) = [4\,m + (2\,m/\tan\phi)]$$

Equating $\Sigma(W\delta) = \Sigma(M\theta)$

$$\frac{wL^2}{2}\left(1 - \frac{\tan\phi}{6}\right) = \left(4m + \frac{2m}{\tan\phi}\right)$$

$$\left(\frac{wL^2}{24\,m}\right) = \left[\frac{2 + (1/\tan\phi)}{(6 - \tan\phi)}\right]$$

For a maximum value of m, differentiating the RHS of Eq. (1), we have the relation:

$$\left[\frac{2 + (1/\tan\phi)}{(6 - \tan\phi)}\right] = \left[\frac{-(1/\tan^2\phi)}{-1}\right] = \left[\frac{1}{\tan^2\phi}\right]$$

Cross multiplying, the quadratic equation is obtained as:
$$\tan^2\phi + \tan\phi - 3 = 0$$

The positive root of this equation is:

$$\tan\phi = \left[\frac{-1 + \sqrt{1 + 12}}{2}\right] = 1.3$$

Substituting the value of $\tan \phi$ in Eq. (1), we have the final relation between collapse load and ultimate moment capacity of the slab as:

$$\left(\frac{wL^2}{24m} \right) = \left[\frac{1}{(1.3)^2} \right]$$

$$w = \left(\frac{14.2\,m}{L^2} \right)$$

REFERENCES

1. Purushothaman P, *Reinforced Concrete Structural Elements, Behaviour, Analysis and Design*, Tata McGraw-Hill Publishing Co. Ltd. and Tor Steel Foundation of India, New Delhi, 1984, pp 377–638.

2. Subramanian N, *Design of Reinforced Concrete Structures*, Oxford University Press, New Delhi, 2013, pp 334–414.

3. Bennett EW, *Structural Concrete Elements*, Chapman & Hall, London, 1973, pp 1–170.

4. Park R and Gamble WL, *Reinforced Concrete Slabs*, 2 edn, John Wiley & Sons, New York, 2000.

5. Rangan BV, Hall AS and Warner, *Reinforced Concrete Structures*, Pitman Publications, London, 1980.

6. Krishna Raju N, *Advanced Reinforced Concrete Design*, 2 edn, CBS Publishers, New Delhi, 2005, pp 250–263.

7. Ingerslav A, The strength of rectangular slabs, *Journal of the Institute of Structural Engineering*, Vol. 1, No. 1, January 1923, pp 3–14.

8. Johanssen KW, Brudlinieteorier, Jul, Gjellarups Forlag, Copenhagen, 1943, Yield Line Theory, Cement and Concrete Association, London, 1962, p 181.

9. Wood RH, *Plastic and Elastic Design of Slabs and Plates*, Thames and Hudson, London, 1961, p 344.

10. Jones LL, *Ultimate Load Analysis of Reinforced and Prestressed Concrete Structures*, Chatto and Windus, London, 1962, p 248.

ASSIGNMENT

1. Design a one-way reinforced concrete slab simply supported at the edges for a public building with a clear span of 4 m supported on 200 mm solid concrete masonry walls. Adopt M20 grade concrete and Fe 415 HYSD bars.

2. Design a two-way RCC slab for a room of size 4 m × 6 m with continuous edges all round at supports. Assume a live load 4 kN/m². Concrete of M20 grade and Fe 415 HYSD bars are available for use. Sketch the details of reinforcements in the slab.

3. Design a cantilever canopy slab projecting 3 m from the lintel beam using M20 grade concrete and Fe 415 HYSD bars. Assume a live load 2 kN/m². Sketch the details of reinforcement in the slab.

4. Design a continuous one-way slab for a school building class room floor. The slab is continuous over T-beams spaced at 4 m intervals. Adopt a live load 4 kN/m^2, M20 grade concrete and Fe 415 HYSD bars.

5. Design the typical interior panel of a flat slab floor for a public bank building with panels of size 4 m × 6 m to support a live load of 4 kN/m^2 using M20 grade concrete and Fe 415 HYSD bars. Sketch the details of reinforcements in the slab.

6. Design the exterior panel of a flat slab floor system for a warehouse 18 m × 18 m divided into panels of 6 m × 6 m by columns. Adopt a live load of 5 kN/m^2 with columns of size 400 mm diameter. Hight of floor = 3 m. Adopt M20 grade concrete and Fe 415 HYSD bars. The design should conform to the specifications of IS:456-2000.

7. A flat slab floor for a garrage is to be designed for loading of 10 kN/m^2. The column grid is 8 m × 8 m. Design an interior panel with drops. Adopt M25 grade concrete and Fe 415 HYSD bars. Sketch the details of reinforcements in the flat slab.

8. A square slab of 5 m side length is simply supported along the edges and supports a uniformly distributed load including its own weight. If the slab is reinforced isotropically to give an ultimate moment of 20 kN·m/m, calculate the magnitude of the uniformly distributed load required to cause collapse from first principles using the method of virtual work.

9. A rectangular slab 6 m × 4.5 m, simply supported along its edges is to be designed as an isotropically reinforced slab to support a uniformly distributed working load of 18 kN/m^2 which includes the self-weight of the slab. Calculate the ultimate moment of resistance required for the slab section from first principles.

10. A triangular reinforced concrete slab ABC has equal sides of length 5 m. The isotropically reinforced slab is simply supported on two sides and carries a uniformly distributed load. If the moment of resistance of the section of the slab is 30 kN·m/m, estimate the ultimate collapse load carried by the slab from first principles.

11. A hexagonal slab, simply supported on all the sides has a side length of 3 m. Find the uniformly distributed load which could cause failure of the isotropically reinforced slab if the ultimate moment of resistance of the slab is 2.7 kN·m/m.

12. A rectangular RC slab 4 m × 5 m is isotropically reinforced with all the edges simply supported. Show that the relation between the ultimate moment capacity of the slab can be expressed as:
$$m = (wL_y^2 \tan^2 \phi)/(24)$$
where, m = ultimate moment capacity of the slab per unit length

   L_y = short span length

   ϕ = angle made by the yield line with the short span.

Also design the slab for a working live load of 4 kN/m^2 using yield line principles. Adopt M20 grade concrete and Fe 415 tor steel.

13. A rectangular slab 4.5 m × 6 m has fixed edges all over the supports. If the slab is isotropically reinforced with equal positive and negative reinforcements, show that the ultimate moment capacity of the slab can be expressed as:

$$m = (wL_y^2 \tan^2 \phi)/(48)$$

where, m = ultimate moment capacity of the slab per unit length

L_y = short span length

ϕ = angle made by the yield line with the short span.

Also design the slab for a working live load of 6 kN/m² using yield line principles. Adopt M20 grade concrete and Fe 415 tor steel.

14. Find the collapse load of a 5.5 m × 3.75 m rectangular slab fixed at all edges for which the support moments are twice of that the corresponding mid-span moment in each direction. Also take $M_x = 0.5 M_y$.

15. An isotropically reinforced square slab of side length L, supports a uniformly distributed ultimate load (including its self-weight) of 'w' kN/m² together with a concentrated load at centre of magnitude 'W' kN. If the ultimate moment of resistance of the slab is 'm' kN·m/m. Show from the first principles that:

$$m = (wL^2/24) + (W/8)$$

16. A square slab of 3 m side length is simply supported along the edges and carries a uniformly distributed load of 30 kN/m² including its own weight. If the slab is reinforced isotropically to give an ultimate moment of resistance of 20 kN·m/m, calculate the magnitude of the additional central point load required to cause collapse. Assume a pattern of simple diagonal yield lines.

REVIEW QUESTIONS

1. How are reinforced concrete slabs classified? List the various types of slabs generally used in the construction industry with examples.

2. Distinguish between one-way and two-way slabs with examples.

3. What is Rankine–Groshoff theory as applied to slabs? How do you design the two-way slabs using this theory.

4. What are the main considerations that generally govern the selection of thickness of two-way slabs?

5. How do you design continuous slabs of multistorey office floors? What method you would use to determine the maximum moment at critical sections?

6. Explain the necessity of using drops in flat slabs? How do you divide the flat slab panels into column and middle strips as recommended in the Indian standard code?

7. How do you determine the design moments in flat slab panels? Sketch the typical arrangement of reinforcements in a flat slab with drops.

8. How do you check for shear stresses in a flat slab? If the shear stresses at critical sections exceed the permissible values, what measures you would suggest to solve the shear stress problem?

9. Explain the principles of yield line analysis of reinforced concrete slabs. Sketch the typical pattern of yield lines developed in slab of various shapes.

10. Distinguish between the virtual work and equilibrium methods as applied to yield line analysis of slabs.

OBJECTIVE QUESTIONS

1. Reinforced concrete slabs can be assumed as one-way slabs when the ratio of long to short span exceeds
   a. 1.0
   b. 1.5
   c. 2.0
2. The basic span/depth ratio for a cantilevered slab specified in IS:456 code is
   a. 20
   b. 25
   c. 7
3. The bending moment coefficients specified in Indian standard code for the design of slabs spanning in two directions at right angles and simply supported on four sides are based on
   a. simple bending theory
   b. Rankine–Groshoff theory
   c. yield line theory
4. In case of two-way slabs spanning in orthogonal directions and continuous over beams at supports, torsion reinforcements should be provided at the
   a. centre of spans
   b. supporting edges
   c. four corners
5. In continuous slabs, reinforcements to resist the negative moments should be provided at the
   a. centre of spans
   b. top of supports
   c. points of contraflexure
6. In flat slabs with drops, the shear stresses in concrete should be checked at the
   a. middle of shorter spans
   b. centre of slab
   c. slab section near the column
7. In case of flat slabs supporting floor loads, the negative bending moments will be maximum in the
   a. middle strip
   b. column strip
   c. centre of slab
8. The minimum reinforcements expressed as a percentage of cross-section to be provided in all slabs when HYSD bars are used should be not less than
   a. 0.12
   b. 0.50
   c. 1.00

9. In simply supported square slabs supporting uniformly distributed loads, the typical yield lines developed at failure stage will be

    a. connecting the middle of edges
    b. connecting the corners
    c. connecting corners to the middle of edges

10. In a rectangular slab with edges restrained at the supports, negative yield lines will develop at the failure stage near the

    a. centre of span
    b. soffit of slab at centres of span
    c. edges at supports

Limit State Design of Columns and Footings

11.1 INTRODUCTION

Reinforced concrete columns in a structure are vertical elements with the primary function of transmitting the dead and live loads from floors and roofs to the foundations. Structural concrete members in compression are generally referred to as columns and struts. The term *column* is associated with members transmitting loads to the ground and the term *strut* is applied to members in a direction such as those in a trussed frame work. The Indian standard code, IS:456-2000[1], Clause 25.1.1 defines the column as a compression member, with an effective length not exceeding three times the least lateral dimension. The term *pedestal* in which the longitudinal reinforcement is not taken into account in strength computations, is also a compression member having an effective length not exceeding three times the least lateral dimension according to the IS:456 code, Clause 26.5.3.1.

The term *column* generally represents all types of compression members and also sometimes the term *column* and *compression member* are used interchangeably to refer to structural concrete members in compression. A structural member subjected to axial compression and bending moment is sometimes referred to as *beam-column*.

11.2 FAILURE MODES OF COLUMNS

Reinforced concrete columns may fail in any of the following three modes[2,3,4]:

1. Pure compression failure
2. Combined compression and bending failure
3. Failure by elastic instability.

The failure modes[5,6] depend primarily on the slenderness ratio of the member which in turn depends upon the cross-sectional dimensions, effective length, and support conditions of the member.

11.3   CLASSIFICATION OF COLUMNS

11.3.1 Classification Based on Type of Loading

Reinforced concrete columns are classified into the three types, based on the type and nature of loading as detailed below:

1. Columns supporting loads applied concentrically (axial loading) as shown in Fig. 11.1a. A typical example being the interior columns of a multistorey building with symmetrical loads from floor slabs from all sides. Also a single column supporting a conical water tank is also axially loaded.
2. Columns subjected to uniaxial eccentric loading as shown in Fig. 11.1b. Typical example being the edge column rigidly connected to beam from one side only.
3. Columns with biaxial loading as shown in Fig. 11.1c, where beams are rigidly connected orthogonally on the top of the column such as in corner columns.

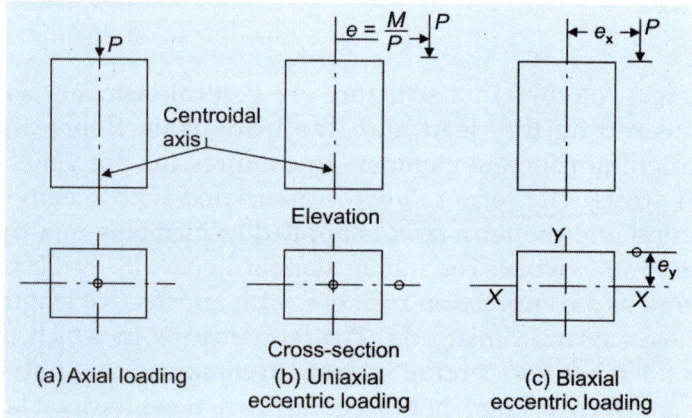

Fig. 11.1: Classification of columns based on type of loading

11.3.2 Classification Based on Type of Reinforcement

Depending on the type of reinforcement used, reinforced concrete columns are classified into the following three groups:

1. *Tied columns* in which the main longitudinal bars are confined within closely spaced lateral ties (Fig. 11.2a).
2. *Spiral columns* having main longitudinal reinforcements enclosed within closely spaced and continuously wound spiral reinforcement (Fig. 11.2b).
3. *Composite columns* in which the longitudinal reinforcement is in the form of structural steel section or pipes with or without longitudinal bars (Fig. 11.2c).

In general tied columns are the most commonly used having different shapes (square, rectangular, T, L, circular etc.).

Spiral columns are generally adopted with circular and octagonal sections.

11.3.3 Classification Based on Slenderness Ratio

Depending on the slenderness ratio defined as the ratio of effective length to least lateral dimension, columns may be classified as:

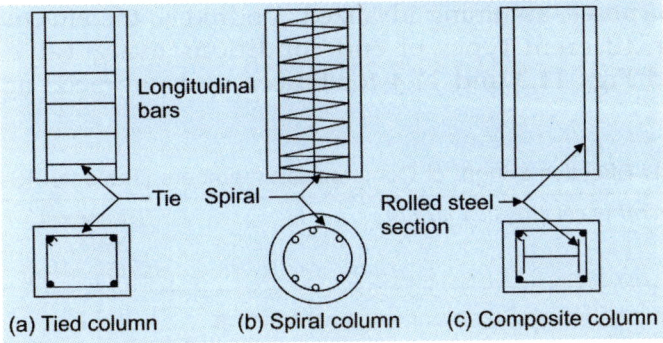

Fig. 11.2: Classification of columns based on type of reinforcement

1. Short columns
2. Slender or long columns

IS:456-2000 code, Clause 25.1.2 classifies a rectangular compression member as short when both the slenderness ratio's (L_{ex}/D) and (L_{ey}/b) are less than 12, where,

L_{ex} = effective length in respect of major axis

D = depth in respect of major axis

L_{ey} = effective length in respect of minor axis and

b = width of the member.

If any of these ratios is equal to or more than 12, then it is termed slender or long column. This definition is not suitable for nonrectangular and noncircular sections, where the slenderness ratio is better defined in terms of the radius of gyration rather than the lateral dimensions.

11.4 EFFECTIVE LENGTH OF COLUMNS

11.4.1 Computation of Effective Length

In the design of columns, the effective length plays an important part and is influenced by the end conditions of the column. The effective length of a column depends upon the unsupported length (distance between lateral connections) and the boundary conditions at the ends of column due to the conditions of the framing beams and other members. The effective length L_{ef} can be expressed in the form:

$$L_{ef} = kL$$

where,

L = unsupported length or clear height of columns

k = effective length ratio or a constant depending upon the degrees of rotational and translational restraints at the ends of column.

The effective length of compression members depends upon the bracing and end conditions. For braced (laterally restrained at ends) columns, the effective length is less than the clear height between the restrains, whereas for unbraced and partially braced columns, the effective length is greater than the clear length between the restraints.

For design purposes, assuming idealized conditions, the effective length L_e may be assessed for different types of end conditions using Table 11.1 (Table 28, IS:456-2000) and Figs 11.3 and 11.4 for braced and unbraced against side sway respectively.

Table 11.1: Effective Length of Compression members (Table-28, IS:456-2000)

Degree of end restraint of compression member	Theoretical value of effective length	Recommended value of effective length
Effectively held in position and restrained against rotation at both ends	0.5 L	0.65 L
Effectively held in position at both ends, restrained, against rotation at one end	0.7 L	0.80 L
Effectively held in position at both ends, but no restrained rotation	1.00 L	1.00 L
Effectively held in position and restrained against rotation at one end, and at the other restrained against rotation but not held in position	1.00 L	1.20 L
Effectively held in position and restrained against rotation at one end, and at the other partially restrained against rotation but not held in position	–	1.50 L
Effectively held in position at one end but not restrained against rotation, and at the other end restrained against rotation but not held in position	2.00 L	2.00 L
Effectively held in position and restrained against rotation at one end but not held in position nor restrained against rotation at the other end	2.00 L	2.00 L

11.4.2 Slenderness Limits

In the design of reinforced concrete columns, the dimensions of the column should be selected in such a way that it fails by material failure only and not by buckling. To ensure this criterion, the code recommends that the clear distance between restraints (unsupported length) should never exceed 60 times the least lateral dimensions of the column (Clause 25.3.1). For unbraced columns, it is recommended that this value is limited to 30. In cantilever columns, in addition to the above restriction ($L \le 60b$), the clear height should also not exceed the value of $L = (100 \, b^2/D)$, where D is the depth of crosssection measured in the plane under consideration and b is the width of crosssection (Clause 25.3.2).

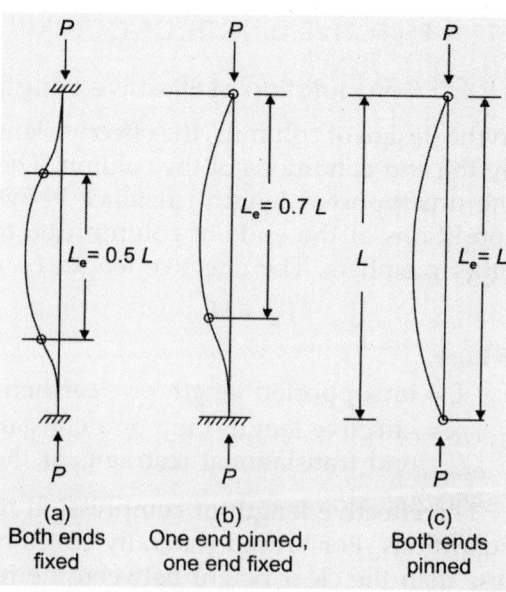

Fig. 11.3: Effective length of columns braced against side sway

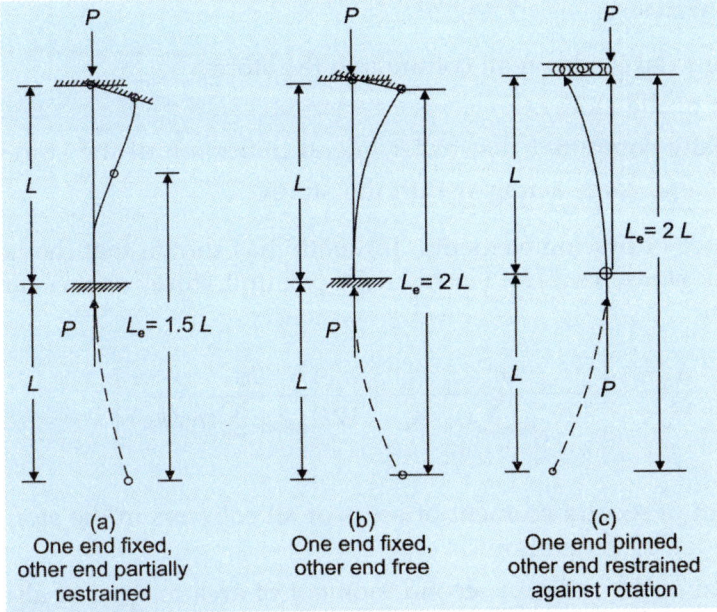

Fig. 11.4: Effective length of columns unbraced against side sways

11.4.3 Minimum Eccentricities

It is mandatory that all columns should be designed for minimum eccentricity according to the Indian standard code (IS:456), Clause 25.4, which takes care of any imperfections in constructions and inaccuracy in loading. The minimum eccentricity is expressed by the relation,

$$e_{min} = \left(\frac{L}{500} + \frac{D}{30} \right)$$

but not less than 20 mm
where,

    L = unsupported length
    D = lateral dimensions in the plane of bending

For non rectangular and non circular cross-sectional shapes, SP:24[7]-1983 recommends the minimum eccentricity as

$$e_{min} = (L_e/300) \text{ or } 20 \text{ mm (whichever is greater)}$$

11.4.4 Braced and Unbraced Columns

In case of framed structures, an approximate method of deciding whether a column is *braced* or *unbraced* is specified in the ACI code[8] and is reproduced in the revised IS:456-2000 code. For this purpose, the *stability index* (Q) of a storey in a framed multistorey structure is defined as

$$Q = \left[\frac{\sum P_u}{h_s} \times \frac{\Delta_u}{H_u} \right]$$

where,

ΣP_u = sum of axial loads on all columns in the storey

h_s = height of the storey

Δ_u = elastically computed first order lateral deflection of the storey

H_u = total lateral force acting within the storey.

In the absence of bracing elements, Taranath[9] has shown that the lateral flexibility measure of the storey (Δ_u/H_u) (storey drift per unit storey shear) can be expressed by the relation

$$\left(\frac{\Delta_u}{H_u}\right) = \left[\frac{h_s^2}{12E_{c,col}\sum(I_c/h_s)} + \frac{h_s^2}{12E_{c,beam}\sum(I_b/L_b)}\right]$$

where,

ΣI_c = sum of second moment of areas of all columns in the storey in the plane under consideration.

$\Sigma(I_b/L_b)$= sum of the ratios of second moment of area to span of all floor members in the storey in the plane under consideration.

E_c = modulus of elasticity of concrete.

The equation for the stability index Q is based on the assumption that the points of contraflexure occurs at the mid heights of all columns and mid-span points of all beams and by applying unit load method to an isolated storey[10]. If bracing elements such as trusses, shear walls and infill walls are used then their beneficial effect will be to reduce the ratio (Δ_u/H_u) significantly.

If the value of $Q \leq 0.04$, then the column may be considered as no sway column (braced), otherwise the column may be treated as sway column (unbraced).

IS:456-2000 codal charts (Figs 11.5 and 11.6) are very useful in determining the effective length ratios of braced and unbraced columns respectively and in terms of β_1 and β_2 which represents the degree of rotational freedom at the top and bottom ends of the column. The values of β_1 and β_2 for braced and unbraced columns are given by the relations,

$$\beta_1 = \left[\frac{\sum I_c/h_s}{\sum I_c/h_s + \sum 0.5(I_b/L_b)}\right] \text{ (for braced columns)}$$

$$\beta_2 = \left[\frac{\sum I_c/h_s}{\sum I_c/h_s + \sum 1.5(I_b/L_b)}\right] \text{ (for unbraced columns)}$$

The limiting values $\beta = 0$ and $\beta = 1$, represent the *fully fixed* and *fully hinged* conditions respectively.

The following numerical example of multistorey building illustrates the method of determining the effective length of columns using IS:456 code charts and to determine whether the columns are braced or unbraced depending upon the stability index.

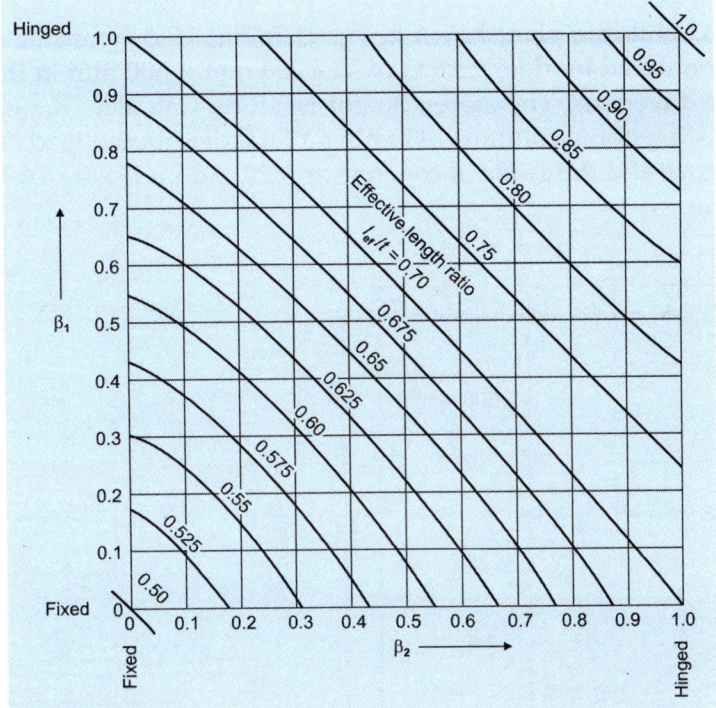

Fig. 11.5: Effective length ratios for columns with no sway (braced columns) (IS:456-2000, Fig. 26)

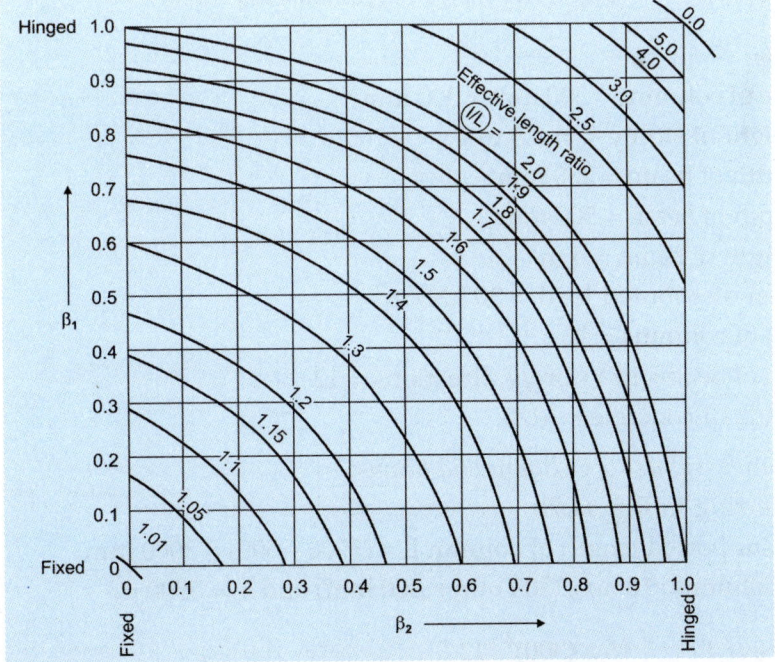

Fig. 11.6: Effective length ratios for columns without restraint against side sway (unbraced columns) (IS:456-2000, Fig. 27)

Example

A multistoreyed building plan shown in Fig 11.7a has 16 columns of size 300 mm × 300 mm interconnected by floor beams of size 250 mm × 500 mm in the longitudinal and transverse directions. The storey height is 3.5 m. Calculate the effective length of the typical lower storey columns assuming a total distributed load 30 kN/m² from all the floors above and the grade of concrete as M20. Adopt IS:456-2000 codal method for computation.

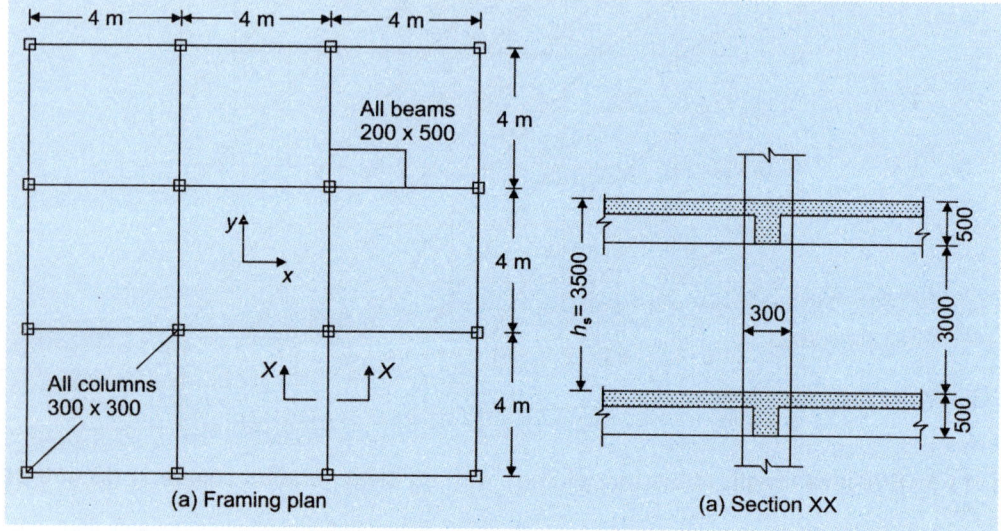

Fig. 11.7: Multistoreyed building frame

i. *Data*:

Size of columns = 300 mm × 300 mm

Height of storey h_s = 3.5 m

Width of beam = 250 mm

Depth of beam = 500 mm

Length of beam = 4 m

Total distributed load = 30 kN/m²

No. of columns = 16

No. of beams in *xx* or *yy* directions = 12

Grade of concrete = M20

ii. *Relative stiffness of Columns and Beams*:

Referring to Fig. 11.7b

Unsupported length of column L = (3500 − 500) = 3000 mm

a. Columns : 16 nos, (300 mm × 300 mm) and h_s = 3500 m

$$\sum\left(\frac{I_c}{h_s}\right) = \left[\frac{16 \times (300)^4/12}{3500}\right] = (3086 \times 10^3)\ \text{mm}^3$$

b. Beams in each direction xx or yy

$$\sum\left(\frac{I_b}{L_b}\right) = \left[\frac{12 \times 250 \times (500)^3/12}{4000}\right] = (7812 \times 10^3)\ \text{mm}^3$$

c. Check for braced or unbraced columns

$$\left(\frac{\Delta_u}{H_u}\right) = \frac{h_s^2}{12E_c}\left[\frac{1}{\sum(I_c/h_s)} + \frac{1}{\sum(I_b/L_b)}\right]$$

$E_c = 5000\sqrt{f_{ck}}$ according to Clause 6.2.3.1, IS:456-2000

Therefore, $E_c = 5000\sqrt{20} = 22360\ \text{N/mm}^2$

$$\left(\frac{\Delta_u}{H_u}\right) = \frac{3500^2}{(12 \times 22360)}\left[\frac{1}{(3086 \times 10^3)} + \frac{1}{(7812 \times 10^3)}\right]$$

$$= (5.991 \times 10^{-6})\ \text{mm/N}$$

Total axial load on all columns = $(12 \times 12 \times 30) = P_u = 4320\ \text{kN}$

Stability index

$$Q = \left[\frac{P_u}{h_s} \times \frac{\Delta_u}{H_u}\right] = \left[\frac{4300 \times 10^3}{3500} \times (5.991 \times 10^{-6})\right] = 0.00739 < 0.04$$

Hence, the columns in the storey can be considered as braced in xx and yy directions.

d. Effective length of columns using IS:456 codal charts

$$\beta_1 = \beta_2 = \left[\frac{\sum(I_c/h_s)}{\sum(I_c/h_s) + \sum 0.05(I_b/L_b)}\right]$$

$$\sum(I_c/h_s) = \left[\frac{(300)^4/12}{3500} \times 2\right] = (385 \times 10^3)\ \text{mm}^3$$

$$\sum(I_b/L_b) = \left[\frac{(250) \times 500^3/12}{4000} \times 2\right] = (1302 \times 10^3)\ \text{mm}^3$$

$\therefore\quad \beta_1 = \beta_2 = \left[\frac{385 \times 10^3}{(385 \times 10^3) + (0.5 \times 1302 \times 10^3)}\right] = 0.371$

Referring to Fig. 11.5 (Fig. 26, IS:456-2000) and interpolating the effective length ratio $k = \left(\frac{L_e}{D}\right) = \left(\frac{1890}{300}\right) = 6.3 < 12.$

$\therefore L_e = (0.630 \times 3000) = 1890\ \text{mm}$

Slenderness ratio of the column $= \left(\frac{L_e}{D}\right) = \left(\frac{1890}{300}\right) = 6.3 < 12.$

Hence, the column should be designed as short column.

11.5 DESIGN OF SHORT AXIALLY LOADED COLUMNS

11.5.1 Design Assumptions

The following salient assumptions are made for the limit state design of columns as specified under Clause 39.1, IS:456-2000.

   a. Plane sections normal to the axis remain plane after bending.
   b. The maximum compressive strain in concrete under axial compression is 0.002.
   c. The maximum compressive strain at the highly compressed extreme fibre in concrete subjected to axial compression and bending and when there is no tension on the section shall be 0.0035 minus 0.75 times the strain at the least compressed extreme fibre.

11.5.2 Design Equations

When the minimum eccentricity does not exceed 0.05 times the lateral dimension, the design ultimate load on the column is computed by the following equation:

$$P_u = 0.4f_{ck} A_c + 0.67f_y A_{sc}$$
$$= 0.4f_{ck} A_g + (0.67f_y - 0.4f_{ck})A_{sc}$$

where,

P_u = ultimate axial load on the member
f_{ck} = characteristic compressive strength of concrete
f_y = characteristic compressive strength of reinforcement
A_c = area of concrete ($A_g - A_{sc}$)
A_{sc} = area of longitudinal reinforcement in the column
A_g = gross area of concrete

11.6 DESIGN OF COMPRESSION MEMBERS WITH HELICAL REINFORCEMENT

The ultimate compressive strength of helically tied columns are given by the relation:

$$P_u = 1.05(0.4f_{ck} A_c + 0.67f_y A_{sc})$$

Clause 39.4.1, IS:456-2000 code specifies that the ratio of the volume of helical reinforcement to the volume of the core shall not be less than $0.36(A_g/A_c - 1) (f_{ck}/f_y)$ where,

A_g = gross area of section
A_c = area of the core of the helically reinforced column measured to the outside diameter of helix
f_{ck} = characteristic compressive strength of concrete
f_y = characteristic strength of helical reinforcement but not exceeding 415 N/mm^2.

11.7 COLUMNS SUBJECTED TO COMBINED AXIAL LOAD AND UNIAXIAL BENDING

External columns of multistoreyed buildings are subjected to direct axial loads and bending moments. The compression members in such cases should be designed for axial load and bending moment based on the assumptions specified in Clauses 39.1 and 39.2, IS:456 code. The analytical design procedure of these members involves lengthy computations using equilibrium equations. Hence, to overcome these difficulties, IS:456 code recommends the use of interaction curves developed by using nondimensional parameters presented in SP:16, design aids for reinforced concrete. The design charts of SP:16 involve the nondimensional parameters of $(P_u/f_{ck}bd)$ and $(M_u/f_{ck}bd^2)$

where,

P_u = axial ultimate load

M_u = limiting moment in association with P_u

f_{ck} = characteristic compressive strength of concrete

b = width of column

D = overall depth of column in the direction of moment.

Using the design charts, the percentage of reinforcement required in the section can easily be determined for given grade of concrete and type of reinforcement.

A typical interaction diagram for f_y = 415 N/mm² and (d'/D) = 0.10 is shown in Fig. 11.8.

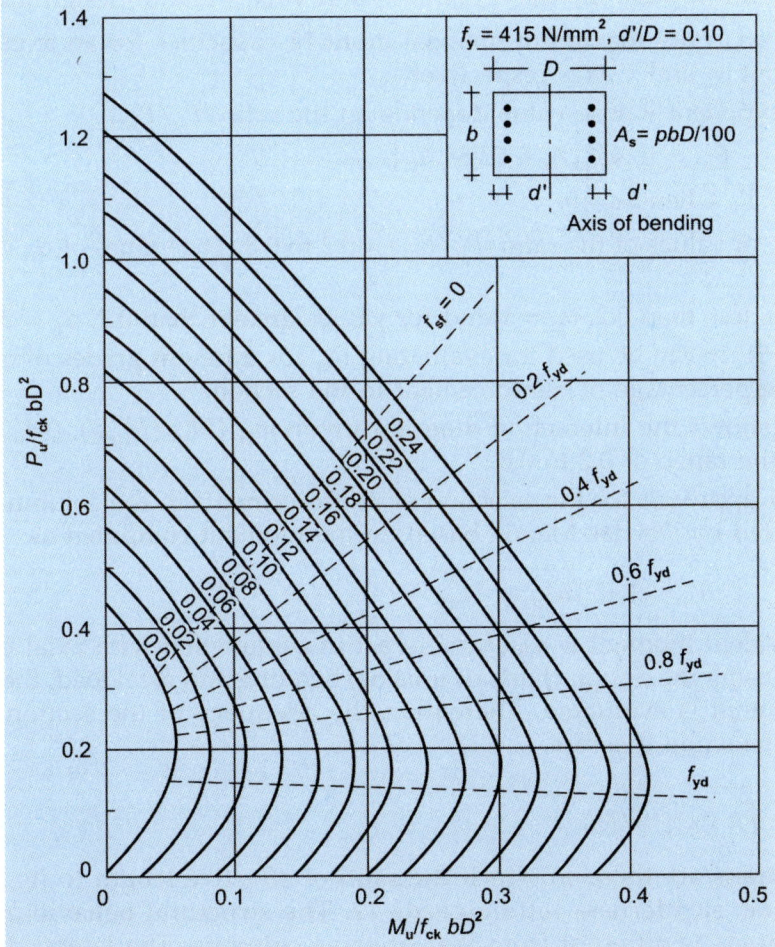

Fig. 11.8: Interaction diagram for column under compression and bending (Chart 32, SP:16)

The entire set of charts for different strengths of steel (f_y = 250, 415 and 500 N/mm^2) and different cover ratio (d'/D) = 0.05, 0.10, 0.15 and 0.20 are presented in SP:16, design aids for reinforced concrete.

11.8 COLUMNS UNDER COMPRESSION AND BIAXIAL BENDING

Columns positioned at the corners of multistoreyed buildings with rigidly connected beams at right angles are subjected to compression and biaxial moments. Analysis and design of such columns involves lengthy computations involving trial and error procedures. To overcome these difficulties, the Indian standard code, IS:456-2000 recommends a simplified procedure based on Bresler's empirical formulation involving the salient design parameters.

The design interaction equation is expressed as

$$\left[\frac{M_{ux}}{M_{ux1}}\right]^{\alpha_n} + \left[\frac{M_{uy}}{M_{uy1}}\right]^{\alpha_n} < 1.0$$

where, M_{ux} and M_{uy} are moments about x- and y-axes due to design loads.

M_{ux1} and M_{uy1} are maximum uniaxial moment capacities for an axial load of P_u bending about x- and y-axes respectively.

α_n is an exponent whose value depends on the ratio (P_u/P_{uz})

where, $P_{uz} = [0.45f_{ck}A_c + 0.75f_yA_{sc}]$,

i.e. values of P_u when $M = 0$.

The range of values of the ratio (P_u/P_{uz}) = 0.2 to 0.8. The values of α_n vary linearly from 1 to 2.

For values less than 0.2, $\alpha_n = 1$ and for values greater than 0.2, $\alpha_n = 2$.

Chart 63, SP:16 can be used for evaluating P_{uz} for different grades of concrete and steel and the percentage of reinforcement in the section.

Chart 64 shows the interaction diagram involving (M_{uy}/M_{uy1}), (M_{ux}/M_{ux1}) and (P_u/P_{uz}) in the range of 0.2 to 0.8.

A simpler approach for the selection of reinforcements in the column section has been suggested by Devdas Menon based on the moment computed as:

$$M_u = 1.15\sqrt{M_{ux}^2 + M_{uy}^2}$$

The equivalent moment is assumed to act in conjunction with axial compressive load P_u and using the design charts of uniaxial bending and axial load, the percentage of reinforcement is evaluated. Thereafter, the adequacy of the section is checked using the interaction diagram.

11.9 SLENDER COLUMNS

Slender columns are those in which the ratio of effective length to its least lateral dimension, i.e. slenderness ratio exceeds 12. The structural behaviour of long or slender columns is different from that of short columns. The lateral deflection of slender columns under loads is significantly greater when compared with that of short columns. When the effect of deflections are not taken into account in the analysis, additional or secondary moment should be considered in the design of slender columns.

The Indian standard code, IS:456-2000 recommends the following formula for the computation of additional moments.

$$M_{ax} = \frac{P_u D}{2000} \left(\frac{L_{ex}}{D}\right)^2$$

$$M_{ay} = \frac{P_u b}{2000} \left(\frac{L_{ey}}{b}\right)^2$$

where,

P_u = axial load on the member
L_{ex} = effective length in respect of the major axis
L_{ey} = effective length in respect of the minor axis
D = depth of cross-section at right angles to the major axis
b = width of the member

The expression for the additional moments can be written in the forms of eccentricities of load as follows:

$$M_{ax} = P_u\, e_{ax}$$
$$M_{ay} = P_u\, e_{ay}$$

where,

$$e_{ax} = \frac{D}{2000} \left(\frac{L_{ex}}{D}\right)^2$$

$$e_{ay} = \frac{b}{2000} \left(\frac{L_{ey}}{b}\right)^2$$

Table 1, SP:16 gives the values of (e_{ax}/D) or (e_{ay}/b) for different values of the slenderness ratio (L_{ex}/D) or (L_{ey}/b) ranging from 12 to 20.

The additional moments computed may be reduced by the multiplying factor k given by the relation:

$$k = \left[\frac{P_{uz} - P_u}{P_{uz} - P_b}\right] \leq 1$$

where, $P_{uz} = [0.45 f_{ck} + 0.75 f_y\, A_{sc}]$.

The values of P_{uz} can be readout from Chart 63, SP:16 and P_b is the axial load corresponding to the condition of maximum compressive strain of 0.0035 in concrete and tensile strain of 0.002 in the outermost layer of tension steel.

According to the IS:456 code, use of multiplying or reduction factor k is optional. However, it is preferred to take advantage of this factor since it can be substantially less than unity resulting in the reduction of the magnitude of additional moments. The value of P_b is influenced by the arrangement of reinforcement at the cover ratio (d/D) and the grades of concrete and steel. The values of the coefficients required for evaluating P_b for various cases are given in Table 60, SP:16.

The expression for k can also be expressed as:

$$k = \left[\frac{1 - (P_u/P_{uz})}{1 + (P_b - P_{uz})} \right] \leq 1$$

Chart 65, SP:16 is very useful in computing k for different ratios of (P_u/P_{uz}) and (P_b/P_{uz}).

11.10 DESIGN OF FOOTINGS

11.10.1 General Principles

Reinforced concrete footings are designed to resist the design factored moments and shear forces due to the imposed loads. The area of footing should be such that the bearing pressure developed at the base of footing does not exceed the safe bearing capacity of the soil.

In plain concrete footings, the thickness at the edge should be at least 150 mm for footings on soils and not less than 300 mm above the tops of piles for footings on piles.

11.10.2 Depth of Foundation

The minimum depth for locating the footing is based on the Rankine theory and is computed using the relation:

$$k = \frac{p}{w} \left\{ \frac{1 - \sin \phi}{1 + \sin \phi} \right\}^2$$

where,

h = minimum depth below the soil surface to the soffit of the footing
p = safe bearing capacity of the soil
w = unit weight of soil

In case of sloped or stepped footings, the effective cross-section at any section should be designed to resist the relevant moments and shear forces acting on the section.

11.10.3 Bending Moments

The bending moment at any section shall be determined by passing through the section of a vertical plane which extends completely across the footing and computing the moment of the forces acting over the entire area of the footing on one side of the section of the plane.

The maximum moment to be used in the design of an isolated concrete footing which supports a column, pedestal or wall is computed at the following sections:

a. At the face of the column, pedestal or wall, for footing supporting a concrete column, pedestal or wall.
b. Halfway between the centre line and the edge of the wall, for footings under masonry walls.
c. Halfway between the face of the column or pedestal and the edge of the gusseted base for footings under gusseted bases.

11.10.4 Shear Forces

The critical section for shear as a measure of diagonal tension shall be assumed as a vertical section located from the face of the column, pedestal or wall at a distance equal to the depth of the footing in case of footings on soils and a distance equal to half the depth of footing for footings on piles.

11.10.5 Transfer of Load at the Base of Column

The compressive stress at the base of the column is transferred by bearing to the top of the footing. The bearing pressure on the loaded area shall not exceed the permissible stress in direct compression multiplied by a value equal to $\sqrt{A_1/A_2}$ but not greater than 2:

where,

A_1 = supporting area for bearing

A_2 = loaded area at the columns base

The design permissible bearing stress in concrete in working stress method of design is $0.25f_{ck}$ and for limit state method of design, the permissible bearing stress is taken as $0.45f_{ck}$.

11.10.6 Tensile Reinforcement

a. In one-way and two-way reinforced square footings, the reinforcement extending in each direction shall be discussed uniformly across the full width of the footing.

b. In two-way reinforced rectangular footing, the reinforcement in the long direction shall be distributed uniformly across the full width of the footing.

For reinforcement in the short direction, a central band equal to the width of the footing shall be marked along the length of the footing and portion of the reinforcement determined in accordance with the equation given below shall be uniformly distributed across the central band.

$$\left[\frac{\text{Reinforcement in central band width}}{\text{Total reinforcement in short direction}}\right] = \left[\frac{2}{\beta + 1}\right]$$

where, β = ratio of the long side to the short side of footing. The remainder of the reinforcement shall be uniformly distributed in the outer portions of the footings.

11.11 DESIGN EXAMPLES

1. A rectangular reinforced concrete column of cross-sectional dimensions 300 mm × 600 mm is to be designed to support an ultimate axial load of 2000 kN. Design suitable reinforcements in the column using M20 grade concrete and Fe 415 HYSD bars.

    i. *Data*:

    $P_u = 2000 \text{ kN}$ $\qquad$ $f_{ck} = 20 \text{ N/mm}^2$

    $b = 300 \text{ mm}$ $\qquad$ $f_y = 415 \text{ N/mm}^2$

    $D = 600 \text{ mm}$

ii. *Longitudinal reinforcements*:

$$P_u = 0.4f_{ck}A_g + (0.67f_y - f_{ck})A_{sc}$$

$$(2000 \times 10^3) = (0.4 \times 20 \times 300 \times 600) + (0.67 \times 415) - (0.4 \times 20)A_{sc}$$

Solving, $A_{sc} = 2073$ mm^2

Provide 6 bars of 22 mm diameter ($A_{sc} = 2280$ mm^2) with 3 bars distributed on each face.

iii. *Lateral ties*:

Tie diameter: $\phi_t \not< \begin{cases} (1/4\,(22) = 5.5 \\ 6 \text{ mm} \end{cases}$ provide 8 mm ties

Tie spacing: $s_t \not> \begin{cases} 300 \text{ mm} \\ (16 \times 22) = 352 \text{ mm} \\ (48 \times 8) = 384 \text{ mm} \end{cases}$ provide spacing of 300 mm

Provide 8 mm diameter ties at 300 mm centres.

iv. *Reinforcement details*:

The details of reinforcements in the column section is shown in Fig. 11.9.

2. Design the reinforcements in a circular column of diameter 300 mm to support a service axial load of 800 kN. The column has an unsupported length of 3 m and is braced against side sway. The column is reinforced with helical ties. Adopt M20 grade concrete and Fe 415 HYSD bars.

i. *Data*:

Length of column $L = 3$ m

Diameter $D = 300$ mm

$f_{ck} = 20$ N/mm^2

$f_y = 415$ N/mm^2

Axial service load = 800 kN

Factored load = $(1.5 \times 800) = 1200$ kN

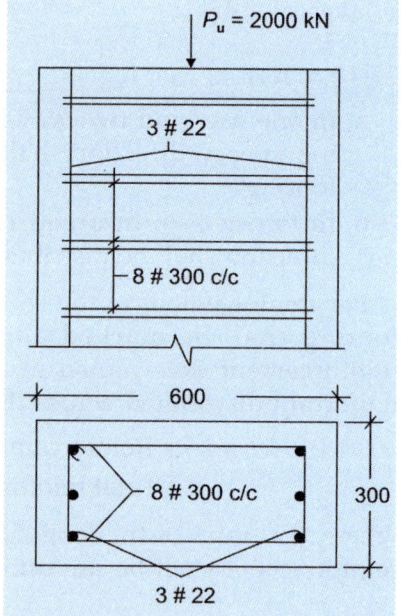

Fig. 11.9: Axially loaded rectangular column

ii. *Slenderness ratio*:

$$(L/D) = (3000/300) = 10.$$

Hence the column is designed as short column.

iii. *Minimum eccentricity*:

$$e_{min} = \left[\frac{L}{500} + \frac{D}{30}\right] = \left[\frac{3000}{500} + \frac{300}{30}\right] = 15 \text{ mm} < 20 \text{ mm}$$

Also $0.05\,D = (0.05 \times 300) = 15$ mm < 20 mm.

iv. *Main reinforcements*:

According to Clause 39.4, IS:456-2000:

$$P_u = 1.05\,[0.4f_{ck}\,A_g + (0.67f_y - 0.4f_{ck})A_{sc}]$$

$$\left(\frac{1200 \times 10^3}{1.05}\right) = \left[\left(\frac{0.4 \times 20 \times \pi \times 300^2}{4}\right) + \{(0.67 \times 415) - (0.4 \times 20)A_{sc}\}\right]$$

Solving, A_{sc} $= 2139\ mm^2$

A_{st} (minimum) = 0.8% of cross-section
$$= (0.008 \times \pi \times 300^2/4) = 565\ mm^2$$

Provide 6 bars of 22 mm diameter ($A_{sc} = 2280\ mm^2$)

v. *Helical reinforcement*:

Adopting clear cover of 50 mm over spirals

Core diameter $= [300 - (2 \times 50)] = 200\ mm$

Area of core $A_c = [\pi \times 200^2)/(4 - 2280)] = 29120\ mm^2$

Volume of core/m $= V_c = (29120 \times 10^3)\ mm^3$

Gross area of section $A_g = (\pi \times 300^2)/4 = 70695\ mm^2$

Using 8 mm diameter helical spirals at a pitch p mm, the volume of helical spiral per metre length is computed as

$$V_{us} = [\pi(300 - 100 - 8)\,50 \times 1000/p]\ mm^3/m$$
$$= (30159.288 \times 10^3)/p\ mm^3/m$$

According to Clause 39.4.1, IS:456

$$(V_{us}/V_e) < 0.36[A_g/A_c) - 1]\,(f_{ck}/f_y)$$

$$\left[\frac{30159.288 \times 10^3}{(29120 \times 10^3)p}\right] < 0.36\{(70685/29120) - 1\}\,(20/415)$$

Solving, $p = 42\ mm$

According to Clause 26.5.3.2(d), IS:456 code, the pitch should comply with the following specifications:

$$p < \left[\frac{75\ mm\ core\ diameter}{6}\right] = \left(\frac{200}{6}\right)$$
$$= 33.6$$

$$p > \begin{bmatrix} 25\ mm \\ 3\ (diameter\ of\ helical\ reinforcement) \\ = (3 \times 8) = 24 \end{bmatrix}$$

Hence provide 8 mm diameter helical spirals at a pitch of 30 mm.

vi. *Reinforcement details*:

The details of reinforcements in the helically reinforced column is shown in Fig. 11.10.

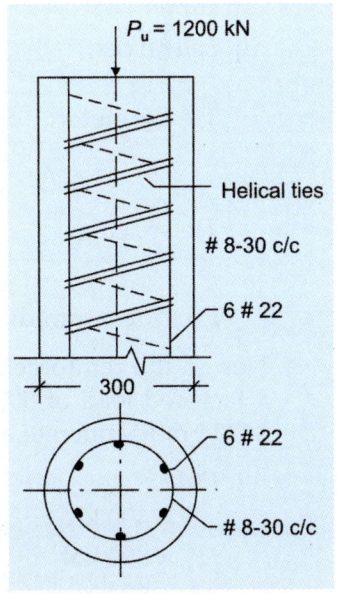

Fig. 11.10: Axially loaded circular column

3. Design the reinforcements in a rectangular column of size 300 mm × 500 mm to support a design ultimate load of 500 kN, together with a factored moment of 200 kN·m. Adopt the value of $f_{ck} = 20$ N/mm^2 and $f_y = 415$ N/mm^2.

   i. *Data*:

   $b = 300$ kN $\qquad\qquad f_{ck} = 20$ N/mm^2

   $D = 500$ mm $\qquad\qquad f_y = 415$ N/mm^2

   $P_u = 500$ kN $\qquad\quad (d'/D) = 0.1$

   $M_u = 200$ kN·m

   ii. *Non dimensional parameters*:

   $$\left(\frac{P_u}{f_{ck}\,b\,d}\right) = \left(\frac{500 \times 10^3}{20 \times 300 \times 500}\right) = 0.166$$

   $$\left(\frac{M_u}{f_{ck}\,bD^2}\right) = \left(\frac{200 \times 10^6}{20 \times 300 \times 500^2}\right) = 0.133$$

   iii. *Longitudinal reinforcement*:

   Refer to chart 32, SP:16 ($d'/D = 0.10$ and $f_y = 415$ N/mm^2) and readout the ratio:

   $(p/f_{ck}) = 0.06$

   $\therefore \qquad p = (0.06 \times 20) = 1.2$

   $\therefore \qquad A_{st} = \left(\frac{pbD}{100}\right) = \left(\frac{1.2 \times 300 \times 500}{100}\right) = 1800$ mm^2

   Provide 4 bars of 25 mm diameter distributed 2 nos. on each face ($A_{sc} = 1964$ mm^2).

   iv. *Lateral ties*:

   Tie diameter: $\phi_t \not< \begin{cases} (1/4) \times 25 = 6.25 \text{ mm} \\ 6 \text{ mm} \end{cases}$ $\qquad$ Provide 8 mm ties

   Tie spacing: $s_t \not> \begin{cases} (16 \times 25) = 400 \text{ mm} \\ (48 \times 8) = 384 \text{ mm} \\ b = 300 \text{ mm} \end{cases}$ $\qquad$ Adopt 300 mm

   Provide 8 mm diameter ties at 300 mm centres.

4. Design the reinforcements in a circular column of diameter 400 mm to support a factored load of 800 kN together with a factored moment of 80 kN·m. Adopt M 20 grade concrete and Fe 415 HYSD bars.

   i. *Data*:

   $D = 400$ kN $\qquad\qquad f_{ck} = 20$ N/mm^2

   $P_u = 800$ kN $\qquad\qquad f_y = 415$ N/mm^2

   $M_u = 80$ kN·m $\qquad (d'/D) = 0.1$

   Assume $d' = 40$ mm.

ii. *Nondimensional parameters*:

$$\left(\frac{P_u}{f_{ck}\,bd}\right) = \left(\frac{800 \times 10^3}{20 \times 400 \times 400}\right) = 0.25$$

$$\left(\frac{M_u}{f_{ck}\,D^3}\right) = \left(\frac{80 \times 10^6}{20 \times 400^3}\right) = 0.0625$$

iii. *Longitudinal reinforcement*:

Refer to Chart 56, SP:16 and readout the ratio

$(p/f_{ck}) = 0.06$

∴ $p = (20 \times 0.06) = 1.2$

∴ $A_{st} = (p\,\pi\,D^2)/(4 \times 100) = (1.2 \times \pi \times 400^2)/400 = 1508 \text{ mm}^2$

Provide 6 bars of 20 mm diameter ($A_{sc} = 1884 \text{ mm}^2$).

iv. *Lateral ties*:

Tie diameter: $\phi_t \not< \begin{cases} (20/4) = 5 \text{ mm} \\ 6 \text{ mm} \end{cases}$ Provide 6 mm ties

Tie spacing: $s_t \not> \begin{cases} 400 \text{ mm} \\ (16 \times 20) = 320 \text{ mm} \\ (48 \times 6) = 288 \text{ mm} \end{cases}$ Adopt 280 mm spacing

Provide 6 mm diameter ties at 300 mm centres.

5. Design the reinforcement in a short column 400 mm × 400 mm at the corner of a multistoreyed building to support an axial factored load of 1500 kN, together with biaxial moments of 50 kN·m acting in perpendicular planes. Adopt M20 grade concrete and Fe 415 HYSD bars.

i. *Data*:

$b = 400 \text{ kN}$ $f_{ck} = 20 \text{ N/mm}^2$

$D = 400 \text{ mm}$ $f_y = 415 \text{ N/mm}^2$

$P_u = 1500 \text{ kN}$ Assume $d' = 400 \text{ mm}$

$M_{ux} = M_{uy} = 50 \text{ kN·m}$ $(d'/D) = 0.10$

ii. *Equivalent moment*:

The reinforcement in section is designed for the axial compressive load P_u and the equivalent moment given by the relation:

$$M_u = 1.15\sqrt{M_{ux}^2 + M_{uy}^2}$$

$$= 1.15\sqrt{50^2 + 50^2} = 81.3 \text{ kN·m}$$

iii. *Nondimensional parameters*:

$$\left(\frac{P_u}{f_{ck}\,bD}\right) = \left(\frac{1500 \times 10^3}{20 \times 400 \times 400}\right) = 0.468$$

$$\left(\frac{M_u}{f_{ck}\,bD^2}\right) = \left(\frac{81.3 \times 10^6}{20 \times 20 \times 400^3}\right) = 0.063$$

iv. *Nondimensional parameters*:

Refer to Chart 44, SP:16 ($f_y = 415$ N/mm^2 and $d'/D = 0.10$) (equal reinforcement on all faces) and readout the value of

$$(p/f_{ck}) = 0.06$$

$$\therefore \qquad p = (20 \times 0.06) = 1.2$$

$$\therefore \quad A_{st} = \left(\frac{pbD}{100}\right) = \left(\frac{1.2 \times 400 \times 400}{100}\right) = 1920 \text{ mm}^2$$

Provide 4 bars of 20 mm diameter and 4 bars of 16 mm diameter ($A_{sc} = 2060$ mm^2) distributed equally on all faces with 3 bars on each face.

$$\therefore \qquad p = (100 \times 2000)/(400 \times 400) = 1.28$$

$$\therefore \quad (p/f_{ck}) = (1.28/20) = 0.064$$

Refer to Chart 44, SP:16 and readout $(M_{ux1}/f_{ck}bD^2)$ corresponding to the values of $(P_u/f_{ck}bD) = 0.468$ and $(p/f_{ck}) = 0.064$.

$$\left(\frac{M_{ux1}}{f_{ck}bD^2}\right) = 0.468$$

$$\therefore \qquad M_{ux1} = (0.068 \times 20 \times 400 \times 400^2)\, 10^{-6} = 87 \text{ kN·m}$$

Due to symmetry $M_{ux1} = M_{uy1} = 87$ kN·m

$$\begin{aligned}
P_{uz} &= [0.45f_{ck}\,A_c + 0.75f_y\,A_s] \\
&= (0.45 \times 20)\,[(400 \times 400) - 2060] + 0.75 \times 415 \times 2060 \\
&= 2062 \times 10^3 \text{ m} \\
&= 2062 \text{ kN}
\end{aligned}$$

$$\left(\frac{P_u}{P_{uz}}\right) = \left(\frac{1500}{2062}\right) = 0.72$$

Refer to Fig. 11.11 and readout the coefficient α_n corresponding to

$$(P_u/P_{uz}) = 0.72 \text{ and}$$

$$\alpha_n = 1.8.$$

iv. *Check for safety under biaxial bending*:

$$\left[\left(\frac{M_{ux}}{M_{ux1}}\right)^{\alpha_n} + \left(\frac{M_{uy}}{M_{uy1}}\right)^{\alpha_n}\right] \leq 1$$

$$\left[\left(\frac{50}{87}\right)^{1.8} + \left(\frac{50}{87}\right)^{1.8}\right] = 0.736 < 1$$

Hence the section is safe under biaxial bending.

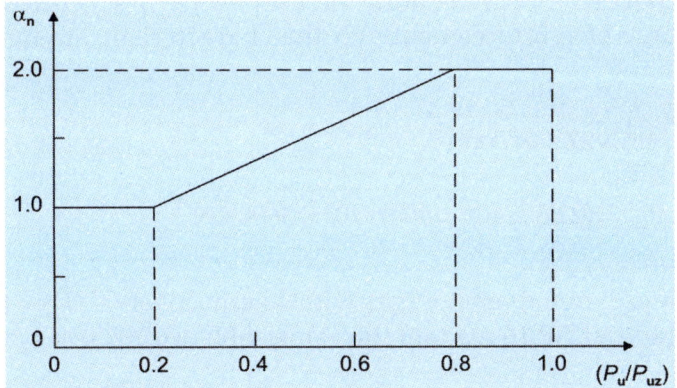

Fig. 11.11: Relation between α_n and (P_u/P_{uz})

v. *Reinforcement details*:

Provide 4 bars of 20 mm and 4 bars of 16 mm diameter as longitudinal reinforcement and 8 mm lateral ties at 300 mm centres.

6. Design the reinforcements in the slender column which is restrained against sway using the following data.

i. *Data*:

Size of column = 400 mm × 400 mm

Concrete grade = M30

Characteristic strength of reinforcement f_y = 415 N/mm^2

Effective length of column $L_{ex} = L_{ey}$ = 6 m

Unsupported length L = 7 m

Factored load P_u = 1500 kN

Factored moment in mutually perpendicular direction = $M_{ux} = M_{uy}$ = 40 kN·m at top and 20 kN·m at bottom.

ii. *Slenderness ratio*:

$$(L_{ex}/D) = (L_{ey}/b) = (6000/400) = 15 > 12.$$

Hence the column is slender about both axes.

From Table 1, SP:16, corresponding to the slenderness ratios of 15, readout the value of:

$$(e_x/D) = (e_y/b) = 0.113$$

iii. *Additional moments*:

$$M_{ax} = M_{ay} = (P_u \, e_x) = (P_u \, e_y) = (1500 \times 0.113)\, 0.4 = 67.8 \text{ kN·m}$$

The above moments will have to be reduced in accordance with the Clause 39.7.1.1 of the (IS:456-1980) code, but multiplication factors can be evaluated only if the reinforcement in the section is known.

iv. *Reinforcement*:

For first trial, assume the percentage reinforcement p = 3 (with reinforcement equally distributed on all four faces).

$$A_g = (400 \times 400) = 16 \times 10^4 \text{ mm}^2$$

From Chart 63, SP:16, readout the ratio of (P_{uz}/A_g) corresponding to the percentage of reinforcement and the characteristic strength of reinforcement as:

$$\left(\frac{P_{uz}}{A_g}\right) = 22.5 \, \text{N/mm}^2$$

$\therefore \qquad P_{uz} = (22.5 \times 400 \times 400)/10^3 = 3600 \, \text{kN}$

v. *Computation of P_b:*

Assuming 25 mm diameter bars with 40 mm cover $(d'/D) = (40/400) = 0.1$. From Table 60, SP:16 readout the values of $k_1 = 0.207$ and $k_2 = 0.328$.

Hence $\quad P_{bx} = P_{by} = \left(k_1 + k_2 \times \dfrac{P}{f_{ck}}\right) f_{ck} bD$

$$= \left[\left(0.207 + 0.328 \times \frac{3}{30}\right) 30 \times 400 \times 400\right] 10^{-3} = 1151 \, \text{kN}$$

vi. *Computation of reduction factor k_x and k_y:*

$$k_x = k_y = \left(\frac{P_{uz} - P_u}{P_{uz} - P_{bx}}\right) = \left(\frac{3600 - 1500}{3600 - 1151}\right) = 0.85$$

Hence, modified additional moments are:

$$M_{ux} = M_{uy} = [(67.8 \times 0.85) = 57.63 \, \text{kN·m}$$

The additional moments due to slenderness effects should be added to the initial moments after modifying the initial moment as per Clause 39.7.1, IS:456-2000.

$$M_{ux} = M_{uy} = [(0.6 \times 40) - (0.4 \times 20)] = 16 \, \text{kN·m}$$

The above actual moments should be compared with those calculated from minimum eccentricity consideration and the greater value is to be taken as the initial moment for adding the additional moments.

$$e_x = e_y = \left[\frac{L}{500} + \frac{D \text{ or } b}{30}\right] = \left[\frac{7000}{500} + \frac{400}{30}\right] = 27.3 \, \text{mm}$$

Both e_x and e_y are greater than 20 mm.

Moments due to minimum eccentricity are computed as:

$$M_{ux} = M_{uy} = \left(1500 \times \frac{27.3}{1000}\right) = 41 \, \text{kN} \cdot \text{m}$$

Total design moments are:

$$M_{ux} = M_{uy} = (41 + 57.63) = 98.63 \, \text{kN·m}$$

vii. *Check for biaxial bending*:

$$\left(\frac{P_u}{f_{ck} bd}\right) = \left(\frac{1500 \times 10^3}{30 \times 400 \times 400}\right) = 0.3125$$

Also $(p/f_{ck}) = (3/30) = 0.10$.

Refer to Chart 44, SP:16 $(f_y = 415 \text{ N/mm}^2)$ and $(d'/D) = 0.10]$ and readout the parameter.

$$\therefore \quad \left(\frac{M_u}{f_{ck}bD^2} \right) = 0.14$$

$$\therefore \quad M_{ux1} = M_{uy1} = (0.14 \times 30 \times 400 \times 400^2) \, 10^{-6} = 268.8 \text{ kN·m}$$

$$(P_u/P_{uz}) = (1500/3600) = 0.416$$

From Fig. 8.4, readout the value of $\alpha_n = 1.35$

Hence, we have the relation:

$$\left(\frac{M_{ux}}{M_{ux1}} \right)^{\alpha_n} + \left(\frac{M_{uy}}{M_{uy1}} \right)^{\alpha_n} \le 1.0$$

$$\left(\frac{98.63}{268.8} \right)^{1.35} + \left(\frac{98.63}{268.8} \right)^{1.35} = 0.50 < 1.0$$

Hence, the assumed reinforcement percentage of 3% is satisfactory.

7. Design a reinforced concrete footing for a rectangular column of section 300 mm × 500 mm supporting an axial factored load of 1500 kN. The safe bearing capacity of the soil at site is 185 kN/m². Adopt M20 grade concrete and Fe 415 HYSD bars.

   i. *Data*:

$P_u = 1500 \text{ mm}$	$f_{ck} = 20 \text{ N/mm}^2$
$b = 300 \text{ mm}$	$f_y = 415 \text{ N/mm}^2$
$D = 500 \text{ mm}$	
$p = 185 \text{ kN·m}^2$	

   ii. *Size of footing*:

   Load of column = 1500 kN
   Self weight of footing (10%) = 150 kN
   Total factored load w_u = 1650 kN

   $$\text{Footing area} = \left(\frac{1650}{1.5 \times 185} \right) = 5.94 \text{ m}^2$$

   Proportion the footing area in the same proportion as the sides of the column.

   Hence $(3x) \times (5x) = 6$

   $\therefore \qquad x = 0.63$

   Short side of fitting $= (3 \times 0.63) = 1.89 \text{ m}$

   Long side of footing $= (5 \times 0.63) = 215 \text{ m}$

   Adopt a rectangular footing of size 2 m × 2 m

   Factored soil soil pressure at base is computed as

   $$p_u = \left(\frac{1500}{2 \times 3} \right) = 230 \text{ kN/m}^2 < (1.5 \times 1.85) = 277.5 \text{ kN/m}^2$$

Hence, the footing area is adequate since the soil pressure developed at the base is less than the factored bearing capacity of soil.

iii. *Factored bending moments*:

Cantilever projection from the short side face of the column

$$= 0.5(3 - 0.5) = 1.25 \text{ m}$$

Cantilever projection from the long side face of the column

$$= 0.5(2 - 0.3) = 0.85 \text{ m}$$

Bending moment at short side face of column

$$= (0.5p_u \cdot L^2) = (0.5 \times 250 \times 1.25^2) = 195 \text{ kN·m}$$

Bending moment at long side face of the column

$$= (0.5p_u \cdot L^2) = (0.5 \times 250 \times 8.85^2) = 90 \text{ kN·m}$$

iv. *Depth of footing*:

(a) From moment considerations, we have:

$$M_u = 0.138 f_{ck} \, b \, d^2$$

$$\therefore \quad d = \sqrt{\frac{M_u}{0.138 f_{ck} \, b}} = \sqrt{\frac{195 \times 10^6}{0.138 \times 20 \times 10^3}} = 266 \text{ mm}$$

(b) From shear stress considerations, we have the critical section for one-way shear is located at a distance d from the face of the column.

Shear force per metre width (longer direction)

$$V_{uL} = 250 \, (1250 - d) \text{ N}$$

Assuming the shear strength $\tau_c = 0.36 \text{ N/mm}^2$ for M20 grade oncrete with nominal percentage of reinforcement, $p_t = 0.25\%$.

Refer to Table 19, IS:456-2000 and readout the permissible shear stress

$$= (k_s \, \tau_c) = (1 \times 0.33) = 0.33 \text{ N/mm}^2$$

Nominal shear stress $\tau_v = (V_u/bd) = (176 \times 10^3)/(10^3 \times 550) = 0.32 \text{ N/mm}^2$

Since $\tau_v < (k_s \, \tau_c)$, shear stresses are within the safe permissible limits.

vii. *Reinforcement details*:

The details of reinforcements in the footing is shown in Fig. 11.12.

8. Design a reinforced concrete circular footing for a circular column of 300 mm diameter supporting a factored axial load of 750 kN. Adopt the safe bearing capacity of the soil as 200 kN/m² and use M20 grade concrete and Fe 415 HYSD bars.

   i. *Data*:

$P_u = 750$ mm	$f_{ck} = 20 \text{ N/mm}^2$
$D = 300$ mm	$f_y = 415 \text{ N/mm}^2$
$p = 200 \text{ kN/m}^2$	
$p_u = (1.5 \times 200) = 300 \text{ kN/m}^2$	

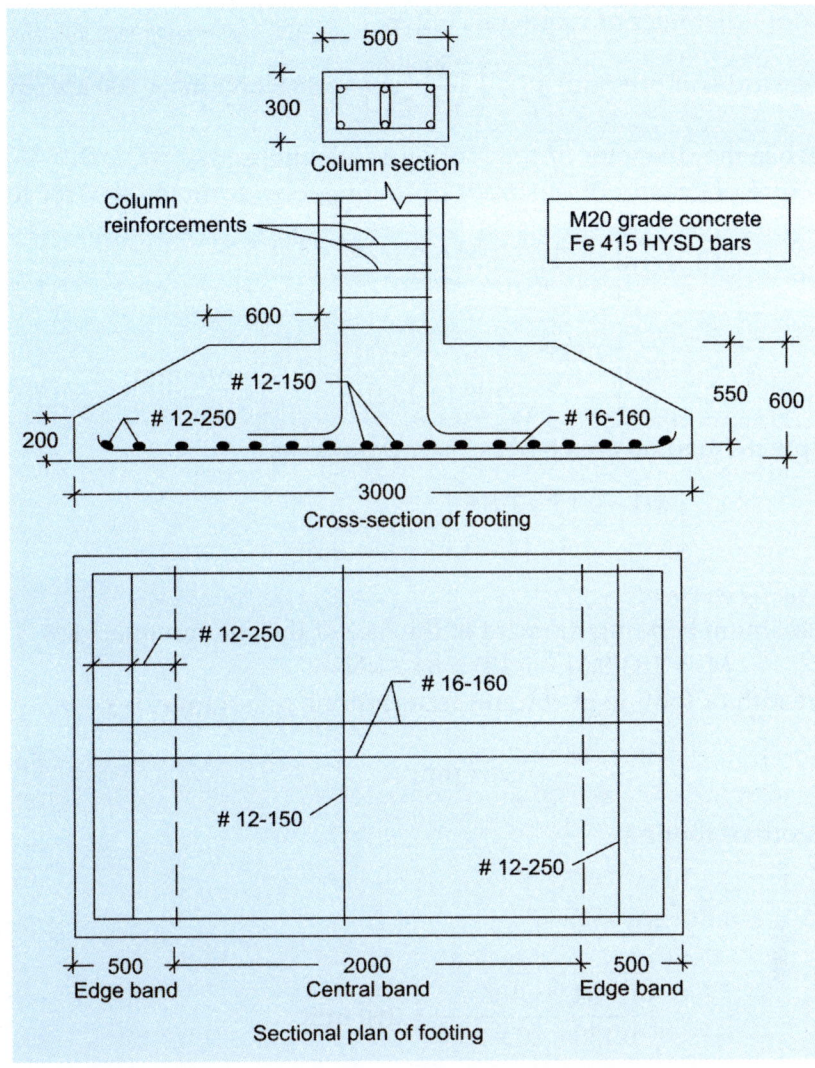

Fig. 11.12. Reinforcement details in rectangular footing

ii. *Size of footing*:

Load of column = 750 kN

Self-weight of footing (10%) = 75 kN

Total load w_u = 825 kN

Let D_f = diameter of the footing

A_f = area of the footing

$$A_f = \left(\frac{\pi D_f^2}{4}\right) = \left(\frac{w_u}{p_u}\right) = \left(\frac{825}{300}\right) = 2.75 \text{ m}^2$$

$$D_f = \sqrt{\frac{4 \times 2.75}{\pi}} = 1.87 \text{ m}$$

Adopt diameter of footing $D_f = 2$ m

Upward soil pressure $p_u = \left(\dfrac{750 \times 4}{\pi \times 2^2}\right) = 238.8$ kN/m^2 < 300 kN/m^2

Hence the diameter of the footing is adequate.

Centre of gravity of quadrant of footing 'obc' from 'o' is (refer to Fig. 8.6)

$$= 0.6\left[\frac{R^2 + r^2 + R \cdot r}{R + r}\right]$$

$$= 0.6\left[\frac{1000^2 + 150^2 + (1000 \times 150)}{(1000 + 150)}\right] = 610 \text{ mm}$$

Upward load on area b b' c c' is computed as

$$\left[\frac{\pi(1 - 0.15^2)238.8}{4}\right] = 183 \text{ kN}$$

iii. *Bending moment*:

Maximum bending moment at the face of the column quadrant
$M_u = 183(0.61 - 0.15) = 84.2$ kN·m

Breadth of footing at column face (for one quadrant c' b')

$$= \left(\frac{\pi \times 300}{4}\right) = 235 \text{ mm}$$

Depth of footing

$$d = \sqrt{\frac{M_u}{0.138 f_{ck}\, b}}$$

$$= \sqrt{\frac{84.2 \times 10^6}{(0.138 \times 20 \times 235)}} = 360 \text{ mm}$$

Depth required for shear considerations will be nearly 1.5 times that for moment considerations.

Hence adopt effective depth $d = 525$ mm and
overall depth $D = 600$ mm

iv. *Reinforcements*:

$$M_u = \left(0.87\, f_y\, A_{st}\, d\right)\left[1 - \left(\frac{A_{st} f_y}{bd f_{ck}}\right)\right]$$

$$(84.2 \times 10^6) = (0.87 \times 415 A_{st} \times 525)\left[1 - \left(\frac{415 A_{st}}{235 \times 525 \times 20}\right)\right]$$

Solving, $A_{st} = 484$ mm^2

Minimum $A_{st} = (0.0012 \times 235 \times 600) = 169.2$ mm^2

Provide 12 mm diameter bars at 150 mm centres ($A_{st} = 754$ mm^2) both ways.

v. *Check for shear stress*:

Ultimate shear force at a distance 0.525 m from the face of the column is computed as

$$V_u = 238.8 \, (2^2 - 1.35^2) \, (\pi/4) = 408 \text{ kN}$$

Shear per metre width of perimeter

$$= \left(\frac{408}{p \times 1.35)} \right) = 96 \text{ kN}$$

$$\tau_v = \left(\frac{V_u}{bd} \right) = \left(\frac{96 \times 10^3}{10^3 \times 525} \right) = 0.18 \text{ N/mm}^2$$

$$= \left(\frac{100 A_{st}}{bd} \right) = \left(\frac{100 \times 754}{10^3 \times 525} \right) = 0.143$$

Refer to Table 19, IS:456-2000 and readout the permissible value of shear strength of concrete

$$(k_s \cdot \tau_c) = (1 \times 0.28) = 0.28 \text{ N/mm}^2 > 0.18 \text{ mm}^2$$

Hence shear stresses are within safe permissible limits.

vi. *Reinforcement details*:

The details of reinforcements in the circular footing is shown in Fig. 11.13.

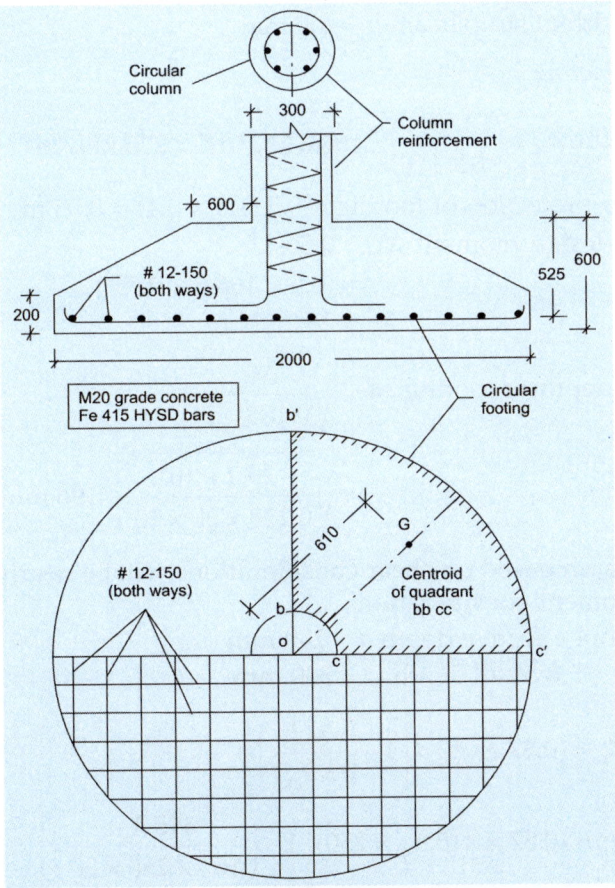

Fig. 11.13. Reinforcement details in circular footing

9. Design a combined column footing with a strap beam for two reinforced concrete columns 300 mm × 300 mm size spaced 4 m apart and each supporting a factored axial load of 750 kN. Assume the ultimate bearing capacity of soil at site as 225 kN/m². Adopt M20 grade concrete and Fe 415 HYSD bars.

i. *Data*:

Size of column = 300 mm × 300 mm
Spacing of column = 4 m
Factored load on each column = 750 kN
Ultimate bearing capacity of soil = 225 kN/m²
f_{ck} = 20 kN/mm² f_y = 415 N/mm²

ii. *Loads on footing*:

Total load on column (2 × 750) = 1500 kN
Self-weight (10%)   =   150 kN
Total ultimate load P_u = 1650 kN

iii. *Size of footing*:

Area of footing $= \left(\dfrac{1650}{225}\right) = 7.33 \text{ m}^2$

Adopt a footing of size 6 m × 1.5 m
Adopt width of strap beam b = 400 mm

iv. *Design of footing*:

Soil pressure $p_u = \left(\dfrac{1500}{6 \times 1.5}\right) = 166.6 \text{ kN/m}^2 < 225 \text{ kN/m}^2$

Cantilever projection of footing = 0.5 (1.5 × 0.4) = 0.55 m
Ultimate design moment $M_u = 0.5 p_u L^2$

$$= (0.5 \times 166.6 \times 0.55^2)$$

$$= 25.2 \text{ kN·m}$$

Effective depth of footing $d = \sqrt{\dfrac{M_u}{0.138 f_{ck} b}}$

$$= \sqrt{\dfrac{25.2 \times 10^6}{0.138 \times 20 \times 10^3}} = 96 \text{ mm}$$

But the depth based on shear considerations will be nearly double than that due to moment considerations.

Hence adopt effective depth d = 250 mm
and   overall depth D = 300 mm

$$M_u = (0.87 f_y A_{st} d \left[1 - \left(\dfrac{A_{st} f_y}{bd f_{ck}}\right)\right]$$

$$(25.2 \times 10^6) = (0.87 \times 415 A_{st} \times 250) \left[1 - \left(\dfrac{415 A_{st}}{10^3 \times 250 \times 20}\right)\right]$$

Solving, $A_{st} = 287$ mm^2

Minimum reinforcement= $(0.0012 \times 1000 \times 300) = 360$ mm^2

Adopt 10 mm diameter bars at 200 mm centres ($A_{st} = 393$ mm^2) as main reinforcement.

v. *Check for shear stress*:

Shear stress at a distance equal to the effective depth

$$V_u = (0.55 - 0.25)166.6 = 50 \text{ kN}$$

$$\tau_v = \left(\frac{V_u}{bd}\right) = \left(\frac{50 \times 10^3}{10^3 \times 250}\right) = 0.2 \text{ N/mm}^2$$

$$= \left(\frac{100 A_{st}}{bd}\right) = \left(\frac{100 \times 393}{10^3 \times 250}\right) = 0.157$$

Refer to Table 19 (IS:456-2000) and readout $\tau_c = 0.28$ N/mm^2

∴ Permissible shear stress $= (k_s \tau_c) = (1 \times 0.28) = 0.28$ N/mm^2

Since $(k_s \tau_c) > \tau_v$, shear stresses are within safe permissible limits.

vi. *Design of strap beam*:

Factored load on beam $w_u = (1.5 \times 166.6) = 250$ kN/m

Neglecting the small cantilever portion of the beam:

$$M_u = 0.125 \, w_u \, L^2 = (0.125 \times 250 \times 4^2) = 500 \text{ kN·m}$$

$$V_u = 0.5 \, w_u \, L = (0.5 \times 250 \times 4) = 500 \text{ kN}$$

Depth of strap beam computed based on amount.

Assuming $\tau_c = 1.2$ N/mm^2

$$d = \left(\frac{V_u}{b\tau_c}\right) = \left(\frac{500 \times 10^3}{400 \times 1.2}\right) = 0.2 \text{ N/mm}^2$$

Adopt effective depth $d = 1150$ mm and

overall depth $D = 1200$ mm

$$M_u = (0.87 \, f_y \, A_{st} \, d\left[1 - \left(\frac{A_{st} f_y}{bd f_{ck}}\right)\right]$$

$$(500 \times 10^6) = (0.87 \times 415 A_{st} \times 1150)\left[1 - \left(\frac{415 A_{st}}{400 \times 1150 \times 20}\right)\right]$$

Solving, $A_{st} = 1290$ mm^2

Provide 4 bars of 22 mm diameter ($A_{st} = 1520$ mm^2)

$$\text{Shear stress } \tau_v = \left(\frac{V_u}{bd}\right) = \left(\frac{500 \times 10^3}{400 \times 1150}\right) = 1.09 \text{ N/mm}^2$$

$$= \left(\frac{100 A_{st}}{bd}\right) = \left(\frac{100 \times 1520}{400 \times 1150}\right) = 0.33$$

Refer to Table 19, IS:456-2000 and readout the permissible shear

$$\tau_c = 0.40 \, \text{N/mm}^2 < \tau_v$$

Hence shear reinforcement are required to resist the balanced shear force computed as

$$V_{us} = [500 - (0.4 \times 400 \times 1150) \, 10^{-3}]$$
$$= 316 \, \text{kN}$$

Using 8 mm diameter 4 legged stirrups,

$$s_v = \left(\frac{0.87 \times 415 \times 4 \times 50 \times 1150)}{316 \times 10^3}\right) = 262 \, \text{mm}$$

Adopt 8 mm diameter 4 legged stirrups at 250 mm centres in the strap beam. Side face reinforcement of 0.1% of web area as specified in the IS:456 code is provided.

vii. *Reinforcement details*:

The details of reinforcements in the combined footing and strap beam is shown in Fig. 11.14.

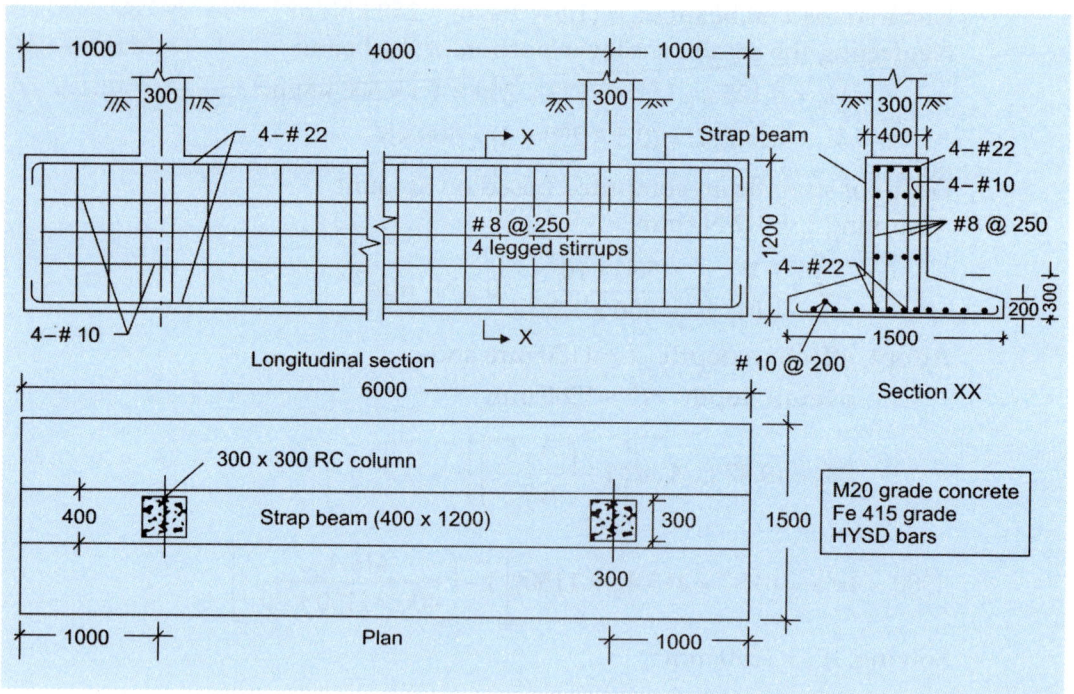

Fig. 11.14. Reinforcement details in combined footing

10. Design a strap footing foundation for two columns C and D spaced 5 m centres. Column C is 350 mm square and supports a factored load of 750 kN and is located on the property line. Column D is 400 mm square and supports a factored load of 1200 kN. The safe bearing capacity of soil at site is 200 kN/m. Adopt M20 grade concrete and Fe 415 HYSD bars.

i. *Data*:

Size of column C = 350 mm × 350 mm

Spacing of column D = 400 mm × 400 mm

SBC of soil = 200 kN/m²

Factored load on column C = 750 kN

Factored load on column D = 1200 kN

Materials: M20 grade concrete (f_{ck} = 20 N/mm²)
Fe 415 HYSD bars f_y = 415 N/mm²

ii. *Footing size*:

Total load on columns (750 + 1200) = 1950 kN
Self-weight of footing (10%)   = 195 kN
Total ultimate load P_u = 2145 kN

$$\text{Area of footing} = \left(\frac{2145}{1.5 \times 200} \right) = 7.15 \text{ m}^2$$

Referring to Fig. 11.15, let L_1 and L_2 be the lengths of the footings under the column C and D respectively.

Consider B = width of the footing,

then $B(L_1 + L_2) = 7.15$, (assuming B = 1.5)

$L_1 = 2.36$ and $L_2 = 2.40$ m

Adopt a footing of size 1.5 m × 2.4 m under both the columns C and D.

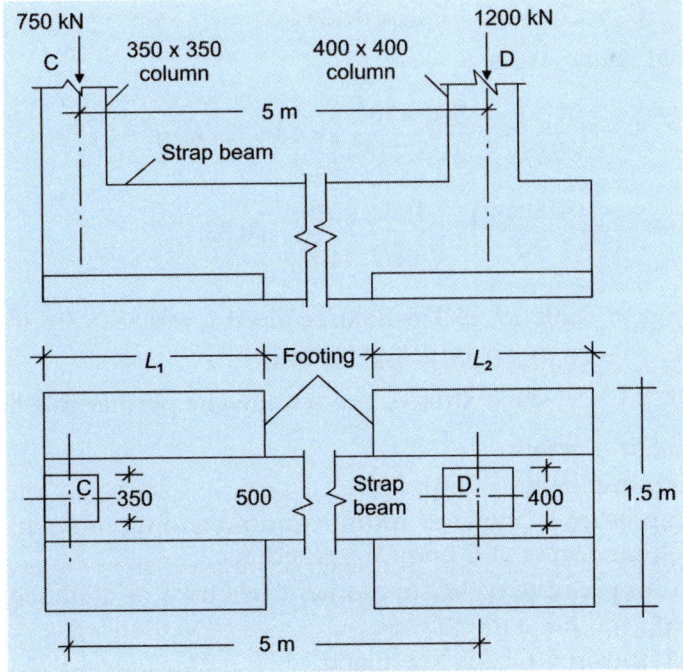

Fig. 11.15: Combined footing (unsymmetrically loaded columns)

iii. *Design of footing*:

Net area of footing = 1.5(2.4 + 2.4) = 7.2 m²

Upward soil pressure = (750 + 1200)/7.2 = 270.8 kN/m²

Assuming width of strap beam = 500 mm

Cantilever projection of slab beyond beam

$$0.5(1.5 - 0.50) = 0.50 \text{ m}$$

Ultimate moment (–ve) $M_u = (0.5 \times 270.8 \times 0.5^2) = 33.85$ kN·m

Effective depth $d = \sqrt{\dfrac{33.85 \times 10^6}{0.138 \times 20 \times 10^3}} = 110$ mm

Depth required from shear consideration is nearly twice that obtained from moment criteria.

Hence, adopt effective depth $d = 200$ mm and overall depth $D = 250$ mm.

$$M_u = (0.87 f_y A_{st} d)\left[1 - \left(\frac{A_{st} f_y}{bd f_{ck}}\right)\right]$$

$$(33.85 \times 10^6) = (0.87 \times 415 A_{st} \times 200)\left[1 - \left(\frac{415 A_{st}}{10^3 \times 200 \times 20}\right)\right]$$

Solving, $A_{st} = 1025$ mm²

Adopt 16 mm diameter bars at 160 mm centres ($A_{st} = 1257$ mm²).

iv. *Check for shear stress*:

Ultimate shear stress at a distance of 200 mm from face of the column

$$V_u = 270.8(0.5 \times 0.2) = 81.24 \text{ kN}$$

Nominal shear stress

$$\tau_v = \left(\frac{V_u}{bd}\right) = \left(\frac{500 \times 10^3}{400 \times 1150}\right) = 1.09 \text{ N/mm}^2$$

$$= \left(\frac{100 A_{st}}{bd}\right) = \left(\frac{100 \times 1520}{400 \times 1150}\right) = 0.33$$

Referring to Table 19, IS:456-2000, readout $\tau_c = 0.38$ N/mm²

∴ $(k_s \cdot \tau_c) = (1.10 \times 0.38) = 0.42$ N/mm²

Since $(k_s \cdot \tau_c) > \tau_v$, shear stresses are within safe permissible limits.

v. *Design of strap beam*:

Load on strap beam = 270.8 kN

The strap beam is analysed for maximum bending moment and shear force with column forces at C and D as reactions.

Maximum positive ultimate moment occurs at a distance of 1.67 m from column C and has a magnitude

M_u (positive) = 561 kN·m and

M_u (negative) at $D = (0.5 \times 270.8 \times 1.2^2) = 293$ kN·m

Adopting width of strap beam $b = 500$ mm

Effective depth $d = \sqrt{\left(\dfrac{561 \times 10^6}{0.138 \times 20 \times 500}\right)} = 638$ mm

Depth required from shear criteria will be larger.

Hence adopt effective depth $d = 950$ mm and overall depth $D = 1000$ mm.

The tension reinforcement for maximum bending moment in the beam is evaluated using the relation,

$$(561 \times 10^6) = (0.87 \times 415 A_{st} \times 950)\left[1 - \left(\frac{415 A_{st}}{500 \times 950 \times 20)}\right)\right]$$

Solving, $A_{st} = 1722$ mm^2

Provide 6 bars of 22 mm diameter ($A_{st} = 2280$ mm^2)

Area of tension reinforcement for negative bending moment at D is computed using the relation:

$$(293 \times 10^6) = (0.87 \times 415 A_{st} \times 950)\left[1 - \left(\frac{415 A_{st}}{500 \times 950 \times 20)}\right)\right]$$

Solving, $A_{st} = 890$ mm^2

Provide 3 bars of 22 mm diameter ($A_{st} = 1140$ mm^2)

vi. *Shear reinforcements*:

Maximum shear force occurs at the left face of beam under column D and is computed as:

$$V_u = 716 \text{ kN}$$

Nominal shear stress $\tau_v = \left(\dfrac{716 \times 10^3}{bd}\right) = 1.5 \text{ N/mm}^2$

$$= \left(\frac{100 A_{st}}{bd}\right) = \left(\frac{100 \times 1140}{500 \times 950}\right) = 0.23$$

Refer to Table 19, IS:456-2000 and readout the permissible shear as

$$\tau_c = 0.32 \text{ N/mm}^2$$

Since $\tau_v > \tau_c$, shear reinforcements are required.

Balance shear force

$$V_{us} = [716 - (0.32 \times 500 \times 950)10^{-3}]$$
$$= 561 \text{ kN}$$

Using 10 mm diameter 4 legged stirrups at a spacing,

$$s_v = \left(\frac{0.87 \times 415 \times 4 \times 79 \times 950)}{561 \times 1000}\right) = 193 \text{ mm}$$

Adopt 10 mm diameter 4 legged stirrups at a spacing of 180 mm centres at supports gradually increasing to 300 mm centres towards the centre of span.

vii. *Reinforcement details*:

The details of reinforcements in the combined footing is shown in Fig. 11.16.

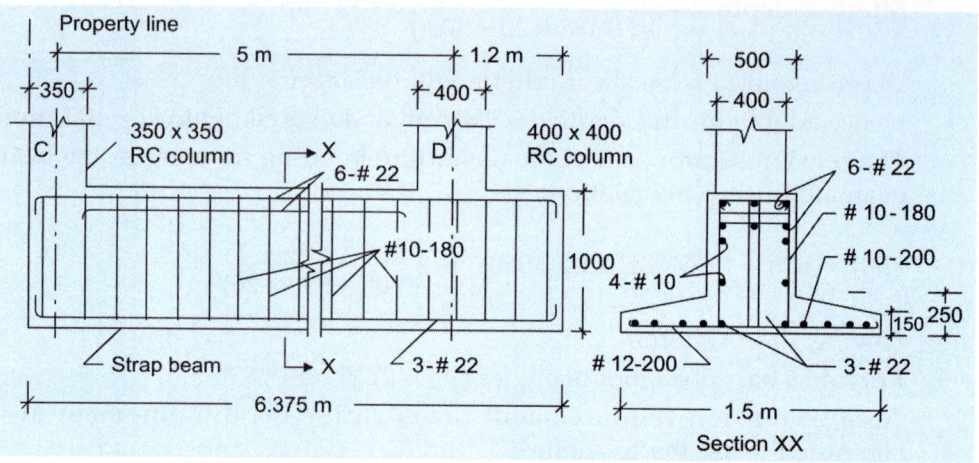

Fig. 11.16. Combined footing (unsymmetrically loaded columns)

REFERENCES

1. IS:456-2000, Indian Standard Code of Practice for Plain and Reinforced Concrete (4th Revision), BIS, New Delhi, 2000, p 37.
2. Rangan BV, Hall AS and Werner, *Reinforced Concrete Structures*, Pitman Publications, London, 1980.
3. Purushothaman P, *Reinforced Concrete Structural Elements, Behaviour, Analysis and Design*, Tata McGraw-Hill, New Delhi, 1984.
4. Winter G and Nilson SH, *Design of Concrete Structures*, McGraw-Hill, New York, 1972.
5. Park & Paulay T, *Reinforced Concrete Structures*, John Wiley and Sons Inc, New York, 1975.
6. Verghese PC, *Limit State Design of Reinforced Concrete*, Prentice Hall of India, New Delhi, 1998.
7. SP:24-1983, Explanatory Handbook on IS:456–2000, BIS, New Delhi, 1983.
8. ACI:318-89, Building code requirements for reinforced concrete, American Concrete Institute, Detroit, Michigan, 1989, pp 119–128.
9. Taranath BS, *Structural Analysis and Design of Tall Buildings* (International Edition), McGraw-Hill, USA, 1988.
10. Macgregor JG and Hage SE, Stability analysis & design of concrete columns, *ASCE Journal of Structural Division*, Vol. 10, October 1977, pp 19–53.

ASSIGNMENT

1. A rectangular reinforced concrete column of cross-sectional dimension 450 mm × 600 mm is subjected to an axial load of 200 kN under service dead and live loads. The column has an unsupported length of 3 m and is braced against side sway in both directions. Adopt M20 grade concrete and Fe 415 HYSD bars.

2. Design the reinforcements in a spirally tied column of 400 mm diameter, supporting an axial factored load of 1500 kN. The column has an unsupported length of 3.4 m and is braced against side sway. Adopt M25 grade concrete and Fe 415 HYSD bars.

3. Design the reinforcements in a rectangular column of size 300 mm × 400 mm subjected to a design factored load of 1200 kN and a factored moment of 200 kN·m with respect to the major axis. Adopt M20 grade concrete and Fe 415 steel.

4. A short reinforced concrete rectangular column of size 300 mm × 500 mm is subjected to a factored load of 1000 kN together with a factored moment of 250 kN·m about the major axis. Adopting M25 grade concrete and Fe 415 HYSD bars, design suitable reinforcements in the column section.

5. Design the reinforcements in a short circular column of diameter 400 mm to support a factored axial load of 1000 kN together with a factored moment of 100 kN·m. Adopt M20 grade concrete and Fe 415 HYSD bars.

6. Design the reinforcements in a short column 400 mm × 600 mm subjected to an ultimate axial load of 1600 kN together with an ultimate moments of 120 kN·m and 90 kN·m about the major and minor axes respectively. Adopt M20 grade concrete and Fe 415 HYSD bars.

7. A short reinforced concrete column located at the corner of a multistoreyed building is subjected to an axial factored load of 200 kN associated with factored moments of 75 and 60 kN·m respectively, acting in perpendicular planes. The column size is fixed as 450 mm × 450 mm. Adopting M20 grade concrete and Fe 415 grade reinforcement, design suitable reinforcements in the column.

8. Design the reinforcements required for a column which is restrained against sway using the following data:
   Size of columns = 450 mm × 530 mm
   Effective length = 6.6 m
   Unsupported length = 7.7 m
   Factored load = 1600 kN
   Factored moment about major axis = 50 kN·m at top and 30 kN·m at bottom
   Factored moment about minor axis = 40 kN·m at top and 20 kN·m at bottom
   Concrete grade = M25
   Steel grade = Fe 500
   Column is bent in double curvature and reinforcement is distributed equally on all the four faces.

9. Design a combined column footing with a strap beam for two reinforced concrete columns 200 mm by 200 mm size spaced 4 m apart and each supporting a service load of 300 kN. The safe bearing capacity of soil is 150 kN/m². Adopt M25 grade concrete and Fe 415 HYSD bars.

10. Design a combined footing with a strap beam for two columns A and B spaced 4 m apart. Column A is 250 mm square and supports a service load of 300 kN and is located on the property line. Column B is 350 mm square and supports a service load of 500 kN. The safe bearing capacity of soil at site is 200 kN/m². Adopt M20 grade concrete and Fe 415 HYSD bars. Sketch the details of reinforcements in the combined footing.

REVIEW QUESTIONS

1. Explain the terms: (a) Short column (b) Long column (c) Braced column and (d) Unbraced column with respect to the design of compression members.

2. What is the significance of the term 'slenderness ratio' with reference to the design of compression members?

3. Explain the function of transverse ties in a reinforced concrete column? What happens if ties are not provided?

4. What is the difference between tied and spirally reinforced columns? In what type reinforced concrete columns do you adopt spirals in preference to lateral ties?

5. Explain the significance of the code clause, specifying that all columns should be designed to resist a minimum eccentricity of loading.

6. What are moment interaction curves? Mention the advantages of using them in the design of eccentrically loaded columns?

7. What are uniaxially and biaxially loaded columns? Explain the basis for the simplified code procedure for the design of biaxially loaded columns.

8. Explain the terms: (a) Isolated footing (b) Combined footing with respect to the design of foundations for columns.

9. Under what circumstances do you prefer trapezoidal footing to a rectangular one while designing foundations for columns.

10. How do you select the depth of a footing? What critical section in a footing do you check for safety against shear?

OBJECTIVE QUESTIONS

1. Short columns have slenderness ratios less than
   a. 30
   b. 20
   c. 12

2. The effective length of a column fixed at both ends is
   a. 1.0 L
   b. 0.5 L
   c. 2.0 L

3. The diameter of main reinforcements in a column should not be less than
   a. 25 mm
   b. 16 mm
   c. 12 mm

4. The minimum cover to reinforcements in columns should not be less than
   a. 25 mm
   b. 30 mm
   c. 40 mm bar whichever is greater

5. The minimum number of bars required in rectangular and circular columns are respectively
   a. 6 and 6
   b. 4 and 6
   c. 8 and 8
6. Corner columns in a multistorey framed building should be designed for
   a. axial loads
   b. lateral loads
   c. combined axial load and moment
7. The minimum and maximum recommended percentages of reinforcements in a column are respectively
   a. 1 and 6
   b. 0.3 and 0.8
   c. 0.8 and 4
8. Reinforced concrete columns should be designed for a minimum eccentricity of
   a. 25 mm
   b. 0.05 times the lateral dimension
   c. 20 mm
9. The size of a footing to support a reinforced concrete column depends upon the
   a. grade of concrete in the column
   b. load transmitted to the footing by the column
   c. safe bearing pressure of the soil
10. The critical section to be considered for checking failure of footing against one way shear is at a distance equal to the
    a. larger dimension of the column from its face
    b. effective depth of the footing
    c. least lateral dimension of the column

Design of Staircases

12.1 GENERAL FEATURES

Reinforced concrete staircases serve an important function of connection between the floors in multistorey buildings. In small buildings they are the only means of access between the floors. In multistorey buildings with several floors, the staircase is generally located around the lift box unit and serves as an emergency exit passage. The staircase unit comprises flight of steps with one or more intermediate landings provided between the floor levels.

The structural components of a flight of stairs consist of:

1. the tread which forms the horizontal portion of the step is generally 250 to 300 mm wide depending upon the type of building.
2. riser which is the vertical distance between the adjacent treads or the vertical projection of the step, generally in the range of 150 to 190 mm depending upon the type of building. The width of stairs varies in the range of 1 to 1.5 m with a minimum value of 850 mm. Generally, public buildings should be provided with larger widths permitting free and uninterrupted passage to users.
3. going forms the horizontal plan projection of an inclined flight of steps between the first and the last riser. A flight of steps consists of two landings and one going with 10 to 12 steps.

The various technical terms generally used in the design of staircases such as flight, landing, going, tread, riser, etc. are shown in Fig. 12.1

12.2 TYPES OF STAIRCASES

Over the years several types of staircases[1,2] have been developed with varying geometrical shapes and structural behaviour under loads. Typical types commonly used in various types of buildings are shown in Fig. 12.2 The nomenclature used to describe these types of stairs according to Subramanian[3] is as follows:

1. Straight flight stairs to negotiate entrances and cellar floors (Fig. 12.2a)
2. Straight flight stairs with landing (Fig. 12.2b)

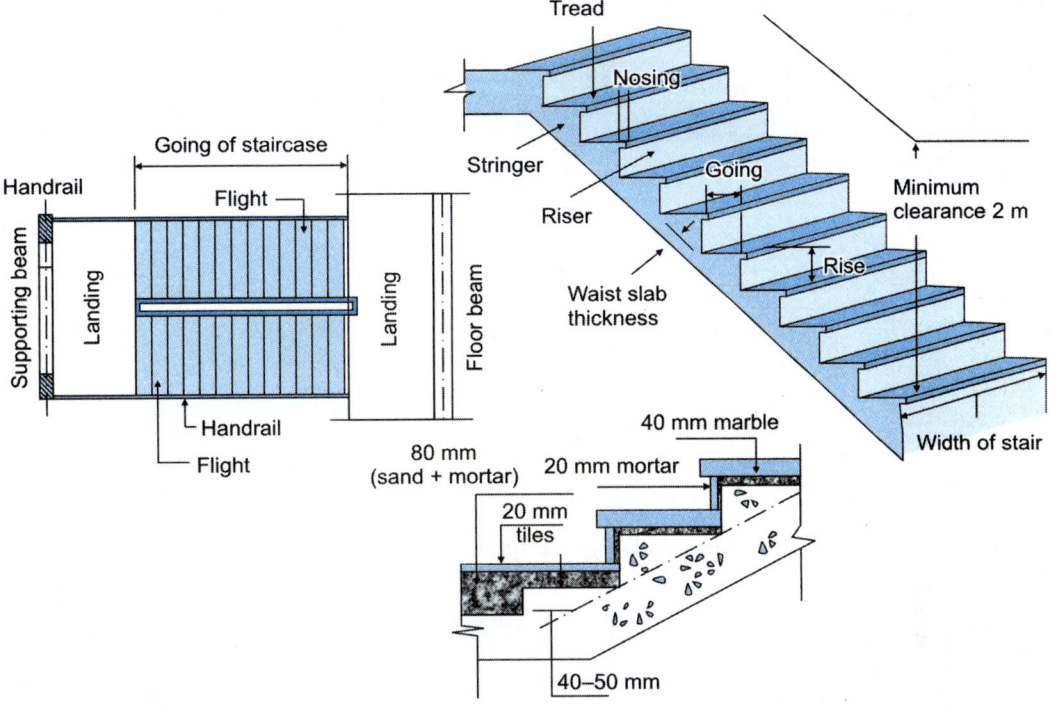

Fig. 12.1: Terminology used in staircases

3. Quarter turn stairs (Fig. 12.2c)
4. Dog-legged stairs or half turn stairs (Fig.12.2d)
5. Branching stairs (Fig. 12.2e)
6. Open-well (half turn) stairs (Fig. 12.2f)
7. Open-well stairs with quarter turn landing (Fig. 12.2g)
8. Part-circular stairs (Fig. 12.2h)
9. Spiral stairs (Fig. 12.2i)
10. Helicoidal stairs (Fig. 12.2j)

The most commonly used types are the dog-legged and open-well staircases. In domestic and office buildings. In multistorey buildings with lift provision, open-well stairs with quarter turn landing are generally used around the lift well for emergency purposes.

In congested locations where there is constraint of space, spiral staircases are ideally suited. It is not user-friendly due to the reduced tread width near the post and suitable only for single person to use the stairs at a time. The stair slabs cantilever out from the main post and functionally act as cantilever in space[4,5].

Helicoidal staircase as shown in Fig. 12.3 is aesthetically superior to other types and generally used in the entrance foyer of cinema theatres and shopping malls to connect the ground and first floors. Helicoidal stair which is built as a ramp following the helicoidal curve with supports at ground & first floor or with intermediate supports involves rigorous structural computations for the determination of design moments and shear forces. The reader may refer to the publications of Bergman[6] and Scordelis[7] for the design aspects of helicoidal stairs.

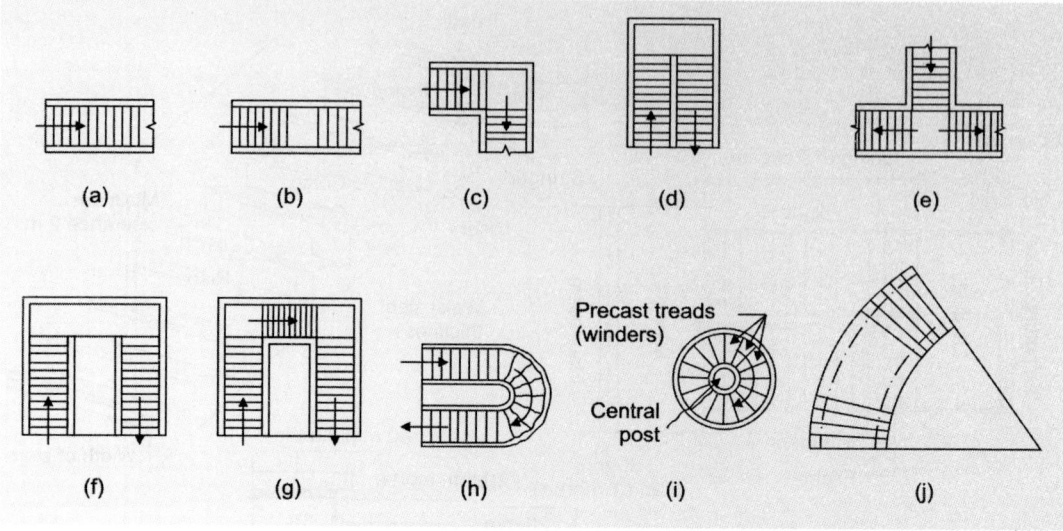

Fig. 12.2: Typical types of staircases

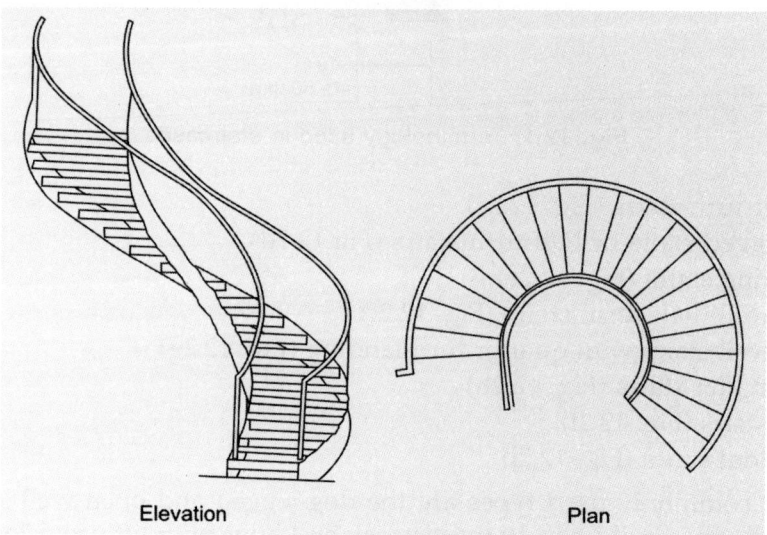

Elevation   Plan

Fig. 12.3: Helicoidal staircase

12.3 STRUCTURAL BEHAVIOUR OF STAIRCASES

The structural behaviour of staircases depends upon the support conditions and the direction of major bending of the slab component under the following categories:

1. Staircase slab spanning longitudinally in the direction of the flight (along the sloping line)
2. Staircase slab spanning transversely (slab widthwise with central or side supports)
3. Staircase slab spanning out as a cantilever from a central pillar from floor to floor

12.3.1 Staircase Slab Spanning in the Longitudinal Direction

In this type, the inclined stair flight together with the landing are supported on walls or beams as shown in Fig. 12.4a. The effective span to be considered in design computations is between the centre to centre of supports.

The slab arrangement may be of the conventional waist slab or the tread–riser type between the supports. The slab thickness depends upon the span and it can be reduced by providing intermediate supports at the junctions of inclined waist slab and horizontal landing slabs. Alternatively, the landing slab may have supports in the transverse direction as shown in Fig. 12.4b.

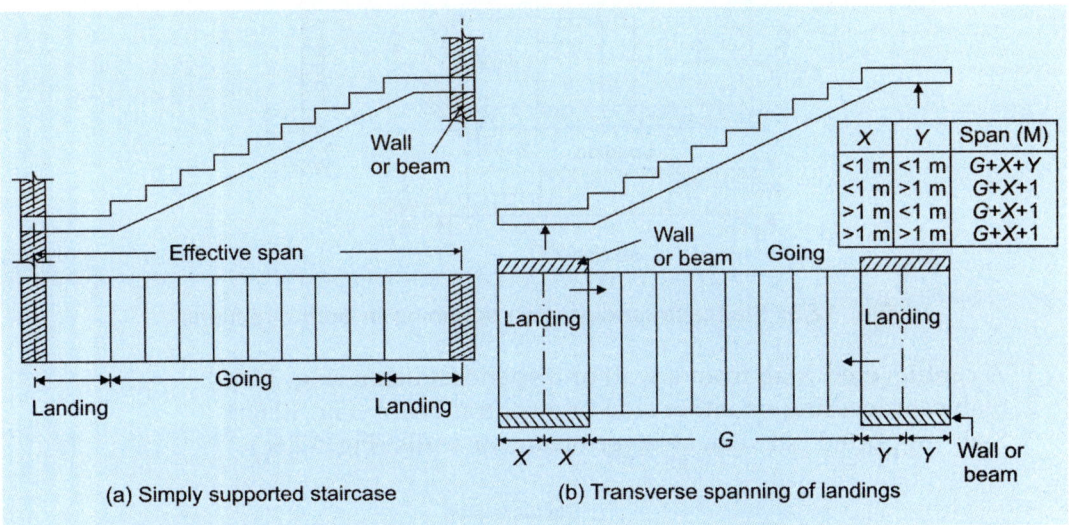

X	Y	Span (M)
<1 m	<1 m	G+X+Y
<1 m	>1 m	G+X+1
>1 m	<1 m	G+X+1
>1 m	>1 m	G+X+1

(a) Simply supported staircase

(b) Transverse spanning of landings

Fig. 12.4: Staircase flight (longitudinal spanning)

In such cases, the effective span to be considered according to the IS:456 code[8], Clause 33.1(b) is given by

$$L = (G + X + Y)$$

where,

G = going

X or Y = half the width of landings

In case of stairs with open wells, where spans partly criss-cross at right angles, the load on areas common to any two such spans may be distributed as one-half in each direction as shown in Fig. 12.5. The IS:456 code also specifies that when flights or landings are embedded into walls for a length of not less than 110 mm and are designed to span in the direction of flight, a length of 150 mm strip may be deducted from the loaded area and the effective breadth of the section increased by 75 mm for purpose of design.

12.3.2 Staircase Slabs Spanning in the Transverse Direction

The most common examples of staircase slabs spanning in the transverse direction are grouped under the following:

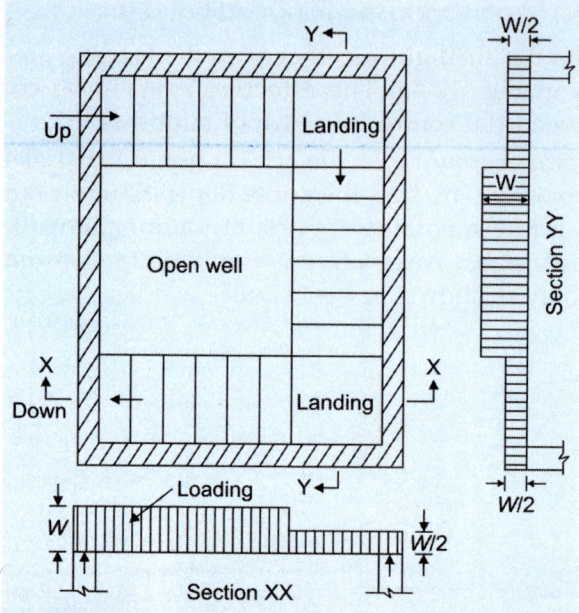

Fig. 12.5: Staircase landing slab spanning in both directions

1. A cantilevered slab from a wall or a spandrel beam (Fig. 12.6a)
2. Slab cantilevering on either side of a central beam (Fig. 12.6b)
3. Slab supported between stringer beams or walls (Fig. 12.6c)

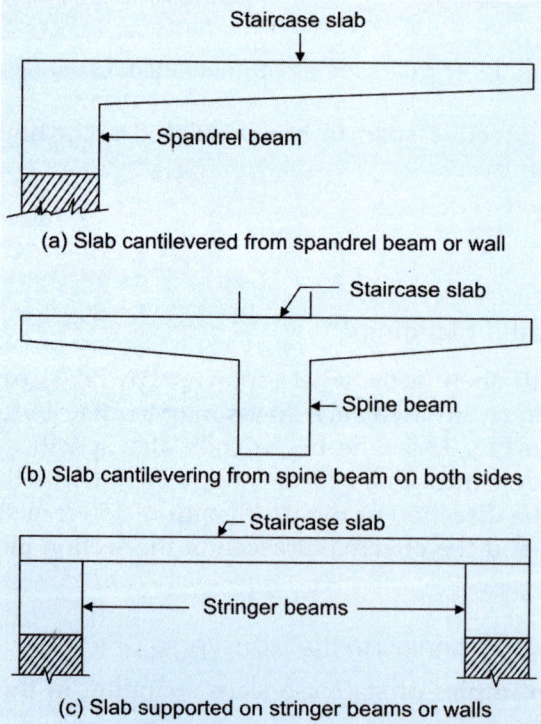

Fig. 12.6: Typical examples of stair slabs spanning transversely

In these types of slabs, the width of flight being small (1 m to 1.5 m), the designed thickness will be very small from structural computations. However, from practical considerations a minimum thickness of 75 mm to 80 mm should be provided and suitable reinforcements to resist the maximum bending moments or the minimum percentage reinforcements (whichever is higher) should be provided in the slab.

12.3.3 Staircase Spanning Out as a Cantilever from a Central Pillar from Floor to Floor

This type referred to as spiral staircase occupies the least space and is aesthetically superior to only tread slab cantilevering out from a massive column and without any risers. It is also referred to as free standing staircase and shown in Fig. 12.2a. This type occupies the least space. The disadvantage of this type of staircase is that it is not user-friendly since the tread width reduces towards the central column.

12.4 LOADS ON STAIRCASES

The various types of loads to be considered in the design of staircases are grouped under dead and live loads.

   a. Dead loads include the self-weight of the stair slab (waist slab), tread and risers along with the self-weight of landing slabs.
   b. The live loads to be considered are specified in IS:875-1987(part II)[9].

   The Indian standard code specifies that imposed or live loads to be considered for the design of stairs, landings should be not less than 3 kN/m^2, provided the structure is not liable to overcrowding. In case of buildings for educational institutions and public offices where overcrowding is expected, the imposed load may be increased to 4 kN/m^2.

12.5 DESIGN EXAMPLES

   1. Design one of the flights of a dog-legged stairs spanning between landing beams using the following data.

      i. *Data*:
         Type of staircase: Dog-legged with waist slab, treads and risers
         Number of steps in the flight = 10
         Tread $T = 300$ mm
         Rise $R = 150$ mm
         Width of landing beams = 300 mm
         M20 grade concrete ($f_{ck} = 20$ N/mm^2)
         Fe 415 HYSD bars ($f_y = 415$ N/mm^2)

      ii. *Effective span*:
         Effective span = (10 × 300) + 300 = 3300 mm = 3.3 m
         Thickness of waist slab = (span/20) = (3300/20) = 165 mm
         Adopt overall depth $D = 165$ mm
         Effective depth $d = 400$ mm

iii. *Loads*:

Dead loads of slab on slope $w_s = (0.165 \times 1 \times 25) = 4.125$ kN/m

Dead load of slab on horizontal span is:

$$w = \left[\frac{w_s \sqrt{R^2 + T^2}}{T} \right]$$

$$= \left[\frac{4.125\sqrt{150^2 + 300^2}}{300} \right] = 4.61 \text{ kN/m}$$

Dead load of one step = $(0.5 \times 0.15 \times 0.3 \times 25) = 0.56$ kN/m

Load of steps per metre length $= \left(\frac{0.56 \times 1000}{300} \right) = 1.86$ kN/m

Finishes etc. = 0.53 kN/m

Total dead load = $(4.61 + 1.86 + 0.53) = 7$ kN/m

Service live load (liable for overcrowding) = 5 kN/m^2

∴ Total service load = $(7 + 5) = 12$ kN/m

Factored load $w_u = (1.5 \times 12) = 18$ kN/m

iv. *Bending moments*:

Maximum bending moment at centre of span is

$M = 0.125 w_u L^2$

$= (0.125 \times 18 \times 3.3^2)$

$= 24.5$ kN·m

v. *Check for depth of waist slab*:

$$d = \sqrt{\frac{M_u}{0.138 f_{ck} \, b}} = \sqrt{\frac{195 \times 10^6}{0.138 \times 20 \times 10^3}} = 266 \text{ mm}$$

$= 94.2$ mm < 140 mm, provided (hence safe).

vi. *Main reinforcements*:

$$M_u = (0.87 f_y A_{st} d \left[1 - \frac{A_{st} f_y}{(bd f_{ck})} \right]$$

$$(24.5 \times 10^6) = (0.87 \times 415 A_{st} \times 140) \left[1 - \frac{415 A_{st}}{(10^3 \times 140 \times 20)} \right]$$

Solving, $A_{st} = 530$ mm^2/m

Provide 12 mm diameter bars at 200 mm centres ($A_{st} = 565$ mm^2) as main reinforcement.

vii. *Distribution reinforcement*:

Distribution reinforcement = 0.12% of cross-section

$= (0.0012 \times 1000 \times 165)$

$= 198$ mm^2/m

Provide 8 mm diameter bars at 200 mm centres ($A_{st} = 251$ mm^2)

The details of reinforcements in the staircase is shown in Fig. 12.7.

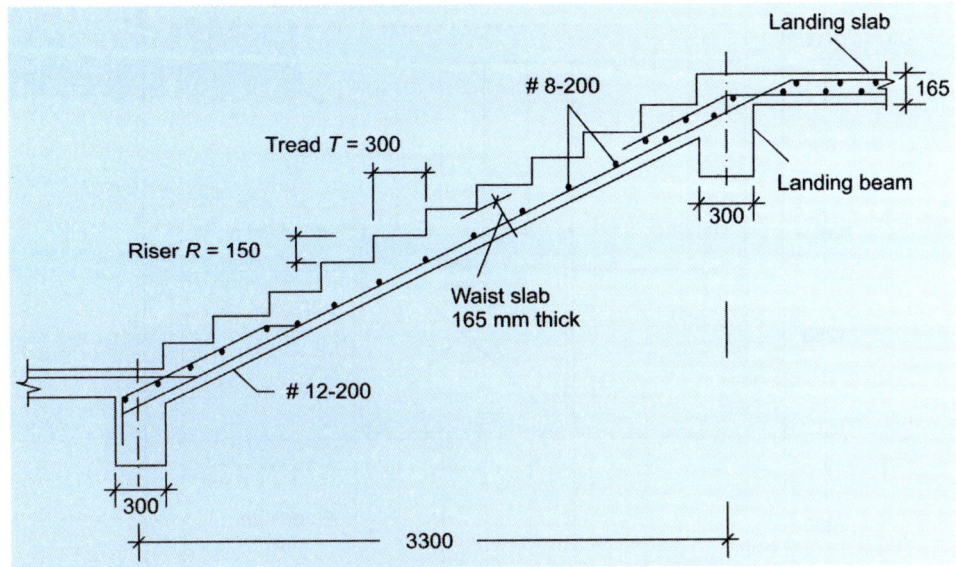

Fig. 12.7: Reinforcement details in a flight of dog-legged staircase

viii. *Design using SP:16 design charts*:

Compute the design parameter

$$\left(\frac{M_u}{bd^2}\right) = \left(\frac{24.5 \times 10^6}{10^3 \times 140^2}\right) = 1.25$$

Refer to Table 2, SP:16 design tables corresponding to $f_{ck} = 20$ N/mm² and readout the percentage of reinforcement as

$$p_t = 0.376$$

∴ $$A_{st} = \left(\frac{pbd}{100}\right) = \left(\frac{0.376 \times 10^3 \times 140}{100}\right) = 527 \text{ mm}^2/\text{m}$$

The reinforcement quantity is the same as that obtained by analytical method.

2. The general arrangement of staircase in a multistorey housing complex is as shown in Fig. 12.8. The risers are 150 mm and treads are 250 mm. The stair slab is embedded into the wall by 200 mm. The height between floors is 3 m. The service load is 3 kN/m². Adopting M20 grade concrete and Fe 415 HYSD bars, design the staircase flight and draw a longitudinal section showing the details of reinforcements in the flight of the staircase.

   i. *Data*:

   Riser $R = 150$ mm
   Tread $T = 250$ mm
   Service live load $q = 3$ kN/m²
   $f_{ck} = 20$ N/mm²
   $f_y = 415$ N/mm²

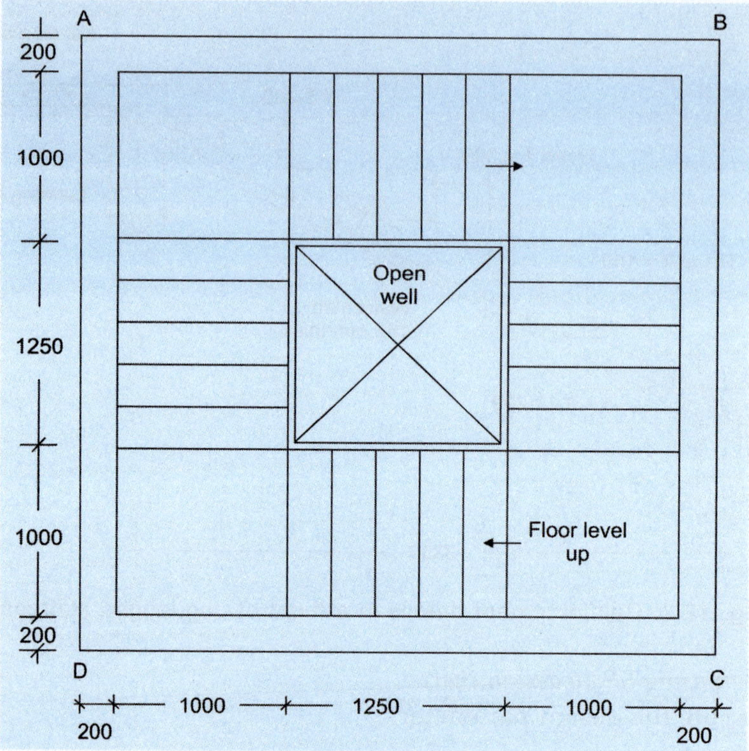

Fig. 12.8: Open well staircase

ii. *Effective span*:

The flights AB, BC, CD and DA are equal in length.

Effective span with 200 mm load bearing wall is computed as

$$L = (1.00 + 1.25 + 1.00 + 0.2) = 3.45 \text{ m}$$

iii. *Thickness of waist slab*:

For simply supported slabs, the span to effective depth ratio is equal to 20. Since a portion of the slab (landing slab) is spanning both ways, the ratio can be increased to 25.

Hence $(L/d) = 25$

$$d = (3450/25) = 138 \text{ mm}$$

Adopt effective depth $d = 135$ mm and overall depth $D = 160$ mm.

iv. *Loads*:

Self-weight of slab on horizontal area is computed as

$$w = \left[(0.16 \times 25)\sqrt{\frac{150^2 + 250^2}{250}}\right] = 4.66 \text{ kN/m}^2$$

Self-weight of steps = $(0.5 \times 0.15 \times 1 \times 25)$ = 1.87 kN/m²
Live load   = 3.00
Finishes etc.   = 0.47
Total load   = 10.00 kN/m²

Self-weight of landing slab = $(0.16 \times 25) = 4.00 \ \text{kN/m}^2$
Weight of finishes etc. $= 1.00$
Live load $= 3.00$
Total load $= 8.00 \ \text{kN/m}^2$

v. *Bending moments and shear forces*:

Figure 12.9 shows the service loads acting on the slab on one of the flights.

$$R_A = [(4 \times 1.1) + (0.5 \times 10 \times 1.25)] = 10.65 \ \text{kN}$$

$$M_{max} = (10.65 \times 1.725) - (4 \times 1.1 \times 1.175) - (0.625 \times 10 \times 0.5 \times 0.625)$$

$$= 11.25 \ \text{kN·m}$$

$$V_{max} = R_A = 10.65 \ \text{kN}$$

Factored moment $M_u = (1.5 \times 11.25) = 16.875 \ \text{kN·m}$

Factored shear $V_u = (1.5 \times 10.65) = 15.975 \ \text{kN}$

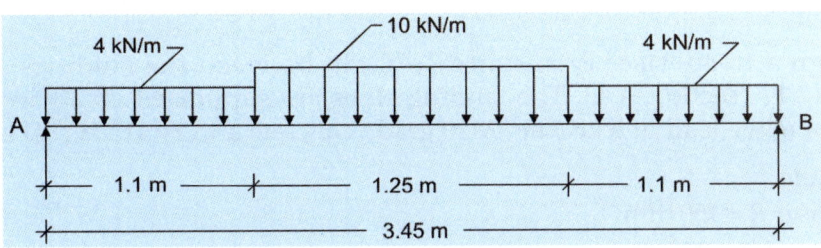

Fig. 12.9: Loads on staircase flight

vi. *Check for depth of waist slab*:

$$d = \sqrt{\frac{M_u}{0.138 f_{ck} b}} = \sqrt{\frac{16.875 \times 10^6}{0.138 \times 20 \times 10^3}} = 78.19 \ \text{mm}$$

Adopted effective depth $d = 135 \ \text{mm}$ (hence safe).

vii. *Reinforcements*:

$$M_u = (0.87 f_y A_{st} d) \left[1 - \left(\frac{A_{st} f_y}{b d f_{ck}} \right) \right]$$

$$(16.875 \times 10^6) = (0.87 \times 415 A_{st} \times 135) \left[1 - \left(\frac{415 A_{st}}{10^3 \times 140 \times 20} \right) \right]$$

Solving, $A_{st} = 530 \ \text{mm}^2/\text{m}$

Adopt 10 mm diameter bars at 200 mm centres ($A_{st} = 393 \ \text{mm}^2$).

Distribution bars = $(0.0012 \times 10^3 \times 160) = 192 \ \text{mm}^2$

Adopt 8 mm diameter bars at 250 mm centres ($A_{st} = 201 \ \text{mm}^2$).

viii. *Reinforcement details*:

The details of reinforcements in one of the flights of staircase is shown in Fig. 12.10.

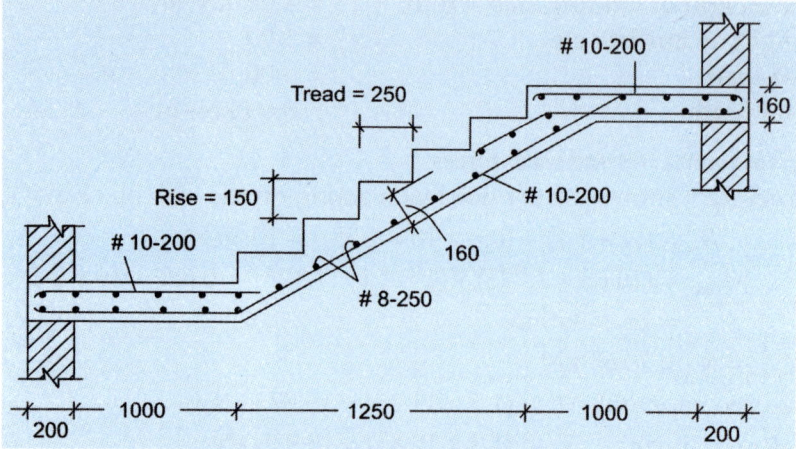

Fig. 12.10: Reinforcement details in staircase flight

3. Design a tread–riser type staircase flight between the landings shown in Fig. 12.11 (section AA). The landing slabs are supported on adjacent edges. Adopt a live load of 5 kN/m², M20 grade concrete and Fe 415 HYSD bars.

    i. *Data*:

    Riser $R = 150$ mm

    Tread $T = 275$ mm

    Width of flight = landing width = 1500 mm

    Materials: M20 grade concrete ($f_{ck} = 20$ N/mm²)

                    Fe 415 HYSD bars ($f_y = 415$ N/mm²)

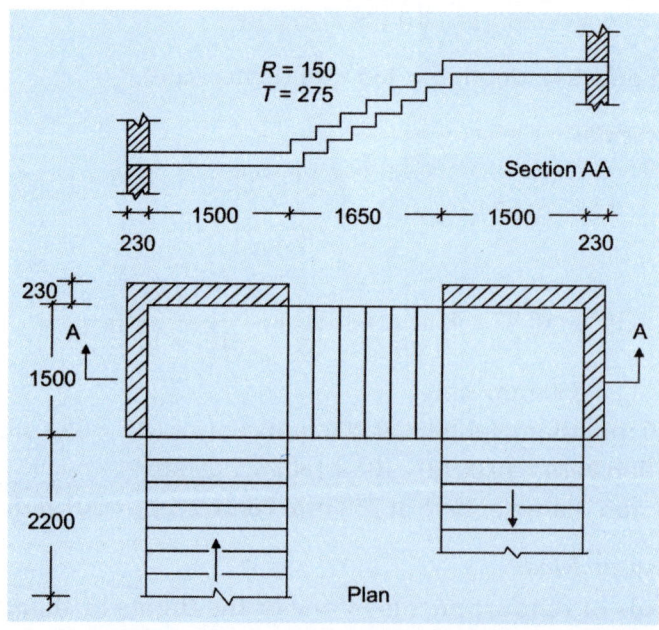

Fig. 12.11: Tread–riser type staircase

ii. *Effective span and thickness of slab*:

$$L = (1650 + 1615 + 1715) = 4.88 \text{ m}$$

Thickness of riser and tread slabs = $(L/25) = (4880/25) = 195$ mm

Adopt effective depth $d = 175$ mm and overall depth $D = 200$ mm

iii. *Loads on going (on projected area)*:

Self-weight of tread–riser slab per steps

$= (0.15 + 0.275) \times 0.20 \times 25$ $= 2.125$ kN

Dead load of steps/m length = $(2.125 \times 10^3)/275$ = 7.72 kN/m²

Weight of finishes   = 0.60

Live load   = 5.00

∴ Total load   = 13.32 kN/m²

Factored load = (1.5×13.32) = 20 kN/m²

iv. *Loads on landing slab (assumed as 200 mm thick)*:

Self-weight of slab = $(0.20 \times 25) = 5.00$ kN/m²

Finishes   = 0.60

Live load   = 5.00

Total load   = 10.60 kN/m²

Factored load = (1.5×10.60) = 16 kN/m²

50% of this load may be assumed to be acting longitudinally in the direction of span = $(0.5 \times 16) = 8$ kN/m².

v. *Design of tread–riser unit*:

Referring to Fig. 12.12, showing the loading on the horizontal span, $L = 4.88$ m, the reaction at left hand support A is

$$R_A = (8 \times 1.615) + (0.5 \times 1.65 \times 20) = 29.42 \text{ kN}$$

Maximum design moment at mid span of tread–riser unit is

$$M_u = (29.42 \times 2.44) - (8 \times 1.615 \times 1.632) - (20 \times 0.825 \times 0.5 \times 0.825)$$
$$= 43.90 \text{ kN·m}$$

Maximum design moment in landing slab at B is

$$M_u = (29.42 \times 1.615) - (8 \times 1.615 \times 0.5 \times 1.615)$$
$$= 37.08 \text{ kN·m}$$

Effective depth required to resist the maximum moment is

$$d = \sqrt{\frac{M_u}{0.138 f_{ck} b}} = \sqrt{\frac{43.90 \times 10^6}{0.138 \times 20 \times 10^3}} = 126 \text{ mm}$$

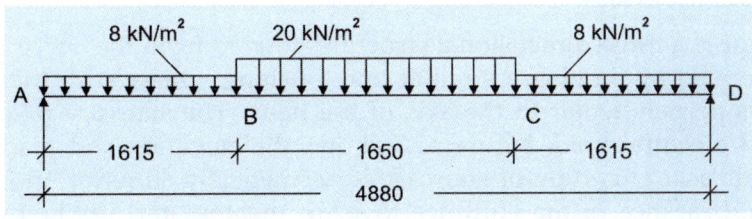

Fig. 12.12: Loading on tread–riser flight

The effective depth provided (175 mm) is more than the required depth of 126 mm. Hence, the section is under-reinforced and safe.

$$M_u = (0.87 f_y A_{st} d) \left[1 - \left(\frac{A_{st} f_y}{bdf_{ck}} \right) \right]$$

$$(43.9 \times 10^6) = (0.87 \times 415 A_{st} \times 175) \left[1 - \frac{415 A_{st}}{(10^3 \times 175 \times 20)} \right]$$

Solving, $A_{st} = 763$ mm^2

Provide 12 mm diameter bars at 140 mm centres ($A_{st} = 808$ mm^2) in the form of closed ties and distribution bars of 8 mm diameter transversely at each bend of the ties.

vi. *Design of landing slab*:

$$(37.08 \times 10^6) = (0.87 \times 415 A_{st} \times 175) \left[1 - \frac{415 A_{st}}{(10^3 \times 175 \times 20)} \right]$$

Solving, $A_{st} = 634$ mm^2

Provide 12 mm diameter bars at 170 mm centres ($A_{st} = 665$ mm^2) in the direction of span and also in the transverse direction.

vii. *Reinforcement details*:

Figure 12.13 shows the details of reinforcements in the tread–riser unit and landing slab.

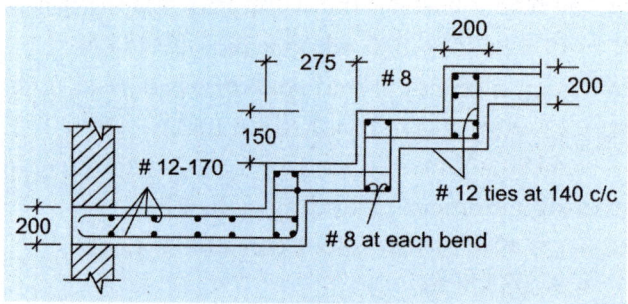

Fig. 12.13: Reinforcement details in tread–riser staircase

12.6 DESIGN OF HELICOIDAL STAIRCASE

12.6.1 Design Procedure

Helicoidal stair is a three-dimensional structure derived from the curve helix and the shape is generated by moving a straight line touching a helix such that the moving line is always perpendicular to the axis of the helix. The staircase follows a helical curve around a central void. Figure 12.3 shows the elevation and plan of a typical helicoidal staircase. This type of staircase is aesthetically superior and provides an impressive appearance in the entrance foyer of theaters and public buildings and hence is increasingly adopted by architects. A rigorous three dimensional structural analysis of this type of staircase is presented by Scordelis[7]. A simplified analysis by considering the horizontal projection of the structure and idealizing it as a curved beam with fixed ends has been attempted by Bergman[6]. This method yields conservative estimate of force components according to Chatterjee[10].

The notations used for the analysis of helicoidal staircase considering it as a curved beam with fixed end boundary conditions shown in Fig. 12.14 are as follows:

R = radius of the centre line of the helicoidal curve

2θ = angle subtended by the helical curve in plan

α = slope of the helicoidal slab

$\tan \alpha$ = ratio of rise to tread for equal size steps

w_u = total uniformly distributed vertical load acting on the curved beam

M_r = radial moment

M_t = torsional moment

ϕ = angle measured from the middle point of the curve of the slab

E = modulus of elasticity of the material

G = modulus of rigidity

C = torsional constant of the cross-section

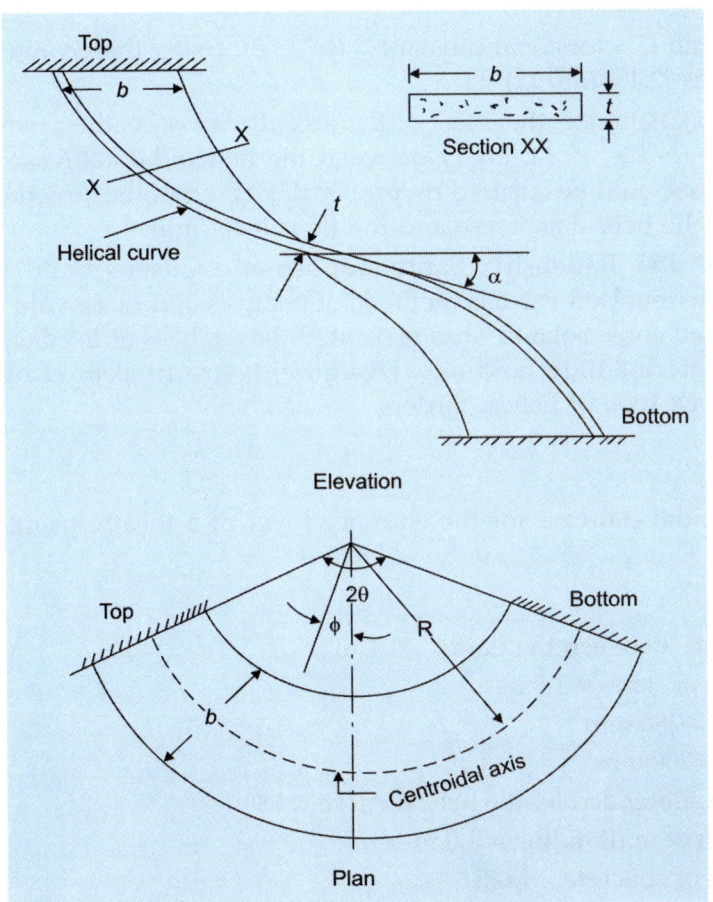

Fig. 12.14: Helicoidal staircase

I = second moment of area of the cross-section

EI = flexural rigidity

GC = torsional rigidity

g = ratio of EI to GC

b = width of the section

t = thickness of the section

e = eccentricity of loading = $[b^2/12R]$

R_c = radius of the centroidal axis of the slab = $(R + e)$

Based on the analysis of Bergman[6] and Chatterjee, we have the relations for the computation of radial and torsional moments as

$$M_r = w_u R_c^2 (c \cos\phi - 1)$$

$$M_t = w_u R_c^2 (c \sin\phi - \phi)$$

The value of c in the above expressions is evaluated by the relation,

$$c = \left[\frac{2(g+1)\sin\theta - 2g \cdot \theta \cdot \cos\theta}{(g+1)\theta - (g-1)\sin\theta \cdot \cos\theta} \right]$$

$I = (bt^3/12)$ and C = torsional constant = $(bt^3/3.5)$. Using these values, the value of g is computed as $(2.4 \times 3.5/12) = 0.7$

According to Chatterjee, the effect of the eccentricity of loading may be neglected when the ratio $(b/R) \leq 0.33$. Fixity between the helicoidal staircase and the floor beams and slabs should be ensured by proper designing of the structural elements at the junction of the helical staircase and the floor level units.

Investigators like Bangash[11], Santhadaporn and Cusens[12] and Reynolds and Steedman[13] have reported various methods of analysis and design aids for helicoidal girders with fixed ends. Solanki[14] has presented the analysis of fixed ended helicoidal girders with intermediate landings. Design aids are presented by Wadud and Ahmed[15] for such type of helical girders.

12.6.2 Design Example

Design a helicoidal staircase for the entrance foyer of a theatre using the following data:

    i. *Data*:

        Distance between the floors = 4.4 m

        Width of stairs = 1.5 m

        Tread = 280 mm

        Riser = 200 mm

        Angle subtended by the helical curve = 180°

        Length of midlanding = 1.0 m

        Grade of concrete = M25

        Type of steel = Fe 415 HYSD bars

ii. *Selection of dimensions*:

Number of risers = [4400/200] = 22

Length of tread at centre line = (22 × 280) = 6160 mm

Total going of the staircase including mid landing of 1.2 m

$$\text{Span} = [6160 + 1000] = 7160 \text{ mm}$$

Included angle =180° and $\pi R = 7160$ mm

Hence, the radius of the helicoids is evaluated as

$$R = [7160/\pi] = 2280 \text{ mm}$$

The gradient of staircase $\alpha = \tan^{-1}$ (rise/tread) = $\tan^{-1}(200/280) = 0.714$

Hence $\alpha = 35.5°$

The thickness of the helical slab is assumed as

(span/25) = (7160/25) = 286.4 mm

Adopt the overall thickness of the slab as 300 mm

Assuming 30 mm as effective cover, effective depth $d = 270$ mm

iii. *Loads*:

Self-weight of the slab = [25 × 0.3 × 1.5] = 11.25 kN/m

Self-weight of steps = [25(0.2/2) × 1.5] = 3.75 kN/m

Live load on slab = (5.0 × 1.5)   = 7.50 kN/m

Finishes   = 0.50 kN/m

Total service load   = 23.00 kN/m

w_u = ultimate load normal to the surface of slab

$$= [1.5 \times 23 \times \cos 35.5°]$$
$$= 28 \text{ kN/m}$$

iv. *Bending and torsional moments*:

Radius of the centroidal axis of the slab is given by the expression

$$R_c = R + (b^2/12R)$$
$$= 2280 + [1500^2/(12 \times 2280)]$$
$$= 2362 \text{ mm} = 2.362 \text{ m}$$

Value of constant $c = \left[\dfrac{2(g+1)\sin\theta - 2g\cdot\theta\cdot\cos\theta}{(g+1)\theta - (g-1)\sin\theta\cdot\cos\theta} \right]$

$$= \left[\dfrac{2(0.7+1)\sin 90° - 2\times 0.7\left(\dfrac{90\pi}{180}\right)\cos 90°}{(0.7+1)\left(\dfrac{90\pi}{180}\right) - (0.7-1)\sin 90°\cdot \cos 90°} \right] = 1.27$$

At mid span, $\phi = 0°$

Bending moment $M_r = w_u R_c^2 (c\cos\phi - 1)$

$$= 28 \times 2.362^2 (1.27 \times 1 - 1)$$
$$= 42 \text{ kN·m}$$

Torsional moment $M_t = w_u R_c^2 (c\sin\phi - \phi) = 0$

At supports, $\phi = 90°$

Bending moment $M_r = w_u R_c^2 (c \cos\phi - 1)$

$\qquad = 28 \times 2.362^2 (1.27 \times 0 - 1)$

$\qquad = -156 \text{ kN·m}$

Torsional moment $M_t = w_u R_c^2 (c \sin\phi - \phi)$

$\qquad = 28 \times 2.362^2 [1.27 \times \sin 90° - (90 \times \pi)/180)]$

$\qquad = -47 \text{ kN·m}$

v. *Check for depth of slab*:

According to Clause 41.4.2 of IS:456 code, equivalent bending moment at support is

$\qquad M_e = M_r + M_t (1 + t/b)/1.7]$

$\qquad = 156 + 47(1 + 300/1500)/1.7]$

$\qquad = 218 \text{ kN·m}$

For Fe 415 grade HYSD bars,

$\qquad M_e = 0.138 f_{ck} bd^2$

$$d = \sqrt{\frac{M_e}{0.138 f_{ck} b}} = \sqrt{\frac{218 \times 10^6}{0.138 \times 25 \times 1500}}$$

$\qquad = 205 \text{ mm} < 270 \text{ mm provided.}$

Hence, the selected overall depth (300 mm) is safe.

vi. *Design of reinforcements*:

$[M_e/bd^2] = [(218 \times 10^6)/(1500 \times 270^2)] = 1.99 \text{ N/mm}^2$

Refer to Table 3, SP:16 design charts and readout the percentage reinforcement corresponding to $f_y = 415 \text{ N/mm}^2$ and the ratio 1.99 N/mm^2 as

$\qquad p_t = [100 A_{st}/bd] = 0.618$

$\qquad A_{st} = [(0.618 \times 1500 \times 270)/100] = 2746 \text{ mm}^2$

Provide 14 bars of 16 mm diameter at the top and 7 bars of 16 mm diameter at soffit since the positive bending moment at centre of span (42 kN·m) is small.

$\qquad A_{st} \text{ (provided)} = 2814 \text{ mm}^2$

Since the moment at the centre of span section is small, provide 7 bars of 16 mm diameter both at the top and the bottom of the central span section.

vii. *Check for shear*:

Ultimate shear at support $V_u = [w_u \cdot L/2] = [28 \times 7.16/2] = 100 \text{ kN}$

Equivalent shear $V_e = [V_u + 1.6 (M_t/b)] = [100 + 1.6 (47/1.5)] = 150 \text{ kN}$

Nominal shear stress $\tau_v = [V_u/bd] = [(150 \times 10^3)/(1500 \times 270)] = 0.37 \text{ N/mm}^2$.

From Table 19, IS:456 code (corresponding to $f_{ck} = 25 \text{ N/mm}^2$) and percentage steel $[100 A_{st}/bd] = [(100 \times 2814)/(1500 \times 270)] = 0.69$

Permissible shear stress $\tau_c = 0.53 \text{ N/mm}^2 > 0.37 \text{ N/mm}^2$

According to Clause 41.4.3, IS:456 code, transverse reinforcement has to be provided having a cross-sectional area given by

$$A_{sv} = \left[\frac{T_u s_v}{b_1 d_1 0.87 f_y}\right] + \left[\frac{V_u s_v}{2.5 d_1 0.87 f_y}\right]$$

Adopting effective cover as 30 mm for top and bottom bars

$$b_1 = (1500 - 60) = 1440 \text{ mm}$$
$$d_1 = (300 - 60) = 240 \text{ mm}$$

$$A_{sv} = \left[\frac{47 \times 10^6 \times s_v}{1440 \times 240 \times 0.87 \times 415}\right] + \left[\frac{150 \times 10^3 \times s_v}{2.5 \times 240 \times 0.87 \times 415}\right] = 1.06 \, s_v$$

Using 10 mm diameter two-legged stirrups, $A_{sv} = (2 \times 78.5) = 157 \text{ mm}^2$

$$s_v = [157/1.06] = 148 \text{ mm}$$

Provide 10 mm diameter two-legged stirrups at 140 mm centres throughout the span length. The details of reinforcements provided in the helicoidal staircase along the span and cross-sections are shown in Fig. 12.15.

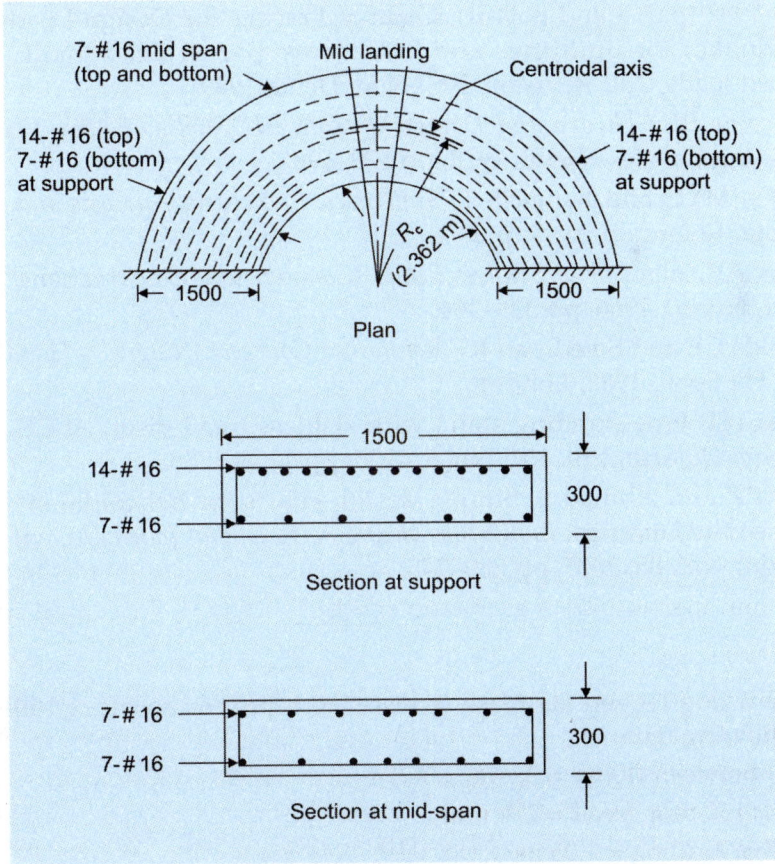

Fig. 12.15: Helicoidal staircase

REFERENCES

1. Jain AK, *Reinforced Concrete Design*, 2 edn, Nem Chand Bros, Roorkee (UP), 1984.
2. Dayaratnam P, *Design of Reinforced Concrete Structures*, Oxford & IBH Publishers, New Delhi, 1983.
3. Subramanian N, *Design of Reinforced Concrete Structures*, Oxford University Press, New Delhi, India, 2003, pp 702–725.
4. Gould PL, Analysis and design of a cantilever staircase, *Journal of the American Concrete Institute*, Vol 60, July 1963, pp 881–889.
5. Ahmed I, Muqtadir A and Ahmad S, Design provisions for stair slabs in Bangladesh building code, *ASCE Journal of Structural Engineering*, Vol. 121, No. 7, 1995, pp 1051–57.
6. Bergman VR, Helicoidal staircases of reinforced concrete, *Journal of the American Concrete Institute*, Vol. 53, Oct. 1956, pp 403–412.
7. Scordelis AC, Internal forces in uniformly distributed loaded helicoidal girders, *Journal of the American Concrete Institute*, Vol. 56, April 1960, pp 1013–1026.
8. IS:456-2000, Indian Standard Code of Practice for Plain and Reinforced Concrete (4th Revision), BIS, New Delhi, 2000, p 63.
9. IS:875-1987 (part-I and part-II), Code of Practice for Design Loads (other than Earthquake) for Buildings and Structures, part-I: dead loads, and part-II: imposed loads, 2nd Revision, BIS, New Delhi, 1989.
10. Chatterjee BK, *Theory and Design of Concrete Shells*, 2 edn, Oxford & IBH Publishing Co, New Delhi, 1978, pp 179–198.
11. Bangash MYH and Bangash T, *Staircases: Structural Analysis and Design*, A.A. Balkema, Roterdam, 1999, p 337.
12. Cusens AR, *Analysis of Slabless Stairs, Concrete and Constructional Engineering*, Vol. 61, No. 10, 1966, pp 359–364.
13. Reynolds CE and Steedman JC, *Reinforced Concrete Designer's Handbook*, 10 edn, E and FN Spon, 1988, London.
14. Solanki HT, Free standing stairs with slabless tread–riser, *ASCE Journal of the Structural Division*, Vol. 101, No. 8, 1975, pp 1733–1738.
15. Wadud Z and Ahmad S, Simple design charts for helicoidal stair slabs with intermediate landings, *Institution of Engineers (India) Journal (Civil Engineering Division)*, Vol. 85, 2005, pp 289–275.

ASSIGNMENT

1. Design a dog-legged staircase (waist slab type) for an office building to suit the following data:

   Height between floors = 3.2 m

   Risers = 160 mm, tread = 270 mm

   Length of landing = 1.25 m

   Width of flight = landing width = 1.25 m

Assume stairs to be supported on 230 mm thick masonry walls at the outer edges of the landing parallel to the risers. Adopt M20 grade concrete and Fe 415 HYSD bars. Assume a live load of 5 kN/m².

2. Design a dog-legged staircase (waist slab type) for an office building assuming floor to floor height of 3 m.

Width of flight = landing width = 1.2 m. Adopt a tread to be 300 mm and rise to be 150 mm. Use M20 grade concrete and Fe 415 HYSD bars. The live load is 5 kN/m². Assume the landings to be supported only on two edges perpendicular to the risers.

3. Design an open-well type staircase (waist slab type) for a multistorey office complex using the following data:

Length of each flight = 1.5 m (5 treads in each flight)

Length of landings = 1.0 m

Tread = 300 mm and riser = 150 mm

The landings are supported all round on 300 mm masonry walls.

Live load = 5 kN/m², f_{ck} = 20 N/mm², f_y = 415 N/mm².

4. A staircase flight is made up of independent tread slabs cantilevered from a reinforced concrete wall. The tread is 300 mm, riser is 150 mm and width of flight is 1.50 m. Design the cantilever slab using M20 grade concrete and Fe 415 HYSD bars. Assume the live load as 5 kN/m².

5. An open-well staircase for an office building is made up of four flights with tread–riser units and landing slabs.

Length and width of tread–riser unit = 1.5 m

Length and width of landing slab = 1.5 m

Tread = 250 mm and riser = 150 mm

Live load = 5 kN/m²

The landing slabs are supported all round on 300 mm masonry walls. Adopting M20 grade concrete and Fe 415 HYSD bars, design one of the flights of staircase and sketch the details of reinforcements in the landing slab and tread–riser unit.

6. Design a dog-legged staircase (tread–riser type) considering the landings to be supported only on two edges perpendicular to the risers using the following data:

Riser = 160 mm and tread = 270 mm

Projected length of tread–riser unit = 2430 mm

Length of landings at each end = 1250 mm

The landings are supported on walls 300 mm thick

Live load = 5 kN/m²

f_{ck} = 20 N/mm² and f_y = 415 N/mm²

Sketch the details of reinforcements in the staircase flight.

7. Design the typical slab of a spiral type staircase cantilevering from a central column. The slab cantilevers over a span of 1.5 m from the edge of a circular reinforced concrete column of diameter 500 mm, imposed load 400 kN/m² and weight of finishes 0.5 kN/m². Adopt M25 grade concrete and Fe 500 HYSD reinforcements.

8. Design a tread–riser type staircase flight between the landings of a staircase using the following data:

Length of landing slabs supported on 230 mm thick walls = 1.2 m

Height between landings = 1200 mm

Length of tread = 275 mm

Riser height = 150 mm

Adopt a live load of 5 kN/m², M25 grade concrete and Fe 415 HYSD bars.

9. Design a stair flight with a central beam supporting cantilevered slab on either side using the following data:

The span length of the beam between supports = 2.24 m

Number of treads in flight = 8

Width of tread = 280 mm

Width of stair flight = 2 m

Height of riser = 150 mm

Adopt M25 grade concrete and Fe 500 HYSD reinforcements.

Sketch the details of reinforcements in the slab and beam.

Design a helicoidal staircase to suit the given data.

10. Design a helicoidal staircase for the entrance foyer of a theatre using the following data:

Distance between the floors = 6 m

Width of stairs = 2 m

Tread = 300 mm

Riser = 160 mm

Angle subtended by the helical curve = 180°

Length of mid landing = 1.5 m

Grade of concrete = M25

Type of steel = Fe 500 HYSD bars

Sketch the details of reinforcements in the stairs at mid span and support sections.

REVIEW QUESTIONS

1. List any three common geometrical configurations of staircases with suitable examples and mention where they are used.

2. Distinguish between stair flights spanning longitudinally and transversely, generally adopted in various types of buildings.

3. Explain with sketches the following terms with reference to staircases:

   a. Going of step and going of stair

   b. Tread and riser

   c. Landing slab and waist slab

4. State the rules for determining the effective span of longitudinally supported stairs as per Indian standard code IS:456.

5. Explain the design features of the following transversely supported stairs:
   a. Stairs with waist slab
   b. Tread–riser stair
   c. Spiral stairs
6. What are open-well staircases? Under what situations would you recommend them and list their advantages over other types.
7. What is meant by "stair slab supported on landings"? Explain the IS code recommendations for the effective span of the stair slab in such cases.
8. How do you design cantilevered step slab supported on a spine beam between landings?
9. What are helicoidal stairs? For what type of buildings would you recommend them for use?
10. What are the design parameters to be considered in the design of helicoidal stairs? Mention the various force components to be considered in the design of helicoidal stairs.

OBJECTIVE QUESTIONS

1. In a dog-legged staircase, the staircase flight spans in the
   a. transverse direction between the edge beams
   b. cantilevers from a spine beam
   c. longitudinal direction
2. In the design of staircases in multistorey buildings, the riser should preferably be
   a. 100 mm
   b. 200 mm
   c. 150 mm
3. The minimum live load to be considered in the design of staircases of public buildings should be
   a. 5 kN/m^2
   b. 3 kN/m^2
   c. 2 kN/m^2
4. When there is limited space to accommodate the staircase, the ideal type to be selected is
   a. dog-legged
   b. open–well
   c. spiral
5. A built in spandrel beam inside the surrounding walls is required in case of
   a. dog-legged staircase
   b. tread-riser staircase
   c. cantilever staircase
6. In case of office buildings, the maximum number of stairs in a flight should be limited to
   a. 4 to 6
   b. 7 to 9
   c. 12 to 15

7. In case of tread-riser type staircase, the reinforcements in the tread and riser units comprise
    a. horizontal bars
    b. vertical bars
    c. horizontal bars with transverse ties

8. In case of free standing staircase, the stair slabs
    a. are supported at both edges
    b. span longitudinally between the floors
    c. cantilever out from a central column

9. In the entrance foyers of theatres and large public buildings, the best type of aesthetically superior staircase is
    a. open-well
    b. tread–riser
    c. helicoidal

10. Helicoidal staircase should be designed for various force components like
    a. bending moments
    b. torsional moments
    c. bending and torsional moments and shear forces

Design of Retaining Walls

13.1 INTRODUCTION

A reinforced concrete retaining wall is defined as a structural element used with the primary purpose of retaining some material on one or both sides of the wall. In some situations the retaining wall may also support vertical loads. In most of the cases in practice retaining walls are used to retain soil at different levels on either side. Retaining walls are widely employed at the ends of bridges in the form of abutments. In urban roadways, retaining walls are invariably used in flyover embankments to retain soil on one side. The retaining wall while retaining earth on one side, is subjected to lateral pressure tending it to bend, overturn and slide the retaining wall.

Retaining walls[1,2] should be designed to resist the lateral earth pressure from the sides and the soil pressure acting vertically on the footing slab integrally built with the vertical slab. Gravity walls of stone masonry were generally used in the earlier days to retain earthen embankments. The thickness of the masonry walls increased with the height of the earthfill. The advent of reinforced concrete has resulted in thinner retaining walls of different types resulting in considerable reduction of costs coupled with durability and improved aesthetics.

13.2 TYPES OF RETAINING WALLS

Reinforced concrete retaining walls are of different types depending upon the height of embankment and the site conditions, and the most common types generally used are discussed below:

13.2.1 Cantilever Retaining Walls

Cantilever retaining wall is the most common type, and universally adopted to retain earth in embankments in various situations like highway, hydraulic, irrigation and marine structures.

Reinforced concrete retaining walls of *cast-in-situ* and precast types have more or less replaced the traditional gravity type masonry retaining walls.

The main structural components of a cantilever retaining wall shown in Fig. 13.1a are identified with the following terminology:

a. Vertical stem or upright slab

The vertical upright slab or stem resists earth pressure from one side and the slab bends like a cantilever. The thickness of the slab is larger at the bottom and gradually decreases towards the top in proportion to the variation in soil pressure.

b. Base slab containing the heel and toe

The base slab forming the foundation comprises the heel slab and the toe slab. The heel slab acts as a horizontal cantilever under the combined action of the weight of retained earth from the top and the soil pressure acting from the soffit. The toe slab also acts as a cantilever under the action of the resulting soil pressure acting upward. The stability of the wall is maintained by the weight of the earthfill on the heel slab together with the self-weight of the structural elements of the retaining wall. Cantilever type retaining walls are adopted for small to medium heights up to 5 m.

13.2.2 Counterfort Retaining Walls

The main structural elements of a counterfort type retaining wall are shown in Fig. 13.1b. When the depth of earthfill exceeds 5 m, it is preferable to adopt counterfort type retaining walls for structural integrity and overall economy. The base slab comprising the heel and toe together with the vertical counterfort and the stem slabs are cast integrally. The stem and heel are effectively fixed to the counterforts so that the stem bends horizontally between the counterforts due to lateral earth pressure. Consequently the thickness of the stem and the heel slab is considerably reduced due to the reduction of moment due to the fixity of these slabs between the counterforts.

13.2.3 Buttressed Retaining Walls

In case of buttressed retaining walls, the counterforts are provided in the front side of the wall and not on the soil side. Buttresses were commonly used in the earlier times for masonry retaining walls. A typical buttressed type retaining wall is shown in Fig. 13.1c. The disadvantage of this type of wall is that it reduces the clearance in front of the wall and its contribution towards the stability is less due to the small toe projection reducing the depth of the buttress at the base.

13.2.4 Anchored Retaining Walls

The stability of cantilever stem of the retaining walls can be improved by anchoring the stem by high strength prestressed guy wires as shown in Fig. 13.1d. The guy wire cable acts as a prop to the cantilever stem and reduces the bending moments at the base of the cantilever slab. The guy wires can be anchored in the rock or soil. This type of wall is used only when high loads are expected or when slender walls are used.

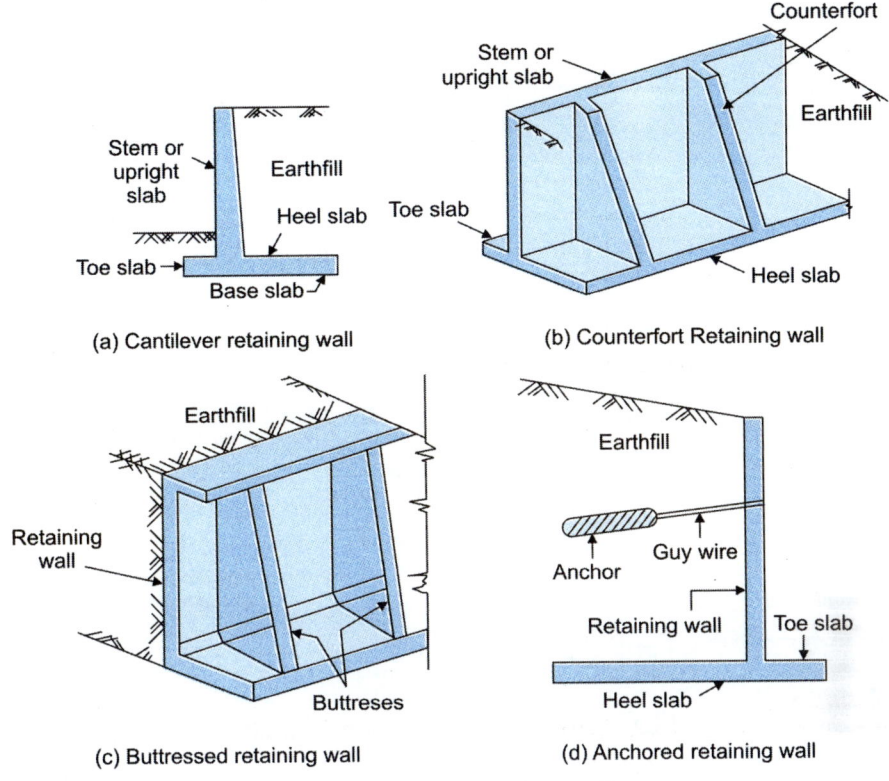

(a) Cantilever retaining wall

(b) Counterfort Retaining wall

(c) Buttressed retaining wall

(d) Anchored retaining wall

13.1: Types of retaining walls

13.3 FORCES ACTING ON RETAINING WALLS

Retaining walls being basically earth retaining structures, are generally subjected to various major types of forces like soil pressure, surcharge loads from the backfill and also small forces due to earthfill on the toe side. The various types of forces acting on a typical cantilever type retaining wall are shown in Fig. 13.2.

13.3.1 Lateral Earth Pressure

The lateral forces due to earth pressure is the major force acting on the retaining wall. The magnitude of the force is expressed by the relation

$$P_a = C_a \gamma_e \, (h')^2 / 2$$

where, C_a = coefficient of active earth pressure

γ_e = density of soil

h = height of the backfill measured vertically above the heel

The coefficient of earth pressure C_a depends upon the angle of shearing resistance (angle of repose) ϕ and the inclination or slope of the backfill to the horizontal expressed as θ.

The general relation for the coefficient of active earth pressure based on Rankine's theory[3,4] is given by the relation:

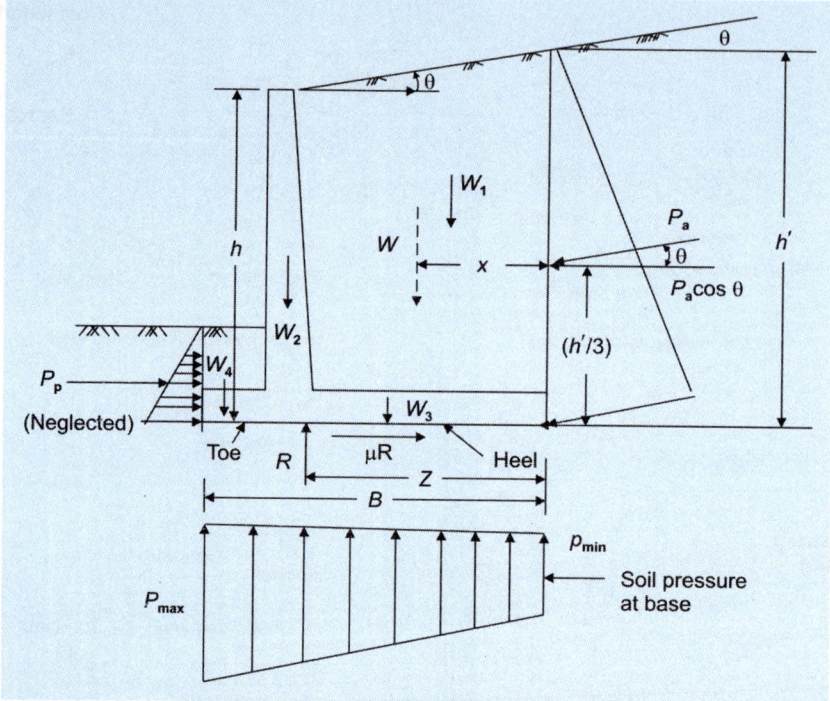

13.2: Forces acting on retaining walls

$$C_a = \left[\frac{\cos\theta - \sqrt{\cos^2\theta - \cos^2\phi}}{\cos\theta + \sqrt{\cos^2\theta - \cos^2\phi}} \right] \cos\theta$$

For the case of a level backfill, $\theta = 0$ and $h' = h$

Hence, $C_a = \left\{ \dfrac{1 - \sin\phi}{1 + \sin\phi} \right\}$

The coefficient of passive earth pressure is given by the relation,

Hence, $C_p = \left\{ \dfrac{1 + \sin\phi}{1 - \sin\phi} \right\}$

The magnitude of the earth pressure P_a acts at one-third the height of the backfill as shown in Fig. 13.2. The force P_p developed due to the passive pressure acts on the toe side of the retaining wall and its magnitude being very small (due to the small height of earthfill on toe slab) is generally neglected in the design computations.

13.3.2 Effect of Surcharge on a Level Backfill

Due to the construction of buildings on a level backfill or due to the movement of vehicles near the top of the retaining wall, gravity loads acting can be considered as uniformly distributed load. This additional load of w_s (kN/m^2) can be treated as statically equivalent to an additional (fictitious) height of soil $h_s = (w_s/\gamma_e)$ acting over the level surface. The force developed due to the effect of surcharge on a level backfill together with the other forces are shown in Fig. 13.3.

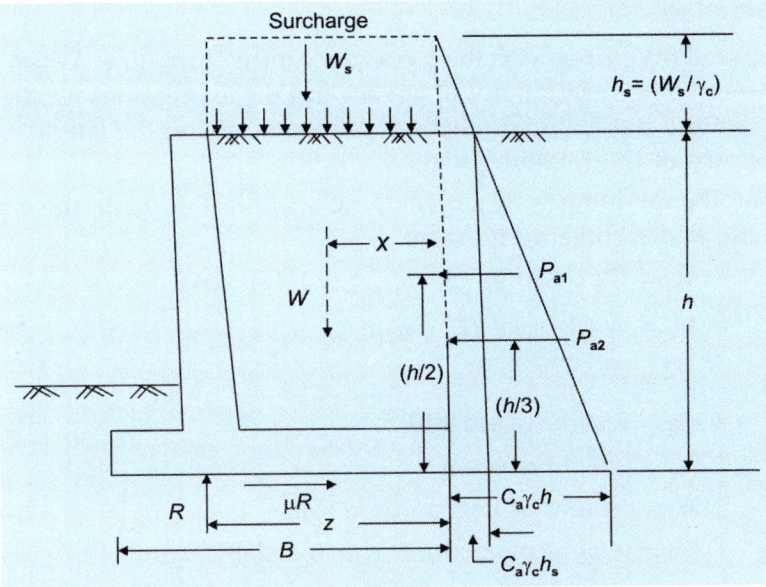

13.3: Effect of surcharge on level backfill

The total force due to active earth pressure is expressed as

$$P_a = P_{a1} + P_{a2}$$

where, $P_{a1} = C_a w_s h = C_a h_s \gamma_e h$ and $P_{a2} = (C_a \gamma_e h^2)/2$

The forces P_{a1} and P_{a2} act at a height of $h/2$ and $h/3$ respectively above the heel.

The vertical forces include the weight of soil, weight of stem, heel, toe slab and the soil fill above toe slab.

The soil pressure developed to resist the earth pressure and other vertical forces acting upwards from heel to toe. The pressure distribution at base is obtained by stability calculations comprising the equilibrium condition of vertical forces and moments.

13.4 STABILITY REQUIREMENTS

The retaining wall should resist the various forces acting on it from the earth backfill, surcharge loads, forces due to earthfill from toe side and in some cases the pressure due to water in the backfill. The design of retaining walls should conform to the stability requirements specified in Clause 20, IS:456-2000[5] which includes overturning and sliding. The factor of safety against overturning and sliding should be not less than 1.4 since the stabilizing forces are due to dead loads. The code also specifies that these stabilizing forces should be factored by a value of 0.9 in calculating the factor of safety.

Hence, the factor of safety can be expressed by the relation,

$$FS = \left[\frac{0.9(\text{stablising force or moment})}{\text{destabilising force or moment}}\right] \geq 1.4$$

13.4.1 Overturning

The retaining wall overturns with the toe as the centre of rotation. When the structure overturns, the upward reaction R will not act and the expressions for the overturning moment M_0 and the stabilizing moment M_s depend only on the lateral earth pressure and the geometry of the retaining wall.

Considering the retaining wall with sloping backfill (Fig. 13.2), the expressions for the overturning and stabilizing moment are

$$M_0 = (P_a \cos \theta)(h'/3) = [C_a \gamma_e (h')^3 / 6] \cos \theta$$

$$M_s = W(B - x) + (P_a \sin q)B$$

where $\quad W = W_1 + W_2 + W_3 + W_4$

and $\quad W_1$ = weight of earthfill

W_2 = weight of stem

W_3 = weight of heel and toe slab

W_4 = weight of earthfill over toe slab

and $\quad x$ = distance of W from the heel

B = Base width of slab

The factor of safety against overturning is expressed as

$$(FS)_{\text{overturning}} = \left(\frac{0.9 M_s}{M_0} \right) \geq 1.4$$

13.4.2 Sliding

The resistance developed against sliding of the retaining wall is mainly due to the frictional forces generated between the base slab and the supporting soil expressed as

$$F = \mu R$$

where,

$R = W$ = resultant soil pressure acting on the base slab and

μ = coefficient of friction between concrete and soil (value of μ varies in the range of 0.35 for silt to about 0.60 for rough rock)

Hence, the factor of safety against sliding is computed by the relation

$$(FS)_{\text{sliding}} = \left(\frac{0.9 \mu W}{P_a \cos \theta} \right) \geq 1.4$$

13.4.3 Shear Key

In the case of backfills with surcharge, the active pressures are relatively high and consequently the required factor of safety against sliding by the frictional forces above will not be sufficient. In such cases, it is advantageous to provide a shear key projecting below the base slab as shown in Fig. 13.4.

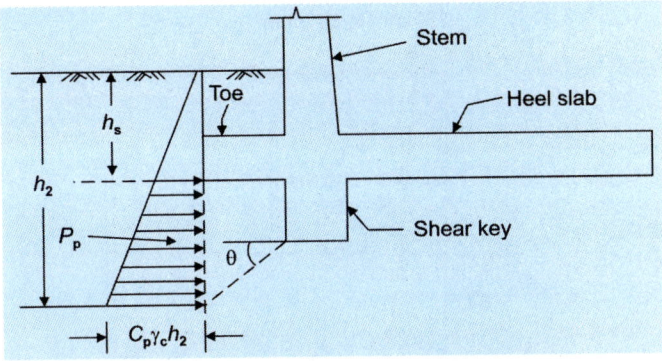

13.4: Passive pressure due to shear key

The passive resistance developed against sliding is computed as

$$P_p = C_p \gamma_e (h_2^2 - h_1^2)/2$$

It is advantageous to provide a shear key just below the stem so that the reinforcements can be extended into the shear key.

The enhanced factor of safety against sliding by the use of the shear key can be expressed as

$$(FS)_{sliding} = \left(\frac{0.9\mu W + P_p}{P_a \cdot \cos\theta} \right) \geq 1.4$$

13.5 DESIGN OF RETAINING WALLS

13.5.1 Preliminary Dimensioning of Structural Components

The design of various structural components of a typical retaining wall comprises fixing the preliminary dimensions of stem, heel and toe slabs of the retaining wall based on empirical rules.

a. Base Slab Dimensions

An economical design of the retaining wall can be obtained by proportioning the base slab so as to align the vertical soil reaction R at the base with the front face of stem or up right slab. Referring to Fig. 13.5,

let h = height of earthfill from the soffit of base slab

B = width of base slab

x_h = width of heel slab

x_t = width of toe slab

C_a = coefficient of active earth pressure.

Assuming the soil pressure distribution as triangular with maximum pressure at the toe and zero at the heel, the resultant vertical pressure will pass through the middle third point. For economical design, the soil pressure resultant should line up with the front face of the stem.

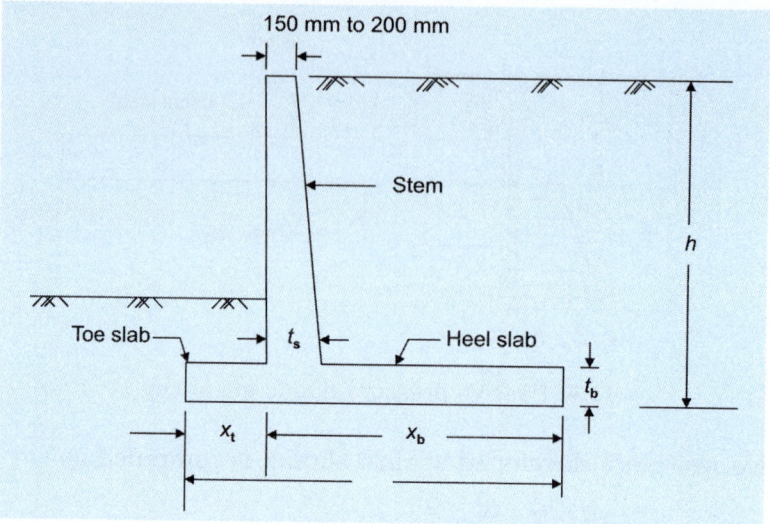

13.5: Dimensioning of retaining wall elements

Adopting this principle, Unnikrishna Pillai and Menon[6] have developed an empirical expression for the minimum width of heel slab as

$$x_h = h\sqrt{C_a/3}$$

The effect of surcharge or sloping backfill can be included by replacing h with $(h + h_s)$ or h' respectively. With known values of h and C_a, the width of heel slab (x_h) can be computed.

Hence, base width $B = 1.5\ x_h$ and $x_t = (1/3)\ B$ so that $x_h = (2/3)\ B$

b. Thickness of Base Slab and Stem

The preliminary computations of the thickness of base slab is expressed as,

$$t_b = (h/2)\ \text{or}\ 0.08h$$

But not less than 300 mm

Thickness of stem at bottom is assumed as

$$t_s = t_b$$

The stem thickness is gradually decreased to a minimum value of 150 mm of 200 mm at top. The front face of the stem is maintained vertical.

c. Design of Stem, Heel, and Toe Slabs

The structural behaviour of stem, heel and toe slabs[7,8] due to earth pressure from the backfill is shown in Fig. 13.6. The stem behaves as a cantilever fixed at the bottom and bends due to the earth pressure from the backfill. The heel slab also bends as a cantilever due to the weight of the earthfill. The deformation characteristics of both the stem and base slabs is shown in the Fig. 13.6. The critical sections XX, YY and ZZ shown in Fig. 13.6 have to be designed to resist the factored moment and shear forces with a load factor of 1.5. Usually shear is not a critical design factor and the flexural reinforcement is provided near the tension face in the slabs with a clear cover of 50 mm. The reinforcements in the stem may be curtailed in stages for economy.

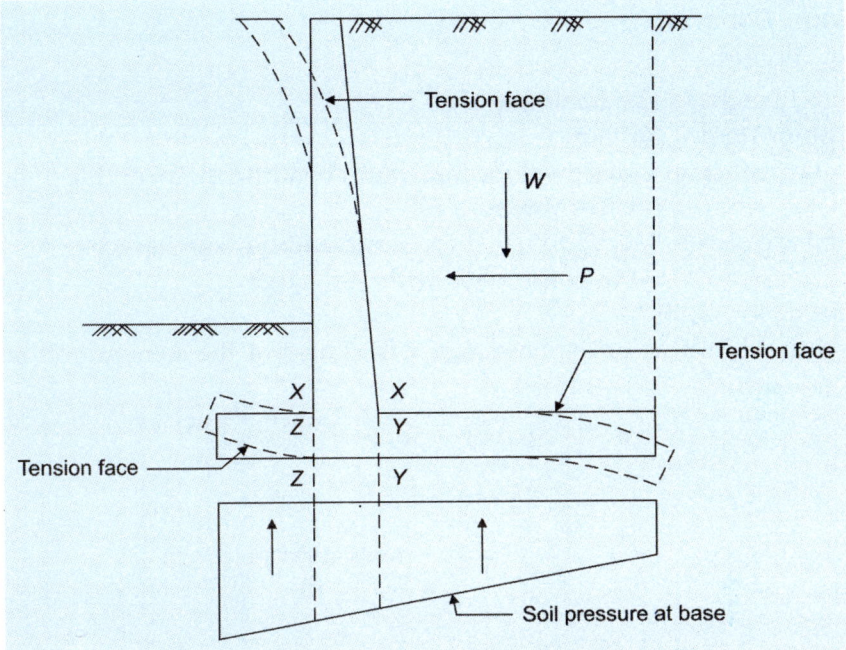

13.6: Deformation characteristics of retaining wall

Temperature and shrinkage reinforcement of 0.12% of the gross cross-section should be provided transverse to the main reinforcement. Normal vertical and horizontal reinforcement should be provided near the front face of stem and also at the bottom face of heel slab and top face of toe slab.

d. Design of Counterforts

The counterforts should be integrally built with proper ties with stem and heel slab so that the horizontal forces due to earthfill is resisted by the tension steel provided in the counterforts. In a similar way the vertical forces on base slab are resisted by the vertical ties in the counterfort.

The counterfort[9,10] is designed as a vertical cantilever, fixed at base. Since the stem acts integrally with the counterfort, the effective section resisting the cantilever moment is a flanged section, with the flange under compression. The counterforts are designed as T-beams with the depth of the section varying linearly from the top to the bottom where the section is maximum to resist the maximum moments.

The stem is designed as a continuous slab[11,12] spanning between the counterforts with negative and positive moments at supports and mid span respectively. The heel and toe slabs are designed for soil pressure as continuous and cantilever slabs respectively.

13.6 BASIC DESIGN STEPS

13.6.1 Cantilever Retaining Wall

The following basic design steps are used to determine the various preliminary dimensions of the stem, heel and toe slabs.

a. Preliminary Dimensions

Referring to Fig. 13.7

H = overall height of the retaining wall

Top width of stem = 200 mm

Bottom width of stem = designed for maximum bending moment due to horizontal earth pressure P

Width of base slab b = 0.5H to 0.6 H for walls without surcharge

= 0.7 H for walls with surcharge

Toe projection = $(b/3)$

Normally the thickness of the base slab = thickness of the stem at bottom.

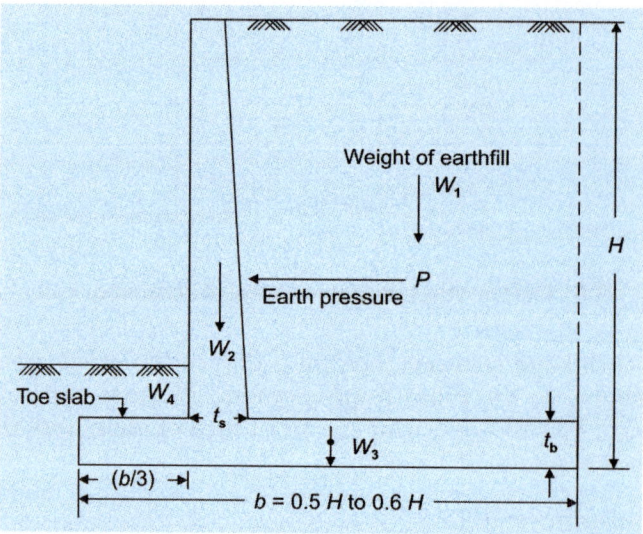

13.7: Preliminary dimensioning of retaining walls

b. Design Principles

1. Heel slab: Designed for maximum bending moment due to earth pressure from bottom and weight of earth and heel slab from top.
2. Toe slab: Designed for maximum bending moment due to earth pressure from bottom and weight of earth on toe slab from top.
3. Check for horizontal sliding

   If W = total vertical load = $(W_1 + W_2 + W_3)$

   μ = coefficient of friction

   P = horizontal earth pressure

   Factor of safety against sliding = $(\mu W/P) > 1.5$

13.6.2 Counterfort Retaining Wall

a. General Features

Retaining walls of height over 5.5 m are usually provided with counterforts

Spacing of counterforts L = 3 m to 3.5 m

Thickness of base slab = 2LH cm

L = spacing of counterforts (m)

H = overall height of the retaining wall (m)

Base width = $0.6H$ to $0.7H$

Toe projection = one-fourth of the width of base

b. Design principles

Stem or upright slab designed as continuous slab spanning between counterforts:

Maximum bending moment $= \left(\dfrac{pL^2}{12} \right)$

where, p = pressure intensity at the base $= wh \left(\dfrac{1 - \sin \phi}{1 + \sin \phi} \right)$

Toe slab is designed for soil pressure and dead weight of slab

Heel slab is designed as continuous slab supported between counterforts to resist the weight of soil from top and upward pressure at base

The thickness of counterfort is the same as that of the base slab and it is designed to resist the lateral earth pressure.

Maximum bending moment in counterfort $= \left[C_p \left(\dfrac{wh^3}{6} \right) L \right]$

where,

h = height of retaining wall above the base

L = spacing of counterforts

C_p = coefficient of active pressure $= \left(\dfrac{1 - \sin \phi}{1 + \sin \phi} \right) = 0.333$

13.7   DESIGN EXAMPLES

13.7.1 Design of Cantilever Type Retaining Wall

Design a cantilever retaining wall to retain earth embankment 4 m high above ground level. The density of earth is 18 kN/m³ and its angle of repose is 30°. The embankment is horizontal at its top. The safe bearing capacity of the soil may be taken as 200 kN/m² and the coefficient of friction between soil and concrete is 0.5. Adopt M20 grade concrete and Fe 415 HYSD bars.

    i. *Data*:

       Height of embankment above GL = 4 m

       Density of soil = 18 kN/m³

       Angle of repose = 30°

       SBC of soil = 200 kN/m²

       Coefficient of friction = 0.5

       Materials: M20 grade concrete (f_{ck} = 20 N/mm²)

                 Fe 415 HYSD bars (f_y = 415 N/mm²)

ii. *Dimensions of retaining wall*:

Minimum depth of foundation $= \left(\dfrac{p}{w}\right)\left(\dfrac{1-\sin\phi}{1+\sin\phi}\right)^2 = \left(\dfrac{200}{18}\right)\left(\dfrac{1}{3}\right)^2 = 1.2\ m$

Provide depth of foundation = 1.2 m
Overall depth of wall $H = (4 + 1.2) = 5.2\ m$
Thickness of base slab $(H/12) = (5200/12) = 433\ mm$
Adopt thickness of base slab = 450 mm
Thickness of stem at base = 450 mm
Height of stem $h = (5.2 - 0.45) = 4.75\ m$
Width of base slab $b = 0.5H$ to $0.6H$
$0.5H = 2.6\ m$
$0.6H = 3.12\ m$ ∴ Adopt $b = 3\ m$
The overall dimensions of retaining wall is shown in Fig. 13.8.

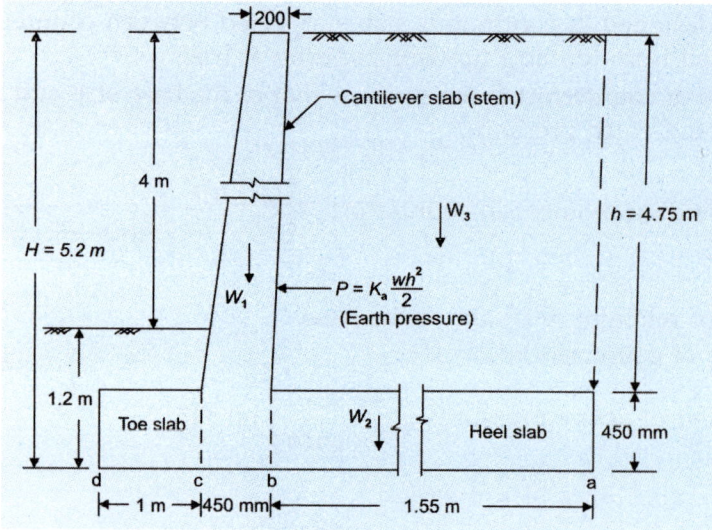

13.8: Dimensions of cantilever retaining wall

iii. *Design of stem*:

Height of stem $h = 4.75\ m$
Maximum working moment in stem

$$M = C_p\left(\dfrac{wh^3}{6}\right)$$

where $C_p = \left(\dfrac{1-\sin\phi}{1+\sin\phi}\right) = (1/3)$

$$M = \left(\dfrac{1}{3} \times \dfrac{18 \times 4.75^3}{6}\right) = 107.17\ kN \cdot m$$

Factored bending moment $M_u = (1.5 \times 107.17) = 161\ kN \cdot m$
Limiting thickness of stem at base

$$d = \sqrt{\frac{M_u}{0.138 f_{ck} b}} = \sqrt{\frac{161 \times 10^6}{0.138 \times 20 \times 10^3}} = 242 \text{ mm}$$

Assumed thickness is more than the limiting value. Hence the section is under-reinforced.

Adopt effective depth of stem $d = 400$ mm at bottom gradually tapering to 200 mm at top. Compute the parameter

$$\left(\frac{M_u}{bd^2}\right) = \left(\frac{161 \times 10^6}{10^3 \times 400^2}\right) = 1.006$$

Refer to Table 2, SP:16 design tables corresponding to $f_{ck} = 20$ N/mm^2 and readout the percentage of reinforcement as,

$$A_{st} = \left(\frac{p_t bd}{100}\right) = \left(\frac{0.3 \times 10^3 \times 400}{100}\right) = 1200 \text{ mm}^2$$

Provide 16 mm diameter bars at 150 mm centres in the vertical direction at the bottom of the stem ($A_{st} = 1341$ mm^2) gradually increasing to 300 mm towards the top.

Distribution bars = $(0.12 \times 1000 \times 450) = 540$ mm^2

Provide 10 mm diameter bars at 250 mm centres on both faces.

iv. *Stability calculations (pressure distribution at base):*

Heel projection = $(2 - 0.45) = 1.55$ m

The overall dimensions of wall is shown in Fig. 13.8.

The stability computations for one metre run of wall is shown in Table 13.1.

Distance of point of application of resultant from end a is

$$Z = \left(\frac{327.45}{204.90}\right) = 1.6 \text{ m}$$

Eccentricity $e = [Z - (b/2)] = [1.6 - (3/2)] = 0.1$ m

$(b/6) = (3/6) = 0.5 \quad \therefore \quad e < (b/6)$

Table 13.1: Stability calculation for one metre run of wall

Loads	Magnitude of load (kN)	Distance from 'a' (m)	Moment (kN·m)
$W_1 = (0.2 \times 4.75 \times 25)$	23.80	1.65	39.27
$(0.5 \times 0.25 \times 4.75 \times 25)$	14.84	1.83	27.15
$W_2 = (3 \times 0.45 \times 25)$	33.75	1.50	50.62
$W_3 = (1.55 \times 4.75 \times 18)$	132.51	0.78	103.35
Moment due to earth pressure = $C_p(wh^3/6)$			
$= (1/3)(18 \times 4.75^3)/6$			107.46
Total	$\Sigma W = 204.90$		$M = 327.45$

Maximum and minimum pressure at base are computed as

$$p_{\substack{\max \\ \min}} = \left(\frac{204.90}{3}\right)\left[1 + \left(\frac{6 \times 0.1}{3}\right)\right]$$

Hence $p_{\max} = 82.00$ kN/m^2 and $p_{\min} = 55$ kN/m^2 (Refer to Fig. 13.9).

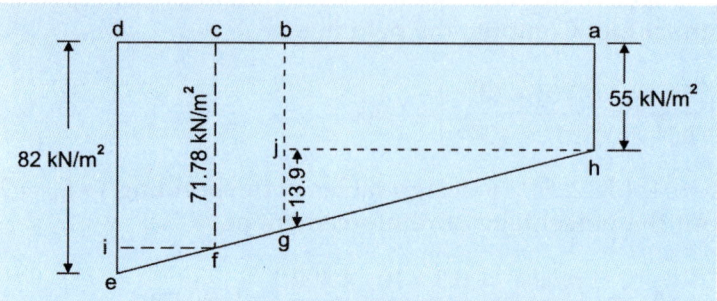

13.9: Pressure distribution at base

v. *Design of heel slab*:

Moment computations are shown in Table 13.2.

Loads	Magnitude of load (kN)	Distance from 'a' (m)	Moment (kN·m)
$W_3 = (1.55 \times 4.75 \times 18)$	132.70	0.775	102.68
Self weight of heel slab $(1.55 \times 0.45 \times 25)$	17.40	0.775	13.51
Total			116.19
Deduct for upward pressure 'abih' (53.84×1.55)	83.45	0.775	64.67
Upward pressure 'ghi' $(0.5 \times 1.55 \times 13.9)$	10.77	0.516	5.55
Total deduction			70.22
Maximum working bending moment in heel slab			46.00

Table 13.2: Moment computation

Maximum design ultimate moment $M_u = (1.5 \times 46) = 69$ kN·m

$$\left(\frac{M_u}{bd^2}\right) = \left(\frac{69 \times 10^6}{10^3 \times 400^2}\right) = 0.43$$

Refer to Table 2, SP:16 and readout the percentage of reinforcement as $p_t = 0.121\%$.

$$\therefore \qquad A_{st} = \left(\frac{p_t bd}{100}\right) = \left(\frac{0.121 \times 10^3 \times 400}{100}\right) = 484 \text{ mm}^2$$

Provide 12 mm diameter bars at 200 mm centres ($A_{st} = 565$ mm^2)

Distribution reinforcement = $(0.0012 \times 1000 \times 450) = 540$ mm^2

Provide 12 mm diameter bars at 200 mm centres ($A_{st} = 565$ mm^2)

vi. *Design of toe slab*:

The maximum bending moment in the toe slab is determined by taking moments of the forces about the point 'c'. The moment computations are shown in Table 13.3.

Loads	Magnitude of load (kN)	Distance from 'a' (m)	Moment about 'c' (kN·m)
Table 13.3: Moment computation			
Upward pressure 'cdif' (71.78 × 1)	71.78	0.5	35.89
Upward pressure 'jfe' (0.5 × 1 × 8.98)	4.49	0.67	3.00
Total			38.89
Deduct self weight of toe slab (1 × 10.45 × 25)	11.2	0.5	5.60
Dead weight of soil over toe slab (0.75 × 1 × 18)	13.5	0.5	6.75
Total deduction			12.35
Maximum service load BM in toe slab			26.54

Maximum design ultimate moment $M_u = (1.5 \times 26.54) = 39.81$ kN·m

Compute the parameter $(M_u/b\,d^2) = (39.81 \times 10^6)/(10^3 \times 400^2) = 0.244$

Refer to Table 2, SP:16 and readout the percentage steel as p_t is less than 0.12%.

Hence, provide maximum reinforcement of 0.12%.

$A_{st} = 0.0012 \times 450 \times 1000 = 540$ mm^2

Provide 12 mm diameter bars at 200 mm centres ($A_{st} = 565$ mm^2)

vii. *Check for safety against sliding*:

Total horizontal earth pressure

$$P = K_a\left(\frac{wH^2}{2}\right) = \left[\frac{1}{3} \times 18 \times \frac{5.2^2}{2}\right] = 81.12 \text{ kN}$$

Assuming coefficient of friction $\mu = 0.5$

Maximum possible frictional force $W = (0.5 \times 204.84) = 102.4$ kN

∴ Factor of safety against sliding = (102.84/81.12) = 1.26 < 1.5

Hence, a shear key has to be designed.

viii. *Design of shear key*:

If p_p = intensity of passive pressure developed just in front of the shear key, the value of p_p is computed as

$$p_p = K_p p$$

where, $K_p = \left(\dfrac{1+\sin\phi}{1-\sin\phi}\right) = (1/K_a) = 3$

and $\quad p = 72.8$ kN/m^2 (Refer to Fig. 10.3)

∴ $\quad p_p = (3 \times 72.8) = 218.6$ kN/m^2

If 'a' is the depth of shear key = 450 mm

Total passive force $p_p = (p_p a) = (218.6 \times 0.45) = 98.3$ kN

∴ Factor of safety (FS) against sliding

$$= \left[\frac{W + p_p}{p}\right] = \left[\frac{102.4 + 98.3}{81.12}\right] = 2.45 > 1.5$$

Hence the retaining wall is safe against failure due to sliding. The reinforcement in stem is extended up the shear key.

ix. *Check for shear stresses at junction of stem and base slab*:

Net working shear force $V = (1.5\,p - mW)$

$$= (1.5 \times 81.12) - 102.4$$
$$= 19.28$$

Factored shear force $\quad V_u = (1.5 \times 19.28) = 28.92$ kN

Nominal shear stress $\quad \tau_v = (28.92 \times 10^3)/(1000 \times 400)$
$$= 0.072 \text{ N/mm}^2$$

$$\left(\frac{100\,A_{st}}{b\,d}\right) = \left(\frac{100 \times 1341}{1000 \times 400}\right) = 1.25$$

From Table 19, IS:456-2000, readout the permissible shear stress as
$$\tau_c = 0.40 \text{ N/mm}^2 > \tau_v$$

Hence shear stresses are within safe permissible limits.

The reinforcement details in the retaining wall are shown in Fig. 13.10.

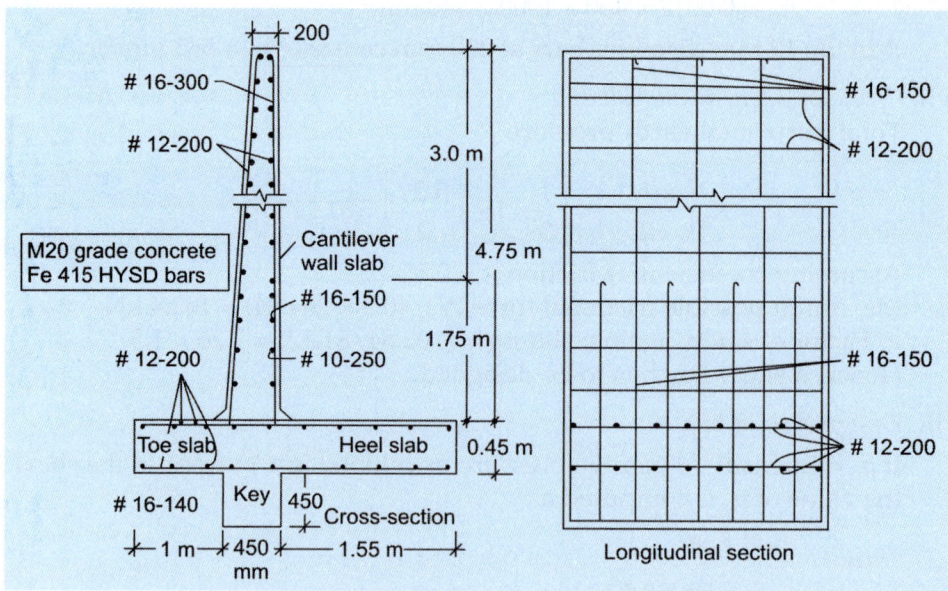

13.10: Reinforcement details in cantilever retaining wall

13.7.2 Design of Counterfort Type Retaining Wall

Design a counterfort type retaining wall to suit the following data:

Height of wall above ground level = 6 m

SBC of soil at site = 160 kN/m²

Angle of internal friction = 33°

Density of soil = 16 kN/m^3

Spacings of counterforts = 3 m c/c

Materials: M20 grade concrete

              Fe 415 HYSD bars

Sketch the details of reinforcements in the wall.

   i. *Dimensions of retaining wall*:

      Minimum depth of foundation $= \dfrac{p}{w}\left(\dfrac{1-\sin\phi}{1+\sin\phi}\right)^2$

$$= \frac{160}{16}\left(\frac{1}{3}\right)^2 = 1.11 \text{ m}$$

      Provide depth of foundation = 1.2 m

      ∴ Overall height of wall $H = (6 + 1.2) = 7.2$ m

      Thickness of slab $= 2\,LH$ cm

                     $= (2 \times 3 \times 7.2) = 43.2$ cm

      Provide 450 mm thick base slab

      Base width $= 0.6H$ to $0.7H$

      (0.6×7.2) $= 4.32$ m

      (0.7×7.2) $= 5.04$ m

      Adopt base width $= 4.5$ m

      Toe projection $= (1/4 \times 4.5) = 1.1$ m

  ii. *Design of stem*:

      Pressure intensity at base

$$= wh\left(\frac{1-\sin\phi}{1+\sin\phi}\right)$$

      where, $h = (7.2 \times 0.45) = 6.75$ m

      ∴ Pressure intensity $= (1.6 \times 6.75 \times 1/3) = 36$ kN/m^2

      Maximum working moment $= \left(\dfrac{36 \times 3^2}{12}\right) = 27$ kN · m

      Factored moment $M_u = (1.5 \times 27) = 40.5$ kN·m

      Effective depth required for balanced section

$$d = \sqrt{\frac{M_u}{(0.138 f_{ck} b)}} = \sqrt{\frac{40.5 \times 10^6}{0.138 \times 20 \times 10^3}} = 121 \text{ mm}$$

Assuming an under-reinforced section and to provide a suitable thickness to resist design shear at base of stem, adopt an overall thickness of 220 mm constant up to the top.

    Effective depth $d = 175$ mm

The reinforcement in the stem is computed using the relation

$$(40.5 \times 10^6) = (0.87 \times 415 A_{st} \times 175) \left[1 - \left(\frac{415 A_{st}}{10^3 \times 175 \times 20} \right) \right]$$

Solving, $A_{st} = 700$ mm^2
Provide 12 mm diameter bars at 150 mm c/c ($A_{st} = 754$ mm^2)
Distribution reinforcement = 0.12% of section

$$= (0.00012 \times 220 \times 1000)$$
$$= 264 \text{ mm}^2/\text{m}$$

Adopt 6 mm diameter bars at 200 mm c/c ($A_{st} = 283$ mm^2)
The dimensions of various structural elements of the counterfort retaining wall are shown in Fig. 13.11a.

iii. *Stability calculations*:
The pressure distribution at base is computed by calculating the various forces acting and taking moments of all the forces about the heel. The various forces acting and their moments about the heel point a are given in Table 13.4.

Table 13.4: Stability computation

Loads	Magnitude of load (kN)	Distance from 'a' (m)	Moment about 'c' (kN·m)
$W_1 = (0.22 \times 6.75 \times 24)$	35.64	3.39	120.80
$W_2 = (0.45 \times 4.5 \times 24)$	48.60	2.25	109.35
$W_3 = (3.28 \times 6.75 \times 16)$	354.24	1.64	580.95
Moment of earth pressure $K_a \times \dfrac{wh^3}{6} = \left(\dfrac{1}{3} \times \dfrac{16 \times 7.2^3}{6} \right)$			331.77
Total	$\Sigma W = 438.49$		$\Sigma M = 1142.87$

Distance of the point of application of the resultant from point a

$$Z = \frac{\Sigma M}{\Sigma W} = \left(\frac{1142.87}{438.49} \right) = 2.66 \text{ m}$$

∴ Eccentricity $e = (Z - b/2) = (2.66 - 4.5/2) = 0.41$ m
But $(h/6) = (4.5/6) = 0.75$ m
∴ $e < (b/6)$
∴ Maximum and minimum pressures at the base are given by:

$$\sigma_{max} = \frac{438.49}{4.5} \left(1 + \frac{6 \times 0.41}{4.5} \right) = 150 \text{ kN/m}^2$$

$$\sigma_{min} = \frac{438.49}{4.5} \left(1 - \frac{6 \times 0.41}{4.5} \right) = 45 \text{ kN/m}^2$$

The maximum intensity of pressure does not exceed the permissible value of 160 kN/m^2.

The pressure distribution at the base of the retaining wall is shown in Fig. 13.11b.

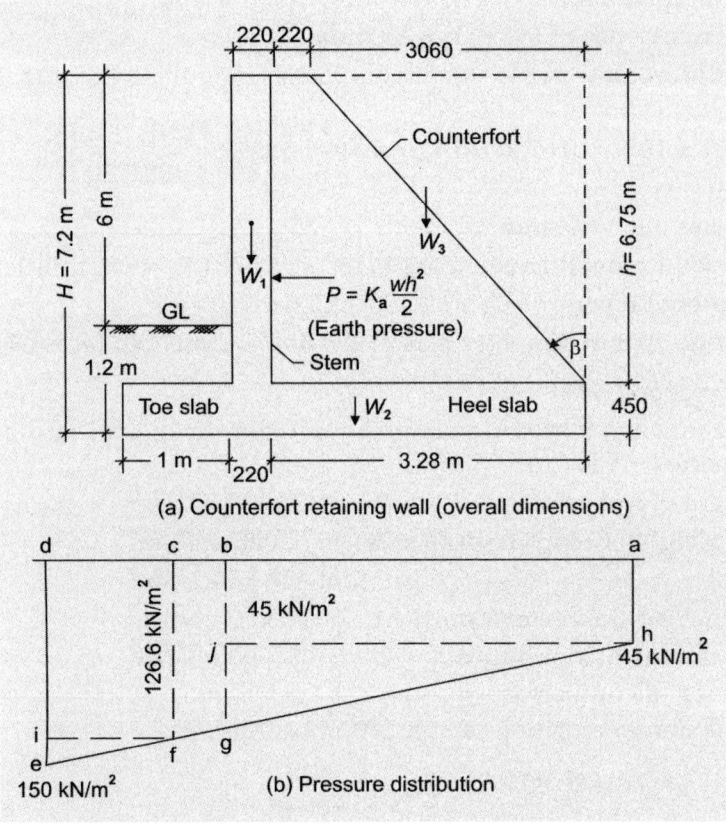

(a) Counterfort retaining wall (overall dimensions)

(b) Pressure distribution

13.11: Counterfort retaining wall

iv. *Design of toe slab*

The maximum bending moment acting on the toe slab is calculated by considering moments of all the forces about the point c. The computations are shown in Table 13.5 for one metre length of the wall.

Loads	Magnitude of load (kN)	Distance from 'c' (m)	Moment about 'c' (kN·m)
Table 13.5: Moment in toe slab			
Upward pressure 'cdif' (126.6 × 1)	126.6	0.5	63.30
Upward pressure 'efi' (0.5 × 1 × 23.4)	11.7	0.67	7.84
Total			71.14
Deduct self-weight of toe slab (1 × 0.45 × 24)	10.8	0.5	5.40
Dead-weight of soil over toe slab (0.75 × 1 × 16)	12.0	0.5	6.00
Total deduction			11.40

Maximum working moment in toe slab
$$M = (71.14 - 11.4) = 59.74 \text{ kN·m}$$
Factored moment $M = (1.5 \times 59.74) = 89.61$ kN·m
Effective depth of toe slab = 400 mm
Reinforcements in toe slab is computed using the following relation

$$(89.61 \times 10^6) = (0.87 \times 415 A_{st} \times 400)\left[1 - \left(\frac{415 A_{st}}{10^3 \times 400 \times 20}\right)\right]$$

Solving, $A_{st} = 644$ mm^2
Provide 12 mm diameter bars at 150 mm c/c ($A_{st} = 754$ mm^2)
Distribution bars = $(0.0012 \times 100 \times 450) = 540$ mm^2
Provide 10 mm diameter bars at 280 mm c/c on both faces ($A_{st} = 561$ mm^2).

v. *Design of heel slab*:
  Considering 1 m wide strip of heel slab near heel end a, upward soil pressure = 45 kN/m^2.
  Weight of soil on strip (16×6.75) =   108.00 kN/m^2
  Self weight of strip $(1 \times 0.45 \times 24)$ =   10.80 kN/m^2
  Total   =   118.80 kN/m^2
  Deduct for downward pressure   = −45.00 kN/m^2
  Net downward pressure   =   73.00 kN/m^2
  Spacing of counterforts = 3 m
  ∴ Maximum negative service BM at counterforts

$$M = \left(\frac{73.80 \times 3^2}{12}\right) = 55.35 \text{ kN} \cdot \text{m}$$

  Factored moment $M_u = (1.5 \times 55.35) = 83$ kN·m
  Reinforcements in heel slab is computed using the relation

$$(83 \times 10^6) = (0.87 \times 415 A_{st} \times 400)\left[1 - \left(\frac{415 A_{st}}{(1000 \times 175 \times 20)}\right)\right]$$

  Solving, $A_{st} = 600$ mm^2
  Provide 12 mm diameter bars at 150 mm c/c ($A_{st} = 754$ mm^2)
  Distribution bars = 0.12% of cross-section
              = $(0.0012 \times 1000 \times 450) = 540$ mm^2
  Provide 10 mm diameter bars at 280 mm centres on both faces ($A_{st} = 561$ mm^2)

vi. *Design of counterforts*:
  Thickness provided at the top = $(220 + 220) = 440$ mm
  Thickness of counterfort = 440 mm
  Maximum working moment in counterfort

$$M = \left(K_a \times \frac{wh^3}{6} \times L\right) = \left(\frac{1}{3} \times \frac{16 \times 6.75^3}{6} \times 3\right) = 820.12 \text{ kN} \cdot \text{m}$$

Factored moment $M_u = (1.5 \times 820.12) = 1230$ kN·m

Reinforcement at the bottom of counterfort is computed by using the following relation

$$(1230 \times 10^6) = (0.87 \times 415 A_{st} \times 4400)\left[1 - \left(\frac{415 A_{st}}{440 \times 4400 \times 20}\right)\right]$$

Solving, $A_{st} = 800$ mm^2

But minimum reinforcement as per IS:456-2000 code is given by

$$A_s = \left(\frac{0.85\,bd}{f_y}\right) = \left[\frac{0.85 \times 440 \times 4400}{415}\right] = 3965 \text{ mm}^2$$

Provide 5 bars of 32 mm diameter ($A_{st} = 4020$ mm^2).

vii. *Curtailment of bars*:

Let h_1 = depth at which 1 bar can be curtailed

Then $\qquad \left(\dfrac{5-1}{5}\right) = \left(\dfrac{h_1}{6.75^2}\right)$

$\therefore h_1 = 6$ m from top

Let h_2 = depth at which 2 bars can be curtailed

Then $\qquad \left(\dfrac{5-2}{5}\right) = \left(\dfrac{h_2}{6.75^2}\right)$

$\therefore h_2 = 5.2$ m from top

Let h_3 = depth at which 3 bars can be curtailed

Then $\qquad \left(\dfrac{5-3}{5}\right) = \left(\dfrac{h_3}{6.75^2}\right)$

$\therefore h_3 = 4.2$ m from top

The remaining two bars are taken right up to the top.

viii. *Connection between counterfort and upright slab*:

Consider the bottom 1 m height of upright slab.

Pressure on this strip = 36 kN/m^2

Total lateral pressure transferred to the counterfort for 1 m height

$\qquad = 36(3 - 0.44) = 91.8$ kN

Factored force = $(1.5 \times 91.8) = 137.7$ kN

Steel required per metre height $= \left(\dfrac{137.7 \times 10^3}{0.87 \times 415}\right) = 381$ mm^2

Provide minimum reinforcement of 10 mm diameter bars in the form of horizontal links at 280 mm centres.

ix. *Connection between counterfort and heel slab*:

Tension transferred in 1 m width of counterfort near the heel end

$\qquad = 73.8(3 - 0.44) = 189$ kN

Factored force = $(1.5 \times 189) = 283.5$ kN

Steel required for 1 m width $= \left(\dfrac{283.5 \times 10^3}{0.87 \times 415} \right) = 785$ mm^2

Spacing of 10 mm diameter bars in the form of two-legged links

$$= \left(\dfrac{2 \times 78.5 \times 10^3}{0.87 \times 415} \right) = 785 \text{ mm}^2$$

Provide 10 mm diameter two-legged vertical links at 200 mm centres.

x. The details of reinforcements in the counterfort retaining wall are shown in Fig. 13.12.

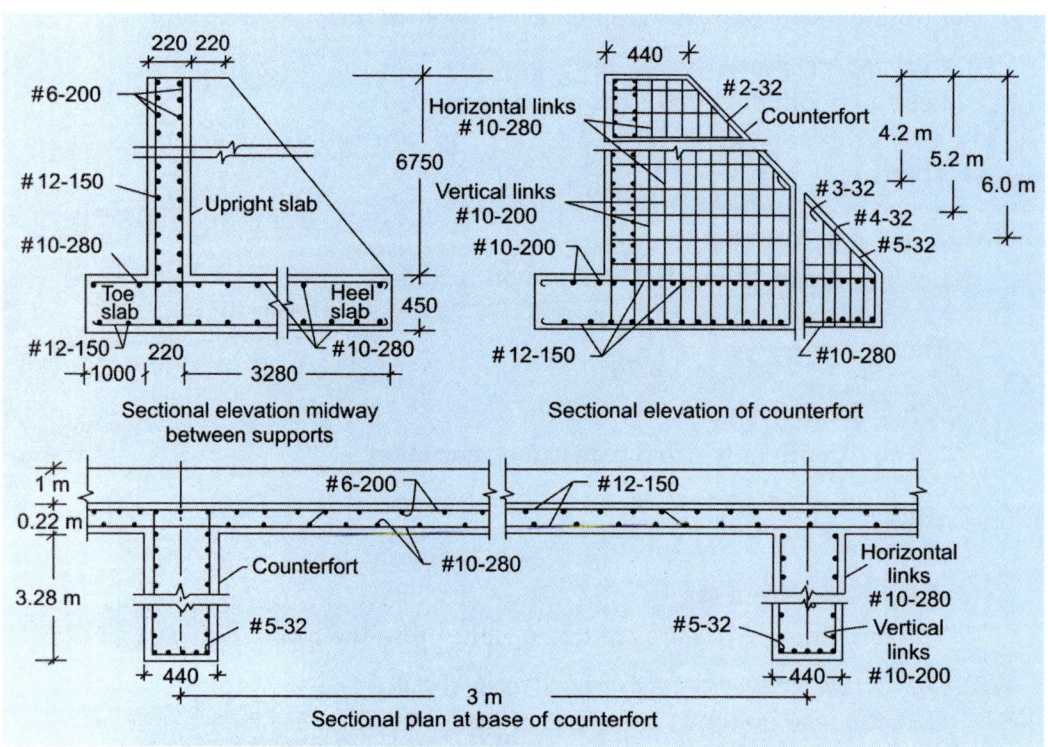

13.12: Reinforcement details in counter fort retaining wall

REFERENCES

1. Bowels JE, *Foundation Analysis and Design*, 3 edn, McGraw-Hill Book Co, New York, 1982.

2. Peck RR, Hanson WE and Thornburn TH, *Foundation Engineering*, 2 edn, John Wiley & Sons Inc, 1974.

3. Budhu M, *Foundations and Earth Retaining Structures*, John Wiley & Sons Inc, New York, 2008, p 500.

4. Huntington WC, *Earth Pressures and Retaining Walls*, John Wiley & Sons Inc, New York, 1968.

5. IS:456-2000, Indian Standard Code of Practice for Plain and Reinforced Concrete (4th Revision), BIS, New Delhi, 2000, p 37.

6. Unnikrishna Pillai S and Menon D, *Reinforced Concrete Design*, 3 edn, Tata McGraw-Hill Education Pvt. Ltd, New Delhi, 2009, pp 716-818.

7. IRC:78-1983, Standard specifications and code of practice for road bridges, Section VII, Foundations and Substructure, Indian Roads Congress, 1994, p 72.

8. Ponnuswamy S, *Bridge Engineering*, Tata McGraw-Hill Publishing Co, New Delhi, 1986, p 544.

9. Clayton CRI, Milititsky J and Woods RI, *Earth Pressure and Earth Retaining Structures*, 2 edn, CRC Press, Boca Raton, 1993, p 408.

10. Park R and Paulay T, *Reinforced Concrete Structures*, John Wiley & Sons, New York, 1975.

11. Rangan BV, Hall AS and Werner, *Reinforced Concrete Structures*, Pitman Publications, London, 1980, p 460.

12. Subramanian N, *Design of Reinforced Concrete Structures*, Oxford University Press, New Delhi, India, 2003, pp 644–701.

ASSIGNMENT

1. Design the stem of a reinforced concrete cantilever type retaining wall to retain earth with the top of the wall to a height of 5 m. The density of soil is 14 kN/m^3 and the angle is repose is 30°. Adopt M20 grade concrete and Fe 415 HYSD bars. Sketch the details of reinforcements in the retaining wall. Assume SBC of soil as 180 kN/m^2.

2. Design a cantilever type retaining wall to support a bank of earth 4 m high above ground level on the toe side of the wall. The backfill surface is inclined at an angle of 15° with the horizontal. Assume that good soil for foundations at a depth of 1.25 m below ground level with a safe bearing capacity of 160 kN/m^2. The granular soil in the backfill has a unit weight of 16 kN/m^3 and an angle of shearing resistance of 30°. Assume coefficient of friction between soil and concrete to be $0.5 f_{ck} = 20$ and $f_y = 415$ N/mm^2.

3. A reinforced concrete retaining wall of cantilever type is required to retain a granular fill to a height of 4 m above ground level and level with the top of the wall. The depth of foundation may be assumed as 1.2 m below ground level. The unit weight to fill is 14 kN/m^3 and angle of internal friction is 30 degree. The safe bearing capacity of soil is 200 kN/m^2. The base slab is to be 400 mm thick. Adopt M20 grade concrete and Fe 415 HYSD bars and design the retaining wall.

4. Design a counterfort type retaining wall to suit the following data:
   Safe bearing capacity of soil = 200 kN/m^2
   Height of soil above ground level = 7 m
   Unit weight of soil = 18 kN/m^2
   Angle of internal friction = 30°
   Spacing of counterforts = 3 m centre

Materials: M20 grade concrete
           Fe 415 HYSD bars.

5. Design a reinforced concrete cantilever retaining wall to retain earth level with the top of the wall to a height of 4.5 m above ground level. The density of soil at site is 16 kN/m³ with a safe bearing capacity of 140 kN/m². Assume the angle of shearing resistance of the soil as 30°. Further assume a coefficient of friction beween soil and concrete as 0.50. Adopt M25 grade concrete and Fe 415 HYSD bars.

6. Design a counterfort type retaining wall to support an earthfill of 6.5 m above ground level. The foundation depth may be taken as 2 m below ground level. The safe bearing capacity of soil at site is 180 kN/m². Unit weight of soil may be taken as 15 kN/m³ and an angle of shearing resistance of 33°. Assume the value of coefficient of friction as 0.50. Adopt M25 grade concrete and Fe 500 HYSD bars. Sketch the details of reinforcements in the retaining wall.

7. Design a cantilever type retaining wall to suit the following data:
Total height of the stem including the base slab = 5.25 m
Height of embankment above base slab = 4.83 m
Height of earthfill above toe slab = 830 mm
Backfill surface is inclined at an angle of 15° with the horizontal
Depth of foundation = 1.25 m
Safe bearing capacity of the soil at site = 160 kN/m²
Unit weight of soil = 16 kN/m³
Coefficient of friction between soil and concrete = 0.5
Angle of shearing resistance of soil = 30°
Grade of concrete = M25
Type of steel = Fe 500 HYSD bars

8. A cantilever type retaining wall is to be designed to contain earth over a depth of 4 m along with a surcharge pressure of 40 kN/m² due to the construction of a building. The density of the soil at site is 15 kN/m³. Angle of repose is 30°. Safe bearing capacity of soil is 160 kN/m². Adopt M20 grade concrete and Fe 415 HYSD bars and design the retaining wall to retain the earth along with the surcharge pressure and sketch the details of reinforcements in various structural elements.

9. Design a suitable counterfort type retaining wall to support a backfill, 7.5 m high above the ground level on the toe side. Assume good soil for foundation at a depth of 1.5 m below ground level with a safe bearing capacity of 170 kN/m². Unit weight of soil 16 kN/m³. Angle of shearing resistance = 30°. Assume coefficient of friction between soil and concrete as 0.5. Adopt M-25 grade concrete and Fe 415 HYSD bars.

10. Design a suitable shear key at the base of the cantilever retaining wall using the following data:
Total height of the stem including the base slab thickness = 4 m
Thickness of stem at base = 400 mm
Thickness of base slab = 400 mm
Total horizontal earth pressure = 80 kN/m

Total weight of stem, base slab and earth = 200 kN/m

Intensity of pressure at base = 70 kN/m^2

Coefficient of friction between soil and concrete = 0.5

Angle of shearing resistance = 30°

Check for factor of safety against sliding and design a suitable shear key.

REVIEW QUESTIONS

1. Explain with sketches the various prominent types of retaining walls used in practice.
2. List the various forces to be considered in the design of reinforced concrete retaining walls.
3. What is the effect of surcharge in the design of retaining walls? Explain the sources developing surcharge forces.
4. What are the stability requirements to be considered in the design of retaining walls?
5. Briefly explain the structural behaviour of various elements of a reinforced concrete retaining wall.
6. When do you resort to a counterfort type retaining wall? Briefly outline the structural behaviour of the various parts of such a wall.
7. Explain with sketches the deformation characteristics of the various structural elements of a cantilever retaining wall.
8. How do you check for the stability against overturning in a cantilever type retaining wall?
9. What is the purpose of using a shear key in a retaining wall? Will the retaining wall be safe without the use of a shear key?
10. What is curtailment of main reinforcement in counterforts of retaining walls? How do you curtail the reinforcements?

OBJECTIVE QUESTIONS

1. Cantilever type retaining walls are used in cases where earth is to be retained over depth of
   a. 6 to 8 m
   b. 2 to 5 m
   c. 1 to 2 m
2. Counterfort type retaining walls are adopted in the case of earthfill
   a. exceeding 5 m
   b. with surcharge
   c. less than 5 m
3. In cantilever retaining walls the earth is retailed on the side of
   a. toe slab
   b. heel slab
   c. base slab

4. In case of checking for stability of retaining walls, passive earth pressure is considered at the
   a. heel slab
   b. stem
   c. toe slab

5. The factor of safety against overturning in a cantilever retaining wall should not be less than
   a. 2.0
   b. 1.4
   c. 1.8

6. In case of heel slab design the main reinforcement is provided at the
   a. soffit of the slab
   b. top of the slab
   c. middle of the slab

7. The shear key below the base slab of a cantilever retaining wall should be designed to resist the
   a. active Earth pressure
   b. passive Earth pressure
   c. earth pressure from the bottom

8. The stem in case of a counterfort retaining wall should be designed as a
   a. cantilever slab
   b. simply supported slab
   c. continuous slab supported at counterforts

9. The main reinforcements in the toe slab is generally provided at the
   a. top of slab
   b. middle of slab
   c. soffit of slab

10. The main reinforcements in the counterforts are
    a. uniformly distributed throughout the depth
    b. curtailed towards the top
    c. curtailed towards the bottom

Pile and Raft Foundations

14.1 GENERAL FEATURES

In designing foundations for different type of structures, reinforced concrete piles and rafts[1,2] are generally used when soil conditions are poor having low bearing capacity. In such cases it is uneconomical to provide spread foundations. In the case of tank beds, black cotton and clayey soils having very low bearing capacity, the foundations for bridges and buildings are invariably built over piles classified under deep foundations[3,4]. Most of the multistorey buildings are built over raft foundations[5] to prevent unequal settlements leading to local distress. In case of highway bridge piers and abutments, pile foundations are widely used to ensure strength, safety and serviceability[6].

Foundations for urban flyover bridge structures are generally built on deep foundations using *cast-in-situ* piles. The metro rail project of Bengaluru has employed precast post-tensioned prestressed concrete girders[7,8] of spans 25 m to 30 m which are supported on massive *cast-in-situ* piers supported on large pile caps combining nine piles of one metre diameter, each cast at site over a depth 10 m to 15 m resting on rocky strata. Piles are extensively used in off shore and marine structures in the construction of foundations for quay walls and for tall structures like chimneys, cooling towers, bins and silos.

14.2 TYPES OF CONCRETE PILES

Reinforced concrete piles are classified under different groups as

 a. *Cast-in-situ*
 b. Precast piles

In case of *cast-in-situ* piles[9], a steel shell is driven first to the required depth and concreting is done after placing the reinforcement cage in the hole. If the shell is left in place, it is called a shell pile. If the shell is removed it is referred to as shell less *cast-in-situ* pile. The world's longest shell pile is provided at Walt Disney World in Florida, USA. The length of the pile is 114 m. During 1950 to 1960, *cast-in-situ* piles were commonly used since the technique of precasting was not well developed. With the introduction of better quality cement and precasting techniques, nowadays precast piles are invariably preferred in place of *cast-in-situ* piles since multistorey buildings and bridge structures generally involve foundations under water or in soils with high water table.

Precast piles[10] can be made with a high degree of quality control regarding dimensions and strength and hence have superior structural properties in comparison with *cast-in-situ* piles. Precast piles can be cast to various shapes such as (i) circular, (ii) square (iii) rectangular (iv) octagonal. Generally circular and square section piles are preferred to other shapes. When more than one pile is used for the foundation, the group of piles are connected at the top by a pile cap to form a single unit. The piles are arranged symmetrically about the axis of the columns so that the loads are distributed uniformly to all the piles. Pile caps are invariably used to support very heavy columns of storied buildings and the piers of bridges and flyovers in metropolitan high ways.

Precast prestressed concrete piles[11,12] have been widely used for foundations of various types of structures. These types of piles are available in standard sizes and different cross-sectional shapes such as circular, square, octagonal and with hollow cores. Precast prestressed piles are stronger and can withstand the driving and handling stresses better than reinforced concrete piles. Due to the high strength concrete used in these type of piles, they are more durable and can withstand shock loads, their strength and serviceability characteristics are superior to the traditional reinforced concrete piles.

Reinforced concrete piles are also classified based upon the load resisting characteristics as

a. friction piles, where the load on the pile is resisted by the friction developed between the concrete pile surface and the surrounding soil (e.g. piles driven in clayey soils).

b. end bearing piles, where the load on the pile is resisted by bearing of concrete pile resting on rocky strata (e.g. piles resting on hard rocky strata).

14.3 DESIGN OF PILE FOUNDATIONS

14.3.1 General Principles

Reinforced concrete piles being the compression members, are designed as columns to resist the loads transmitted from the structure. Structural design of reinforced concrete piles is influenced by the loads acting on the pile, the depth of the pile below the ground level, type of soil, the grade of concrete and quality and type of steel used as reinforcements. The precast piles are designed for handling and driving stresses together with loads to be sustained under service conditions. The design of reinforced concrete piles should conform to the Indian standard code provisons[9,10].

The minimum longitudinal reinforcement in the pile should not be less than the following values:

1. 1.25% of the cross-sectional area of the pile for piles having length upto 30 times their least lateral dimension.

2. 1.5% of the cross-sectional area of the pile for piles having length between 30 to 40 times their least lateral dimension.

3. 2% of the cross-sectional area of the pile for piles having length greater than 40 times their least lateral dimension.

The lateral reinforcement comprises ties or links of not less than 6 mm diameter and the spacing of the links or spirals shall not be greater than 150 mm. Also the spacing of the ties should not exceed half the least lateral dimension.

The minimum steel requirements of a typical precast concrete pile are shown in Fig. 14.1 based on the guidelines specified in SP:34-1987[13].

14.3.2 Structural Design of Piles

Reinforced concrete piles are designed by the stepwise procedure given below:

1. The number of piles is decided based on the total load transmitted to the foundation. The service load on each pile is evaluated and the factored load is computed.

2. The size of the pile is selected depending upon the service load. For loads in the range of 400 kN to 600 kN, 300 mm square piles will be sufficient. Depending upon the increase in the load, the size of the pile is increased.

3. The length of the pile depends upon the depth of hard strata below ground level for bearing piles and the friction developed in case of cohesive soils.
   The length of the pile above ground is generally around 0.6 m to cast the pile cap and the columns.

4. If the slenderness ratio of the pile is greater than 12, it is designed as a long column, considering the reduction coefficient applied to the permissible stresses.

5. The longitudinal reinforcement designed should be more than the minimum percentage of steel specified in Section 14.3.1.

6. The lateral reinforcement consisting of ties or links and spirals and their percentages expressed as percent of volume of the pile should not be less than the values specified in Fig. 14.1. The detailing of the larger percentage volume

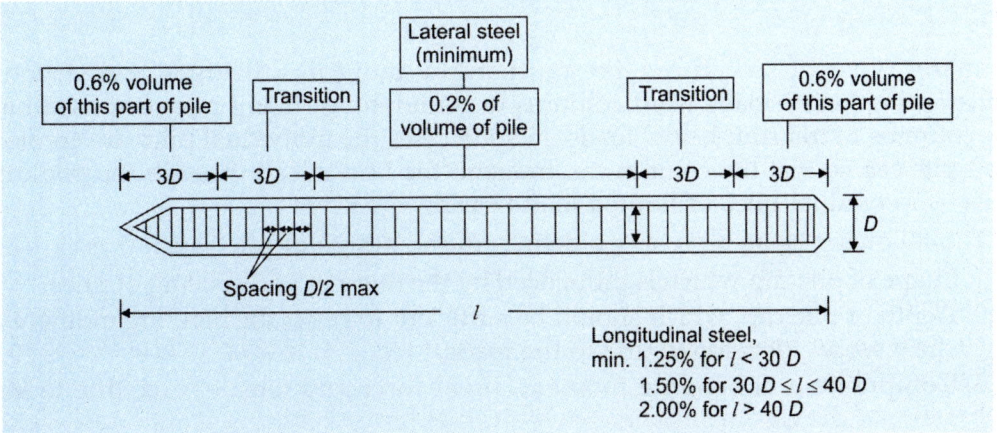

14.1: Minimum steel requirements in precast concrete piles

of lateral reinforcement near the pile head and pile end is of particular significance due to the driving stresses developed at the pile shoe end and pile head.

7. Clear cover to all main reinforcements in pile shall be not less than 50 mm.

8. Steel forks (spacer bars) in pairs are provided at regular intervals to hold the main reinforcement in position. A steel shoe made up of mild steel plates is embedded at the pile end to facilitate easy driving of the pile into the soil strata.

Typical details of reinforcements requirements in a precast concrete pile are shown in Fig. 14.2.

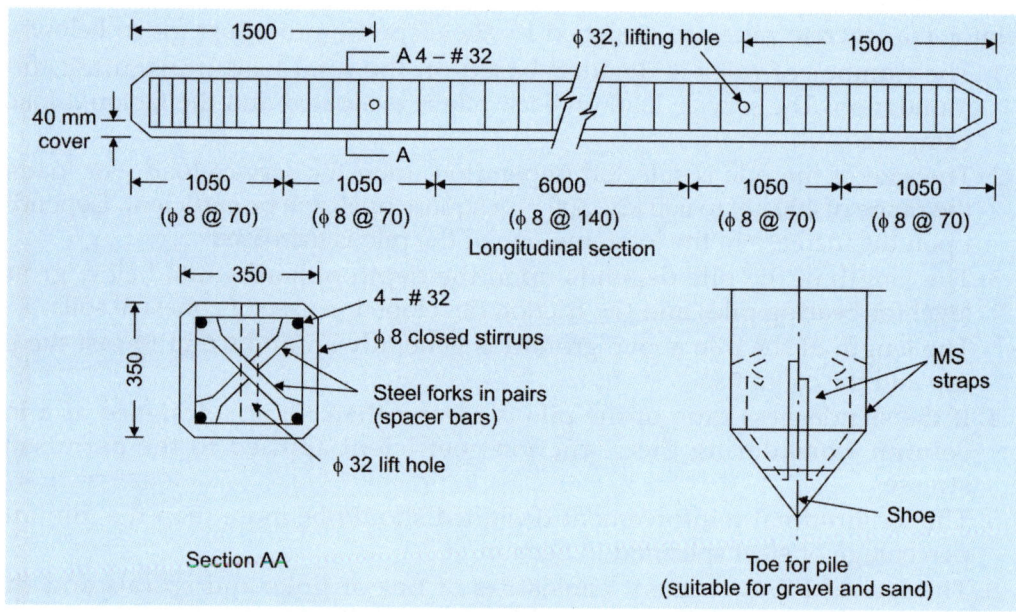

14.2: Typical Details of Reinforcements in a precast concrete pile

14.4 DESIGN OF PILE CAPS

In many types of structures like multistorey buildings, bridges and structures transmitting heavy loads from columns to foundations, many piles are used below the columns to transmit heavy loads. In such cases the individual piles are connected by a pile cap so that the column can transmit the heavy load through the pile cap to various individual piles of limited load bearing capacity.

The following steps serve as guidelines in the design of pile caps:

1. Shape of pile cap which is influenced by the number and spacing of piles.

2. Depth of pile cap, which should be sufficient to resist the bending moment and shear forces, developed due to the loads.

3. Computation of bending moment, shear force and tensile force due to struts action.

4. Amount of reinforcement and its arrangement.

The following guidelines based on the British practice and Indian standard code (IS:2911)[10] recommendations are useful in the design of pile and pile caps.

14.4.1 Shape of Pile Cap

Whenever number of piles are used symmetrically, square or rectangular shaped pile caps are commonly employed. When odd numbers of three piles spaced asymmetrically are used, triangular shaped pile caps are used.

Minimum spacing of piles = 2.5 to $3d_p$

where, d_p = diameter of the piles

For accommodating deviations in driving of piles, the size of pile cap is made 300 mm more than the outer distance of the exterior piles.

Minimum cover = 60 mm to 80 mm

Another criterion in arriving at the shape of the pile cap is to arrange the center of gravity of all the piles to coincide with the centroid of the pile cap. Based on these principles, the common shapes of pile caps used for two to nine piles are as shown in Fig. 14.3.

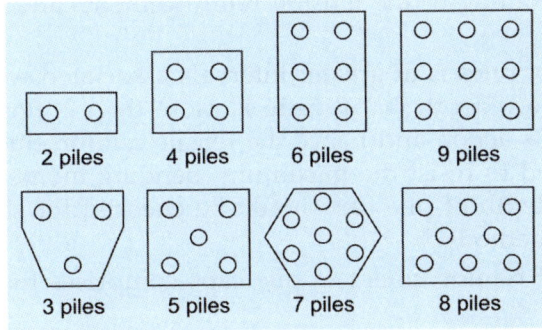

14.3: Typical shapes of pile caps

14.4.2 Depth of Pile Cap

Based on cost analysis, an empirical relation has been recommended by Varghese expressing the thickness of the pile cap as a function of diameter of the pile given by

$$D = (2d_p + 100) \text{ mm for } d_p \text{ not greater than 550 mm}$$
$$D = (1/3)(8d_p + 600) \text{ mm for } d_p \text{ greater than or equal to 550 mm}$$

where, D = overall thickness of pile cap (mm) and d_p = diameter of pile (mm)

14.4.3 Design of Reinforcements in Pile Caps

The transfer of loads from the column to the pile cap and the piles depends upon the structural behaviour of the pile cap under the system of column loads and pile reactions.

The theories that are commonly used in the design of reinforcements in pile caps are grouped as,

a. Truss theory, and
b. Beam theory

Referring to Fig. 14.4a, when the angle of dispersion of load θ is less than 30° (tan 30° = 0.58), the value of shear span/depth ratio (a_v/d) is less than 0.6. Under these conditions, the load is transferred to the piles by strut action shown in Fig.14.4b, where AB is in compression and BC in tension. Experiments have shown that the truss action similar to deep beams and corbels is significant for ratios of (a_v/d) = 2.

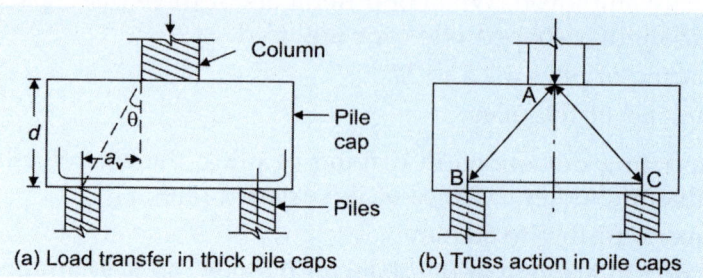

(a) Load transfer in thick pile caps   (b) Truss action in pile caps

14.4: Truss theory of design of pile caps

In truss theory, the tensile force between the pile heads is assumed to be resisted by the reinforcements similar to the tie member of a truss and hence special care should be taken in detailing of the tension reinforcements and its anchorage at the ends.

When the spacing of piles is at greater intervals associated with thinner pile caps in which the shear span/depth (a_v/d) ratio is more than 2, flexural action is more predominant than truss action and hence the tensile reinforcement at the bottom of the pile cap is designed to resist the maximum bending moment as in an ordinary beam. However, the depth of pile cap should be checked for shear when designed by either of the two methods.

The arrangement of reinforcement in pile caps comprises the following types of bars as shown in Fig. 14.5.

a. Main reinforcements located at the bottom of the pile cap in the direction xx bent up at the ends to provide adequate anchorage.

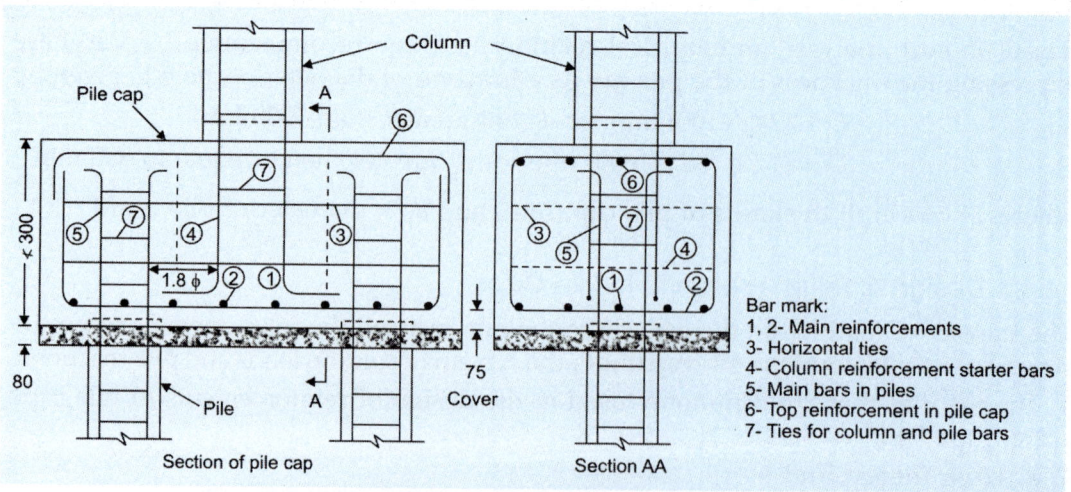

14.5: Reinforcement details in pile caps

b. Main reinforcements placed at the bottom in the direction of yy also bent up at their ends.

c. Horizontal ties comprising of 2 to 3 layers of 16 mm diameter bars as secondary reinforcement to resist bursting.

d. Vertical column starter bars which are L-shaped, located at the level of the bottom main reinforcements.

e. Reinforcements of the pile are extended into the pile cap to provide the required development length in compression.

f. Reinforcements provided as compression steel in the pile cap at the top if required as per computations. They are tied to the bent up bottom bars to form a rigid cage before casting the pile cap.

g. Ties are provided to the pile reinforcements extended into the pile cap and column bars. The reinforcements detailing in pile caps are shown in Fig. 14.5. The plan arrangement of reinforcements used in pile caps with different number of piles is shown in Fig. 14.6.

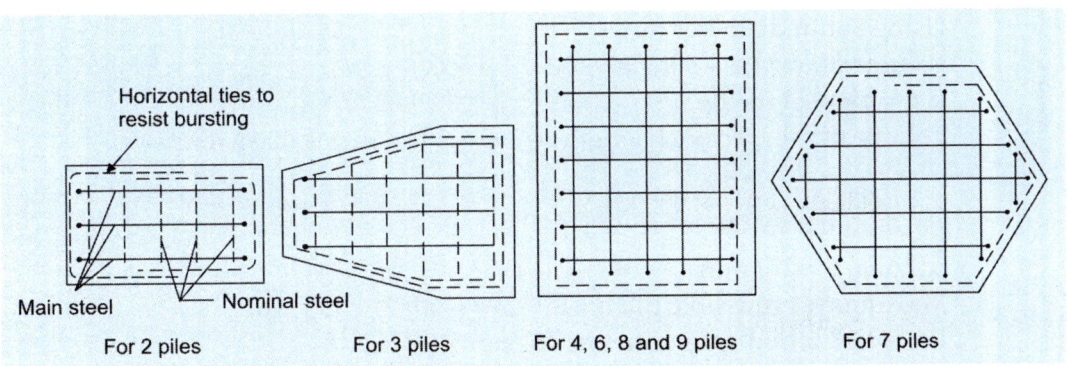

For 2 piles   For 3 piles   For 4, 6, 8 and 9 piles   For 7 piles

14.6: Plan arrangement of reinforcement in pile caps

14.5 DESIGN EXAMPLES OF PILES AND PILE CAPS

1. The foundation for a structure consists of 10 piles to carry a load of 6000 kN. The piles are spaced 1.5 m centres. They are driven through a hard stratum available at a depth of 6 m. Design one of the piles and sketch the details of reinforcements. Adopt M20 grade concrete and Fe 415 HYSD bars.

   i. *Data*:
   Service load on each pile = $(6000/10)$ = 600 kN
   Design ultimate or factored load $P_u = (1.5 \times 600)$ = 900 kN
   Depth of foundation = 6 m
   Materials: M20 grade concrete (f_{ck} = 20 N/mm^2)
   Fe 415 HYSD bars (f_y = 415 N/mm^2)

   ii. *Dimensions of pile*:
   Length of pile above ground level = 0.6 m
   Total length of pile = $(6 + 0.6)$ = 6.6 m
   Assume the cross-sectional dimensions of pile as 300 mm × 300 mm

iii. *Longitudinal reinforcement*:

The pile is designed as a compression member with axial loads.

$$P_u = 0.4f_{ck} A_{st} + (0.67f_y - 0.4f_{ck}) A_{sc}$$

$$(900 \times 10^3) = (0.4 \times 20 \times 300^2) + [(0.67 \times 415) - (0.4 \times 20)] A_{sc}$$

Solving, $A_{sc} = 668$ mm^2

For piles of length, $L = 30D = (30 \times 300) = 9000$ mm

Reinforcement is 1.25% of the cross-sectional area of pile.

$$\therefore \quad A_{sc} = \left(\frac{1.25}{100} \times 300 \times 300 \right) = 1125 \text{ mm}^2$$

Hence provide 4 bars of 20 mm diameter ($A_{sc} = 1256$ mm^2) with a clear cover of 50 mm.

iv. *Lateral reinforcement (ties)*:

Lateral reinforcement in the central portion of the pile = 0.2% of gross volume.

Using 8 mm diameter ties,

volume of one tie = $50[4(300 - 100)] = 40000$ mm^3

If p = pitch of the tie,

volume of pile per pitch length = $(300 \times 300 \times p) = 90000\,p$ mm^3

$$\therefore \quad 40000 = \left[\frac{0.2}{100} \times 90000\,p \right]$$

Solving, $p = 222$ mm

Maximum permissible pitch = $(D/2) = (300/2) = 150$ mm

Provide 8 mm diameter ties at 150 mm centres.

v. *Lateral reinforcement near pile head*:

Spiral reinforcement is provided in the core of the pile for a length of

$$3D = 3 \times 300 = 900 \text{ mm}.$$

Volume of spiral = 0.6% of gross volume

Using 8 mm diameter helical ties ($A_s = 50$ mm^2),

Volume of spiral per mm length = $(0.006 \times 300 \times 300 \times 1) = 540$ mm^3

If p = pitch of spiral, diameter of the spiral

$$d = (300 - 100 - 40) = 160 \text{ mm}$$

$$p = \left(\frac{\text{Circumference of spiral} \times A_s}{540} \right) = \left(\frac{\pi \times 160 \times 50}{540} \right) = 46.51 \text{ mm}$$

Provide 8 mm diameter spiral at a pitch of 45 mm for a length of 900 mm near the pile head.

The spiral is enclosed inside of the main reinforcements.

vi. *Lateral reinforcements near pile ends*:

Volume of ties = 0.6% of gross volume of concrete for a length of

$$3D = (3 \times 300) = 900 \text{ mm}$$

Using 8 mm diameter ties, $A_s = 50$ mm^2.

Volume of each tie = $50[4(300 - 100)] = 40000$ mm^3

If p = pitch of the ties

or $\quad 40000 = (0.006 \times 300 \times 300 \times p)$

Solving, $p = 64$ mm

Provide 8 mm diameter ties at 70 mm centres for a distance of 900 mm from the ends of the pile both at top and bottom.

vii. *Spacer and lifting holes*:

Provide spacer forks in pairs of steel using 25 mm diameter but spacers at 1500 mm centres.

Provide 32 mm diameter holes at 1500 mm from ends.

viii. *Reinforcement details*:

The details of reinforcements in the pile is shown in Fig. 14.7.

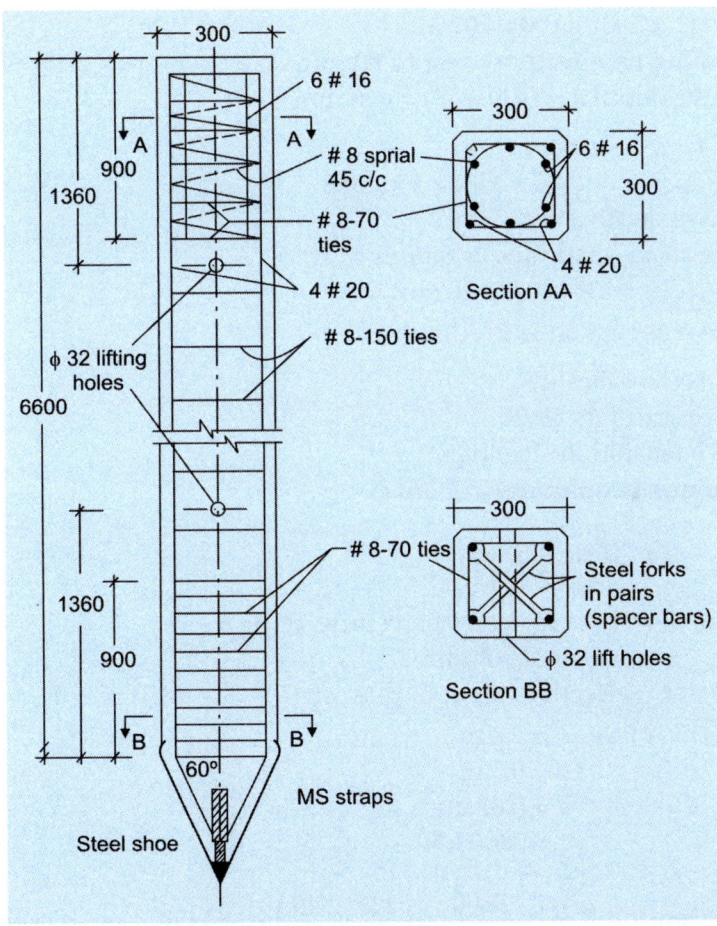

14.7: Details of reinforcements in pile

2. A pile cap consisting of 4 piles of 300 mm × 300 mm is to be designed to support a reinforced concrete 500 mm × 500 mm carrying a load of 200 kN. The piles are located parallel to the column faces with their centres located 600 mm from the centre to the column. Using M20 grade concrete and Fe 415 HYSD bars, design the pile cap and sketch the details of reinforcements.

i. *Data*:

Size of RC column = 500 mm × 500 mm
Size of piles = 300 mm × 300 mm
Service load on column = 200 kN
Factored load = (1.5 × 2000) = 3000 kN
Spacing of piles = 1200 mm centres
f_{ck} = 20 N/mm^2
f_y = 415 N/mm^2
The piles are arranged as shown in Fig. 11.2a.

ii. *Depth of pile cap*:

Total depth of pile cap $D = (2d_p + 100)$ mm
where, d_p = diameter or width of pile = 300 mm
∴ $D = (2 × 300) + 100 = 700$ mm
Assuming an effective cover of 80 mm,
effective depth $d = (700 - 80) = 620$ mm

iii. *Check for truss action*:

Shear span $a_v = (600 - 250) = 350$ mm
Effective depth $d = 620$ mm
Hence shear span/depth ratio is
$(a_v/d) = (350/620) = 0.56 < 0.6$
Hence trust action is predominant.

iv. *Design of tension steel*:

Referring to Fig. 11.2b
Let T = tension in steel bars
for moment equilibrium about A

$$Td = \frac{P_u}{4}\left[\frac{L}{2} - \frac{a}{4}\right]$$

where, L = distance between centres of pile
a = width of column
d = effective depth of pile cap

In this example, $L = 1.2$ m
$a = 0.5$ m
$d = 0.62$ m
$P_u = 3000$ kN

Hence tension $T = \left(\dfrac{3000}{4 × 0.62}\right)\left(\dfrac{1.2}{2} - \dfrac{0.5}{4}\right) = 575$ kN

Area steel $A_{st} = \left(\dfrac{575 × 10^3}{0.87 × 415}\right) = 1592$ mm^2

Adopt 4 bars of 25 mm diameter in *xx* and *yy* directions ($A_{st} = 1964$ mm^2)
within a width of 500 mm (width of column).

Percentage of steel provided

$$p_t = \left(\frac{100\, A_{st}}{b\, d}\right) = \left(\frac{100 \times 1964}{500 \times 620}\right) = 0.63\%$$

v. *Check for moment action*:

Maximum moment at the centre of pile cap

$$M_u = \frac{P_u}{4}\left[\frac{L}{2} - \frac{a}{4}\right]$$

$$= \left(\frac{3000 \times 10^3}{4}\right)\left[\frac{1200}{2} - \frac{500}{4}\right] = 356 \times 10^6 \text{ N·mm}$$

Compute parameter $(M_u/bd^2) = (356 \times 10^6)/(500 \times 620^2) = 1.85$

Refer to Table 2, SP-16 and readout the percentage of reinforcement as $p_t = 0.585$

$$\therefore \qquad A_{st} = \left(\frac{0.585 \times 500 \times 620}{100}\right) = 1814 \text{ mm}^2 < 1964 \text{ mm}^2$$

vi. *Check for shear*:

Nominal shear $V_u = (P_u/4) = (3000/4) = 750$ kN

Nominal shear stress $\tau_v = (V_u/bd) = (750 \times 10^3)/(500 \times 620)$
$$= 0.24 \text{ N/mm}^2$$

Neglected enhanced shear stress, refer to Table 19, IS:456-2000 and readout the permissible shear stress in concrete for M20 grade concrete.

$$\tau_c = 0.52 \text{ N/mm}^2 < \tau_c, \text{ hence safe.}$$

vii. *Reinforcement details*:

The details of reinforcements in the pile is shown in Fig. 14.8.

14.6 DESIGN OF RAFT FOUNDATIONS

Tall structures like multistorey buildings, chimneys and water towers transmit very heavy loads to the foundations through columns. Individual spread footings used for each column in a row are likely to overlap indicating the requirement of larger area to transmit the loads due to limited safe bearing capacity of the soil at site. In such situations raft foundations are generally adopted to support a number of heavily loaded columns on soils with low bearing capacity. The raft foundation comprising of interconnecting beams between the columns with an inverted slab in contact with the soil provides a larger area. The interconnected sill beams with an inverted slab forming the raft foundation provides a solid foundation medium to transmit the loads uniformly to the soil. The elevation and cross-section of a typical raft foundation is shown in Fig. 14.9.

The Indian standard code, IS: 2950-1981(part-1)[14] specifies guidelines for the design of the structural components of raft foundations. The design of a typical raft foundation for a multistory building is illustrated by the following example.

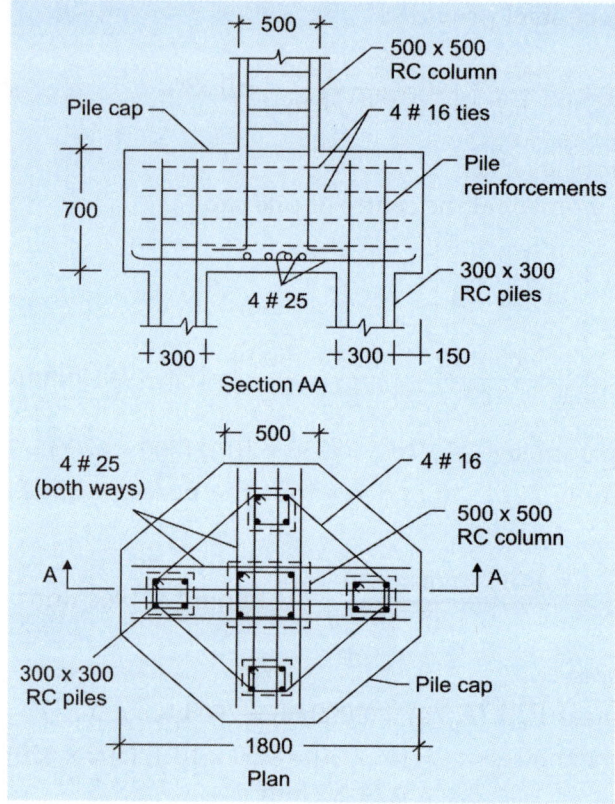

14.8: Reinforcement details in pile cap

Design Example of Raft Foundation

A multistorey building of overall size 12 m × 12 m has 16 reinforced concrete columns of size 300 mm × 300 mm spaced at intervals of 4 m on each side forming a square grid as shown in Fig. 14.9a. Each column transmits a service load of 500 kN at the base. The safe bearing capacity of the soil at site is 100 kN/m². Adopting M20 grade concrete and Fe 415 HYSD reinforcements, design a raft foundation comprising the interconnecting beam between the columns and the inverted slab and sketch the details of reinforcements in the structural elements of the raft. The design should conform to the specifications of Indian standard codes IS:456-2000 and IS:2950-1981.

   i. *Data*:

   Size of building = 12 m × 12 m

   Spacing of columns all round = 4 m intervals

   Service load transmitted = 500 kN on each column

   Size of column = 300 mm × 300 mm

   Safe bearing capacity of soil = 100 kN/m²

   Materials: M20 grade concrete (f_{ck} = 20 N/mm²)

   Fe 415 HYSD bars (f_y = 415 N/mm²)

ii. *Design of raft slab*:

Total working load on all columns = (12 × 500) = 6000 kN

Saft weight of slab and beam (10%)   =   600 kN

Total service load   = 6000 kN

Factored moment in cantilever slab is

$$M_u = (0.5 \times 1.5 \times 83.3 \times 0.6^2) = 22.5 \text{ kN·m}$$
$$V_u = (0.6 \times 83.3) = 50 \text{ kN}$$

Effective depth (required)

$$d = \sqrt{\frac{M_u}{0.138 f_{ck}\, b}} = \sqrt{\frac{22.5 \times 10^6}{0.138 \times 20 \times 10^3}} = 90.3 \text{ mm}$$

Adopt $d = 100$
      $D = 130 \text{ mm}$

Compute parameter $\left(\dfrac{M_u}{b d^2}\right) = \left(\dfrac{22.5 \times 10^6}{10^3 \times 100^2}\right) = 2.25$

Refer to Table 2 (SP:16) and readout the percentage steel as

$$p_t = 0.737 = (100 A_{st})/(b\, d)$$

∴ $A_{st} = \left(\dfrac{0.737 \times 10^3 \times 100}{100}\right) = 737 \text{ mm}^2/\text{m}$

Providing 12 mm diameter bars at 150 mm centres (A_{st} = 754 mm^2)
Distribution bars = (0.0012 × 10^2 × 130) = 156 mm^2
Provide 8 mm diameter bars at 200 mm centres (A_{st} = 251 mm^2)
Check for shear

$$\tau_v = \left(\frac{V_u}{b d}\right) = \left(\frac{50 \times 10^3}{10^3 \times 100}\right) = 0.5 \text{ N/mm}^2$$

$$= \left(\frac{100 A_{st}}{b d}\right) = \left(\frac{100 \times 754}{10^3 \times 100}\right) = 0.754$$

Refer to Table 19, IS:456 and readout τ_c = 0.56 N/mm^2

$$(k_s \cdot \tau_c) = (1.3 \times 0.56) = 0.728 \text{ N/mm}^2 > \tau_v$$

Hence, the slab is safe against shear forces.

iii. *Design of continuous beam over raft slab*:

Maximum service load on beam

$$w = (83.3 \times 15 \times 1) = 125 \text{ kN/m}$$
$$M_u = (1.5 \times 0.1 \times wL^2) = (1.5 \times 0.1 \times 125 \times 4^2) = 300 \text{ kN·m}$$
$$V_u = 1.5(0.6\, wL) = 1.5(0.6 \times 125 \times 4) = 450 \text{ kN}$$

Assuming width of beam $b = 300$ mm

Effective depth $d = \left(\dfrac{M_u}{0.138 f_{ck} b}\right) = \left(\dfrac{300 \times 10^6}{0.138 \times 20 \times 300}\right) = 602$ mm

Adopt $d = 650$ mm and overall depth $D = 700$ mm

Compute $(M_u/bd^2) = (300 \times 10^6)/(300 \times 650^2) = 2.36$

Refer to Table 2 (SP:16) and readout the percentage reinforcement

$\qquad p_t = 0.781 = (100 A_{st})/(bd)$

$\therefore \qquad A_{st} = \left(\dfrac{0.781 \times 300 \times 650}{100}\right) = 1523$ mm^2

Provide 4 bars of 25 mm diameter both at top and bottom to resist the negative moment at supports ($A_{st} = 1963$ mm^2).

iv. *Shear reinforcements*:

$\qquad V_u = 450$ kN

Nominal shear stress

$$\tau_v = \left(\frac{V_u}{bd}\right) = \left(\frac{450 \times 10^3}{300 \times 650}\right)$$

$$= 2.30 \text{ N/mm}^2 < \tau_{c,max} = 2.8 \text{ N/mm}^2$$

$$\left(\frac{100 A_{st}}{bd}\right) = \left(\frac{100 \times 1963}{300 \times 650}\right) = 1.006$$

Refer to Table 19 (IS:456-2000) and readout the permissible shear stress in concrete

$$\tau_c = 0.62 \text{ N/mm}^2 < \tau_v$$

Hence shear reinforcements are to be designed to resist the balance shear given by

$$V_s = [V_u - \tau_c bd]$$
$$= [450 - (0.62 \times 300 \times 650) \, 10^{-3}] = 330 \text{ kN}$$

Using 10 mm diameter four-legged stirrups, spacing

$$S_v = \left[\frac{0.87 f_y A_{sv} \, d}{V_s}\right]$$

$$= \left[\frac{0.87 \times 415 \times 4 \times 78.5 \times 650}{330 \times 10^3}\right] = 223 \text{ mm}$$

Provide 10 mm diameter four-legged stirrups at 200 mm centres.

v. *Reinforcement details*:

The reinforcement details of longitudinal and cross-section of the raft is shown in Figs 14.9b and c.

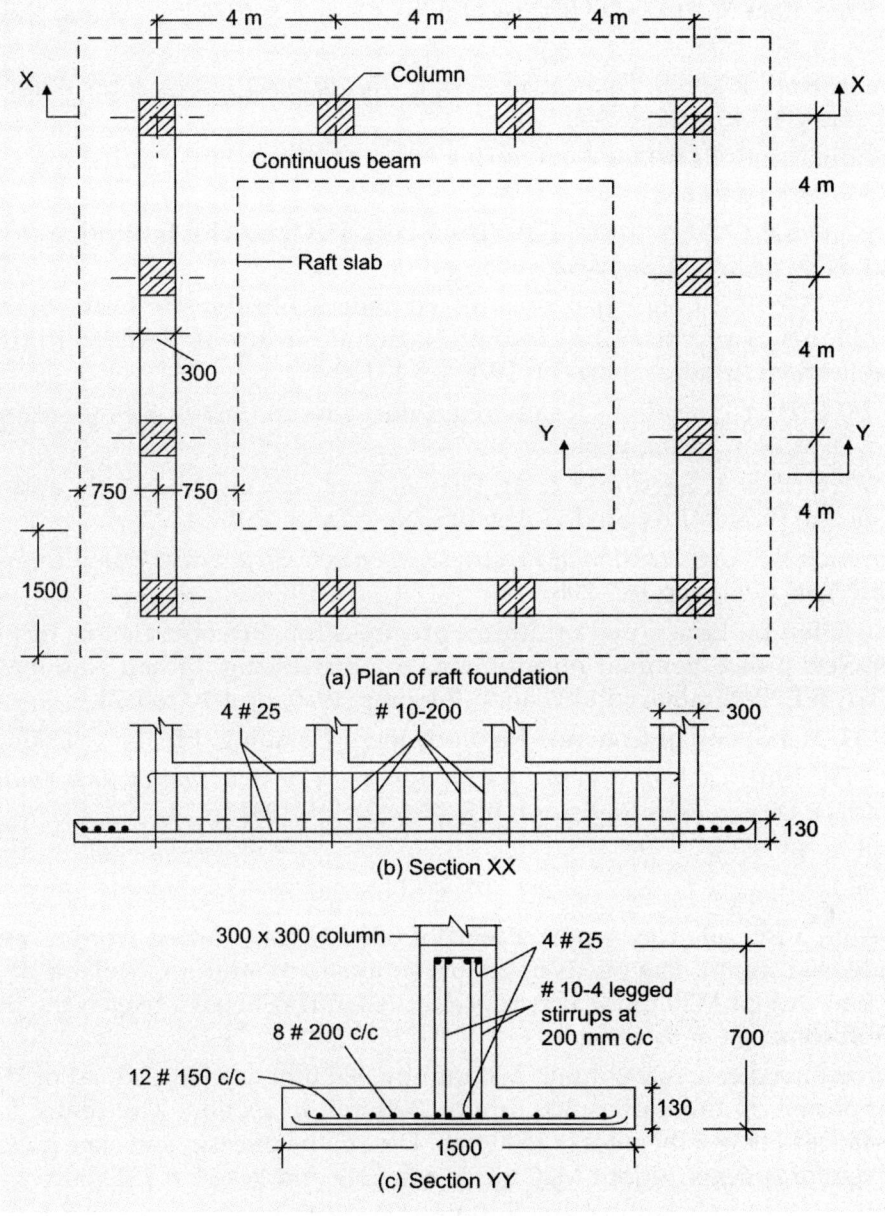

(a) Plan of raft foundation

(b) Section XX

(c) Section YY

14.9: Reinforcement details in raft

REFERENCES

1. Dunham CW, *Foundations of Structures*, 2 edn, McGraw-Hill Book Co, New York, 1962, p 722.

2. Bowles JE, *Foundation Analysis & Design* 5 edn, McGraw-Hill Publishing Company, New York, 1996, p 1175.

3. Peck RB, Hanson WE and Thornburn TH, *Foundation Engineering*, Asia Publishing House, Mumbai, First Indian Edition, 1959, p 410.

4. Krishna Raju N, *Design of Bridges* (4 edn), Oxford & IBH Publishers, New Delhi, 2009, pp 489–506.

5. Kameswara Rao NS, *Foundation Design, Theory and Practice*, John Wiley & Sons Inc, New York, 2011, p 544.

6. Ponnuswamy S, *Bridge Engineering*, Tata McGraw-Hill Publishing Co, New Delhi, 1986, p 544.

7. Krishna Raju N, *Prestressed Concrete*, 5 edn, McGraw-Hill Education (India) Pvt Ltd, New Delhi, 2012, pp 625–678.

8. Subba Rao TN, Long span prestressed concrete bridges in India, Seminar on problems of prestressing, Indian National Group of IABSE, Preliminary Publications, Madras, Jan/Feb 1970, pp 1–130.

9. IS:2911 (Part-1, section-1), Concrete Piles, Driven *Cast-in-Situ* Concrete Piles, BIS, 2010, New Delhi, p 44.

10. IS:2911 (Part-1, Section-3 & 4), Driven Precast Concrete Poles and Precast Concrete Piles in Prebored Holes, BIS, New Delhi, 2010, p 23.

11. Gerwick BC, *Construction of Prestressed Concrete Structures*, Wiley Interscience, New York, 1971, pp 167–208.

12. Shu-T'Ien Li, Functional optimum prestress for different classes of prestress concrete piling, Seminar on problems of prestressing, Indian National Group of IABSE, Preliminary Publication, Madras, 1970, pp I-10 to I-21.

13. SP:34, Handbook of Concrete Reinforcement Detailing, BIS, New Delhi, 1987.

14. IS:2950-1981, *Code of Practice for Design and Construction of Raft Foundations* (part-1), Design (2nd Revision), BIS, New Delhi, 1981, p 24.

ASSIGNMENT

1. Design a RC pile to support a load of 600 kN transmitted from a reinforced concrete column. The pile is to be driven to a hard stratum available at a depth of 9 m. Adopt M20 grade concrete and Fe 415 HYSD bars. Sketch the details of reinforcements in the pile.

2. A reinforced concrete column 500 mm by 500 mm carrying a load of 750 kN is supported on three piles 300 mm by 300 mm in section. The centre to centre distance between the piles is 1500 mm. Design the pile cap and sketch the details of reinforcements. Adopt M20 grade concrete and Fe 415 HYSD bars.

3. The columns of a multistorey building 7 m × 14 m are arranged all round the periphery such that the spacing between the columns is 3.5 m. The columns are 300 mm by 300 mm in cross-section and transmit a load of 400 kN each to the foundations. Design a suitable raft foundation connecting all the columns. Adopt M15 grade concrete and Fe 415 HYSD bars. The safe bearing capacity of soil at site is 80 kN/m². Sketch the details of reinforcements in the raft slab and continuous beam.

4. A typical column of a multistoreyed building transmits a load of 3200 kN to the foundations. This load has to be supported by 4 piles having a square cross-section. The piles are driven through hard stratum and rest on hard rock, 300 mm by 300 mm size precast piles are proposed to be used for the

foundations. Design the reinforcements required for a typical pile assuming the pile to be 8 m long. Adopt M30 grade concrete and Fe 415 grade high yield strength reinforcement. Sketch the typical details of reinforcements in the pile.

5. A reinforced concrete column 400 mm × 600 mm carrying a factored load of 2400 kN is to be supported by six precast piles of length 6 m. The piles are driven through hard gravelly soil and resting on hard strata. Using M20 grade concrete and Fe 415 HYSD bars design the reinforcements required in a typical pile and sketch the details.

6. A reinforced concrete column 400 mm × 400 mm carrying a service load of 800 kN is supported on three piles 300 × 300 mm in section. The centre to centre distance between the piles is 1500 mm. Design the reinforcements in the pile and the pile cap. The length of the piles may be assumed as 6 m bearing on hard rock. Adopt M20 grade concrete and Fe 415 HYSD reinforcement.

7. A pile cap connecting 4 reinforced concrete piles of 300 × 300 mm is to be designed to support an reinforced concrete column 400 mm × 400 mm carrying a service load of 2000 kN. The piles are located parallel to the column faces with their centres located 800 mm from the centre of the column. Using M30 grade concrete and Fe 500 grade reinforcement, design the pile cap and sketch the details of reinforcements.

8. Design a reinforced concrete raft foundation connecting the columns of a multistoreyed building. The columns are arranged in square grid 16 m × 16 m with their spacings 4 m apart. The safe bearing capacity of the soil at site is 120 kN/m². The total service load on all the columns is 4800 kN. The columns are 400 mm × 400 mm in section. Adopt M20 grade concrete and Fe 415 HYSD bars. Sketch the details of reinforcements in the raft foundation.

9. The columns of a multistoreyed building with their centre lines forming a rectangular grid of 10.5 m × 14 m has the columns spaced at 3.5 m centres in the grid. The columns are 300 mm × 300 mm in cross-section and transmit a factored load of 800 kN each to the foundation. The safe bearing capacity of soil at site is 80 kN/m². Adopting M25 grade concrete and Fe 415 HYSD bars, design a suitable raft foundation for the columns and sketch the details of reinforcements in the raft beam and slab.

10. A raft foundation has to be designed to support a mulitistoreyed building with a symmetrical column grid of 5 m × 5 m in perpendicular directions. The overall size of the building is 20 m × 20 m with columns of size 350 mm × 350 mm spaced at 5 m intervals. The total number of columns at ground level is 20. The columns support a total service load of 800 kN each at ground level. Adopting M25 grade concrete and Fe 500 HYSD reinforcements, design a suitable raft foundation for the building assuming the safe bearing capacity of the soil at site as 150 kN/m².

REVIEW QUESTIONS

1. Under what situations you would use pile foundations? What is the advantage of using pile foundations?

2. Discuss briefly the advantages and disadvantages of *cast-in-situ* piles. Under what situations you would prefer *cast-in-situ* piles?

3. What type of piles are used for multistory buildings constructed in tank beds? Explain with sketches the different types piles used in practice.

4. Distinguish between *cast-in-situ* and precast piles. What are the advantages and disadvantages of these types of piles?

5. Explain with sketches the different shapes of precast piles used for foundations.

6. Sketch a typical precast pile showing the various reinforcements used and explain the function of these reinforcements.

7. What are the salient points to be followed in the structural design of a pile? What are the code specifications regarding the longitudinal and lateral reinforcements?

8. What are piles caps and where do you use them? Sketch a typical pile cap to connect nine precast piles of a building foundation showing the various reinforcements.

9. What are the design considerations to be followed in designing pile caps? How do you fix the depth of a pile cap?

10. Explain the necessity of using raft foundations? Under what situations you would resort to the use of raft foundations?

11. Sketch the details of a typical raft foundation connecting 16 reinforced concrete columns spaced 4 m apart arranged in a square grid.

OBJECTIVE QUESTIONS

1. Pile foundations are used to transmit heavy loads from buildings and bridges when the foundation soil at site is classified as
   a. hard gravelly soil
   b. tank beds with clayey soil
   c. soft laterite rock

2. Piles driven into soils in tank beds and bearing on hard strata resist the loads by
   a. bearing only
   b. friction only
   c. combined bearing and friction

3. The codified minimum percentage of longitudinal reinforcement in precast concrete pile having a length greater than 40 times the least lateral dimension should be not less than
   a. 1.5%
   b. 3%
   c. 2%

4. The clear cover to all main reinforcements in a pile should not be less than
   a. 20 mm
   b. 50 mm
   c. 75 mm

5. When groups of piles are used, the minimum spacing between the piles expressed as diameter of the pile d_p should be in the range of
   a. 1.5 to 2 d_p
   b. 3 to 4 d_p
   c. 2.5 to 3 d_p

6. Precast concrete piles should be designed with suitable spiral reinforcement near the pile head with the volume of the spiral steel not less than
   a. 0.15%
   b. 0.4%
   c. 0.6%

7. Pile caps connecting a group of piles should be designed to resist
   a. bending moment
   b. shear force
   c. bending moment and shear force

8. In designing reinforcements in pile caps, truss action will be predominant if the ratio of shear span to depth is less than
   a. 0.7
   b. 0.6
   c. 0.9

9. Precast concrete piles have to be designed for stresses developed due to
   a. flexure
   b. shear
   c. flexure, shear and handling

10. Raft foundations are preferred in situations where
    a. number of columns transmit the load to the foundations
    b. the soil has very low bearing capacity
    c. the bearing area required for each column overlaps that of the adjacent column

Design of Tension Members

15.1 INTRODUCTION

Concrete being weak in tension, requires reinforcements in zones where tensile stresses occur in structural elements. The tensile strength of concrete is about 7% to 10% of the compressive strength. In case of flexural members like beams and eccentrically loaded columns, the cracked concrete in the tension zone is normally neglected. In case of shear and bond, cracked concrete can resist a small amount of tension between the cracks and this factor is considered in most of the national codes. Also in deflection computations, the tension stiffening effect of concrete between the cracks is considered in the estimation of effective flexural rigidity.

In many structural applications like cylindrical water tanks[1], trusses[2], tie members of bow and Vierendeel girders[3] and hopper bottoms of bunkers[4] and silos[5], the concrete section will be entirely subjected to tension or combined bending and tension. Small magnitudes of tension can be managed by the use of steel reinforcements. However, if the magnitude of tension is high, prestressing[6] of such members is commonly adopted as in large water tanks, pipes, trusses and tie members of girders.

The Indian[7], British[8] and American[9] codes have specified guidelines for the design of concrete members subjected to direct tension. The specifications regarding the design of concrete members with direct tension are presented in the following section.

15.2 INDIAN AND BRITISH STANDARD CODE SPECIFICATIONS

Normally in concrete members subjected to direct tension, it is preferable to use concrete of grade M25 and above to achieve better resistance to cracking, although the IS:456-2000 code allows the use of concrete of grades M10 to M20. The permissible stresses in concrete and steel under direct tension and lap length according to IS:456 and British code BS:8007[10] are given in Tables 15.1 and 15.2 respectively.

Table 15.1: Permissible stresses in concrete under direct tension (IS:456 and IS:3370) (values in N/mm²)

Cracking condition	M25	M30	M35	M40	M45	M50
Direct tension						
Cracking not permitted (IS:3370, Part 2:2009)	1.3	1.5	1.6	1.8	2.0	2.1
Cracking permitted (IS:456, Clause B-2.1.1)	3.2	3.6	4.0	4.4	4.8	5.2
Tension due to bending (IS:3370, Part 2:2009)	1.8	2.0	2.2	2.4	2.6	2.8

Table 15.2: Permissible stresses in steel under direct tension and lap length (BS:8007, IS:3370 and IS:456)

Code	Exposure category (minimum grade of concrete and design crack width)	Permissible stress (N/mm²) (lap length)	
		Plain bars Fe 250	High yield strength deformed bars Fe 415
BS:8007-1987	Very severe (M40 and 0.1 mm)	$85(19d_b)$	$100(14d_b)$
BS:8007 and IS:3370 Part 2:2009	Moderate to severe (M30 and 0.3 mm)	$115 (26d_b)$	$130(18.5d_b)$
IS:456-2000 (Clause 35.3.2 and Tables 5 and 22)	Mild (M25 and 0.3 mm)	140 for $d_b \leq 20$ mm $(35d_b)$ 130 for $d_b > 20$ mm $(32.5d_b)$	$230(36d_b)$

15.3 DESIGN OF CONCRETE MEMBERS UNDER DIRECT TENSION

15.3.1 Design Principles

Concrete being weak in tension cannot withstand tensile loads without cracking. Tests of reinforced concrete members in tension, carried out by Morsch[11] in 1908 indicated that uncracked concrete between two adjacent cracks was able to decrease the extent to which steel reinforcement is stretched as compared to only steel bars. This effect was referred to as *tension stiffening* and was attributed to the bond between the steel and concrete. The load deformation characteristics of a tension member as reported by Subramanian[12] are shown in Fig. 15.1.

When a reinforced concrete member is subjected to direct tension, the member is initially uncracked and the load deformation characteristic is linear as indicated by the line OX. Beyond this point for increasing loads, the concrete suffers initial and continued cracking in the range XY and the stabilised crack development in the zone YZ which is non linear. The values N_s and N_y in Fig. 15.1 represent the average applied load shared by steel and concrete. In this way after cracking, the response of the member depends upon the stiffness of the reinforcing bar and there is a gradual transition of the full load being shared by the steel with more cracks developing in concrete. Once the cracking stage has stabilised, the load shared by concrete decreases as secondary cracks develop between the primary cracks as reported by Bischoff[13] and Beeby and Scott[14].

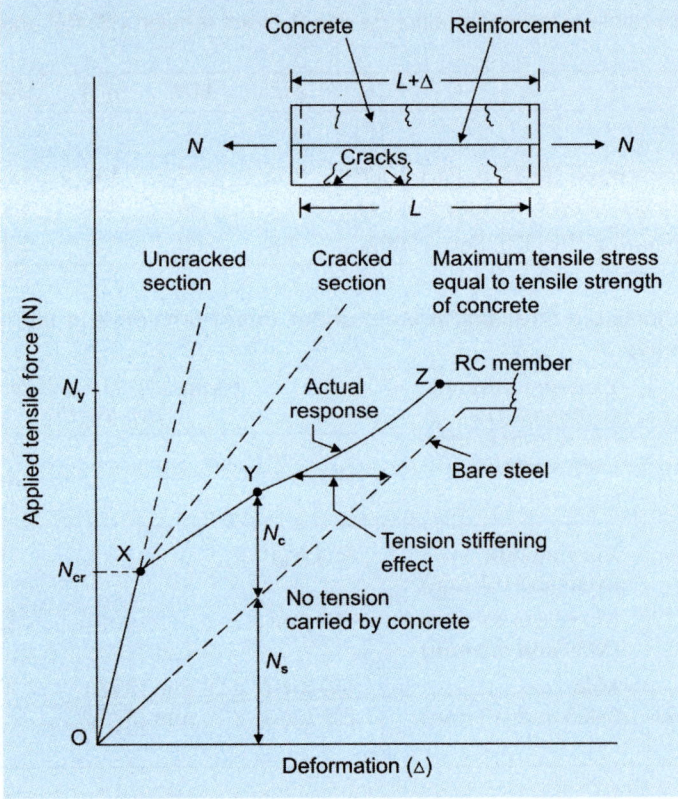

Fig. 15.1: Load deformation characteristics of tension members

The tension stiffening effect shown in figure was found to be time dependent and it decreases with increase in the degree of cracking. Wu and Gilbert[15] have studied the tension stiffening effect in reinforced concrete members under short and long term loads.

The design of tension members is governed by the following criteria as specified by the various standard codes. In general, the design is governed by the limit state of serviceability rather than strength.

1. The total tensile force on the member is considered to be resisted by the reinforcement and the force resisted by concrete is ignored being very negligible due to the low tensile strength of concrete.

2. The stress in the reinforcement is curtailed to be less than the allowable stress in the serviceability limit state.

3. The allowable stress in concrete based on the transformed section should be less than the allowable tension in concrete in serviceability limit state to prevent excessive cracking.

Based on the above criteria, the design equations are expressed as follows:

$$A_{st} \geq \left[\frac{F_u}{0.87 f_y}\right] \text{ for limit state design} \qquad (15.1)$$

$$A_{st} \geq \left[\frac{F_t}{\sigma_{st}}\right] \text{ for working stress design} \qquad (15.2)$$

$$\left[\frac{F}{A_c + (m-1)A_{st}}\right] \leq \sigma_{ct} \qquad (15.3)$$

where,

F_t = axial service load force

F_u = factored axial force

A_{st} = area of tensile reinforcement

A_c = area of concrete section

f_y = yield strength of the reinforcement

σ_{st} = allowable tensile stress in steel

σ_{ct} = allowable direct tensile stress in concrete

m = modular ratio = $(280/3\sigma_{cbc})$ or (E_s/E_c)

15.3.2 Design Procedure

The following steps are useful in the design of tension members:

1. Compute the area of tensile reinforcement required by using Eqs (15.1 and 15.2). In general, the latter equation governs the design. It is important to note that this area of steel should be greater than the minimum area of steel based on Eq.(15.3).
2. The area of concrete section required is determined by Eq. (15.3) using the value of f_{act} from Table 15.1 depending upon the exposure conditions and extent of permissible cracking.
3. Minimum area of steel and cover as prescribed in the codes should be ensured.
4. If limit state design is adopted for sizing of the member, crack width should be evaluated and checked with permissible values.

15.4 DESIGN EXAMPLES

1. Ring beam of a water tank

Design the ring beam of spherical dome of a circular water tank using the following data:

Hoop tension in the ring beam at service loads = 110 kN

Grade of concrete = M20

Type of reinforcement = Fe 415 HYSD bars

Permissible stresses should comply with the values recommended in IS:456-2000 code.

    i. *Data*:

        Tension in ring beam F_t = 110 kN

        For M20 grade concrete and Fe 415 HYSD bars

        Permissible tensile stress in concrete σ_{ct} = 2.8 N/mm²

        Permissible stress in reinforcements σ_{st} = 150 N/mm²

ii. *Area of steel and concrete*:

$$A_{st} = [(110 \times 1000)/150] = 734 \text{ mm}^2$$

Provide 4 bars of 16 mm diameter ($A_{st} = 804 \text{ mm}^2$)

Let A_c be the cross-sectional area of ring beam.

Allowing a tensile stress of 2.8 N/mm^2 in concrete, we have the relation

$$\left[\frac{F_t}{A_c + (m-1)A_{st}} \right] = \left[\frac{110 \times 10^3}{A_c + (13-1)804} \right] = 2.8$$

Solving, $A_c = 29638 \text{ mm}^2$

Adopt a ring beam of size 200 mm × 200 mm with 4 bars of 16 mm diameter as hoop reinforcement and stirrups of 6 mm diameter at 150 mm centres.

2. Tie member of a bow string girder

The vertical tie member of a bow string girder used for a highway bridge crossing is subjected to a working tensile force of 200 kN. Design a suitable cross-section and reinforcements for the tie using M25 grade concrete and Fe 415 HYSD bars. Sketch the details of reinforcements in the tie and its connection to the main horizontal girder of the bow string truss.

i. *Data*:

Tension in tie member $F_t = 200 \text{ kN}$

For M25 grade concrete and Fe 415 HYSD bars

Permissible tensile stress in concrete $\sigma_{ct} = 3.2 \text{ N/mm}^2$

Permissible stress in reinforcements $\sigma_{st} = 150 \text{ N/mm}^2$

Modular ratio $m = (280/3\sigma_{cbc}) = (280/3 \times 8.5) = 11$

ii. *Area of steel and concrete*:

$$A_{st} = [(200 \times 1000)/150] = 1333 \text{ mm}^2$$

Provide 8 bars of 16 mm diameter ($A_{st} = 1608 \text{ mm}^2$)

Let A_c be the cross-sectional area of ring beam.

Allowing a tensile stress of 2.8 N/mm^2 in concrete, we have the relation

$$\left[\frac{F_t}{A_c + (m-1)A_{st}} \right] = \left[\frac{200 \times 10^3}{A_c + (11-1)1608} \right] = 3.2$$

Solving, $A_c = 46420 \text{ mm}^2$

Adopt a tie member of size 200 mm × 250 mm with 8 bars of 16 mm diameter and stirrups of 6 mm diameter at 150 mm centres.

iii. *Detailing of reinforcements*:

The tie member of size 200 mm by 250 mm is reinforced with 8 bars of 16 mm diameter arranged symmetrically around the periphery. The main reinforcements in the tie are anchored into the main girder of the bow string truss by either of the two methods shown in Fig. 15.2.

Anchorage for the tension member reinforcement can be ensured using rebars embedded into the main horizontal girder or by hooked anchor bars.

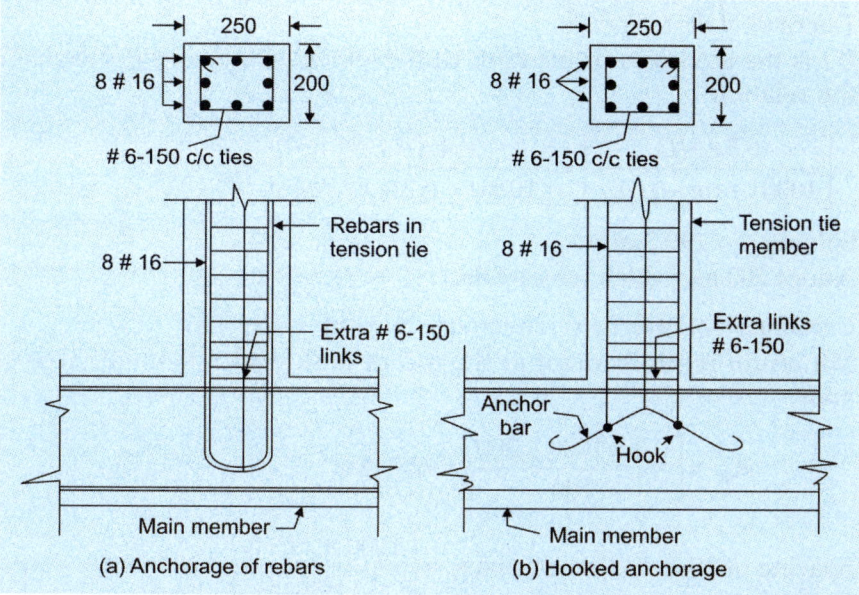

Fig. 15.2: Methods of anchorage used in tension tie members of bow string girder

3. Circular wall of water tank

Design the reinforcements for the circular wall of a water tank resting on ground with a flexible base. The diameter of the water tank is 12.6 m and depth of water storage is 4 m. Adopt M20 grade concrete and Fe 415 HYSD bars. The design should conform to the specifications of the Indian standard codes IS:3370 and IS:456-2000.

    i. *Data*:
      Diameter of water tank = 12.6 m
      Height of water storage = 4 m
      Free board = 200 mm
      Materials: M20 grade concrete
              Fe 415 HYSD bars

   ii. *Permissible stresses*:
              σ_{ct} = 1.2 N/mm^2 (for tank walls)
              σ_{st} = 150 N/mm^2
      Modular ratio m = 13

  iii. *Reinforcements in tank walls*:
      Total height of tank wall H = (4 + 0.2) = 4.2 m
      Maximum hoop tension = (0.5 wHD) = (0.5 × 10 × 4.2 × 12.60) = 264.6 kN/m
      Tension reinforcement per metre height of wall is given by

$$A_{st} = \left[\frac{264.6 \times 10^3}{150}\right] = 1764 \text{ mm}^2$$

Using 12 mm diameter bars on both faces of the wall

$$\text{Spacing} = \left[\frac{1000 \times 113 \times 2}{1764}\right] = 128 \text{ mm}$$

Provide 12 mm bars at 120 mm centres at the base of the tank on either face.

iv. *Thickness of the tank wall*:

If t is the thickness of the tank wall, from cracking considerations, we have the relation

$$\left[\frac{0.5\,wHD}{1000t + (m-1)\,A_{st}}\right] = \left[\frac{264.6 \times 10^3}{1000t + (13-1)1764}\right] = \sigma_{ct} = 1.2$$

Solving, $t = 199.3$ mm

Adopt 200 mm thick tank walls.

v. *Curtailment of main hoop reinforcement in tank walls*:

Minimum reinforcement at the top of tank wall = 0.3% of cross-sectional area

$$A_{st} = \left[\frac{0.3 \times 1000 \times 200}{100}\right] = 600 \text{ mm}^2$$

Spacing of 10 mm diameter bars $= \left[\dfrac{1000 \times 79 \times 2}{600}\right] = 263$ mm

Provide 10 mm diameter bars at 250 mm centres on both faces for a height of 1 m from the top of the tank.

Similarly at mid height (2 m below top), provide 12 mm diameter bars at 250 mm centres.

vi. *Distribution and temperature reinforcements in tank wall*:

Area of vertical distribution reinforcement = 0.3% = 600 mm²

Provide 10 mm diameter bars at 250 mm centres on both faces.

The details of reinforcements in the tank wall are shown in Fig. 15.3.

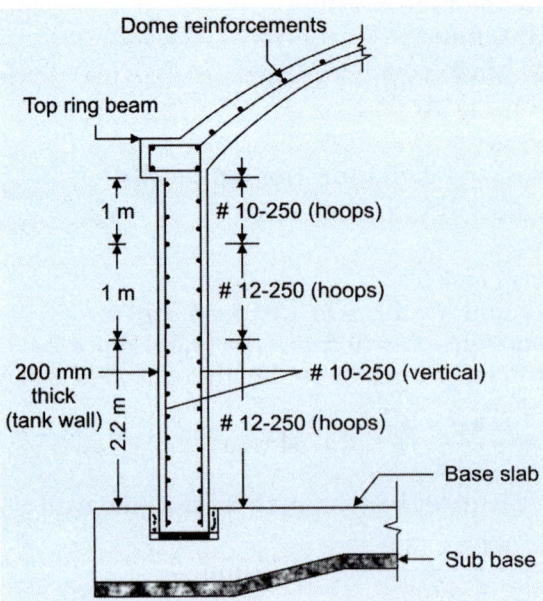

Fig. 15.3: Reinforcement details in tank wall

4. Reinforced concrete water pressure pipe

A reinforced concrete pipe has to be designed to carry water under a pressure intensity of 75 kN/m². The internal diameter of the pipe is 1 m. Design a suitable section and hoop reinforcements for the water main adopting M30 grade concrete and Fe 415 HYSD bars. The design should comply with the specifications of the Indian standard codes IS:3370-1965 and IS:456-2000. Sketch the details of reinforcements in the cross-section of the pipe.

i. *Data*:

Intensity of water pressure $p = 75$ kN/m²
Diameter of the pipe, $D = 1$ m
Hoop tension in the pipe $F_t = (0.5pD) = (0.5 \times 75 \times 1) = 37.5$ kN/m
Materials: M30 grade concrete
            Fe 415 HYSD bars
Modular ratio $m = [280/(3 \times 10)] = 9.3$
Permissible direct tensile stress in concrete $= 1.5$ N/mm² (for M30 grade concrete)

ii. *Area of steel reinforcement*:

$$A_{st} = \left[\frac{37.5 \times 10^3}{150}\right] = 250 \text{ mm}^2/\text{m}$$

Provide 8 mm diameter hoop bars at 200 mm centres ($A_{st} = 250$ mm²).

iii. *Thickness of concrete pipe*:

$$\left[\frac{F_t}{1000t + (m-1)A_{st}}\right] = \left[\frac{37.5 \times 10^3}{1000t + (9.3-1)250}\right] = \sigma_{ct} = 1.5$$

Solving, $t = 22.925$ mm which is impracticable.
Provide a minimum thickness of 75 mm with 30 mm cover for reinforcement.

iv. *Distribution and temperature reinforcements in pipe*:

Area of vertical distribution reinforcement = 0.3% of the cross sectional area

$$A_{st} = (0.003 \times \pi \times 1000 \times 75) = 706.5 \text{ mm}^2$$

Provide 10 bars of 10 mm diameter distributed around the circumference. The details of reinforcements in the pipe are shown in Fig. 15.4.

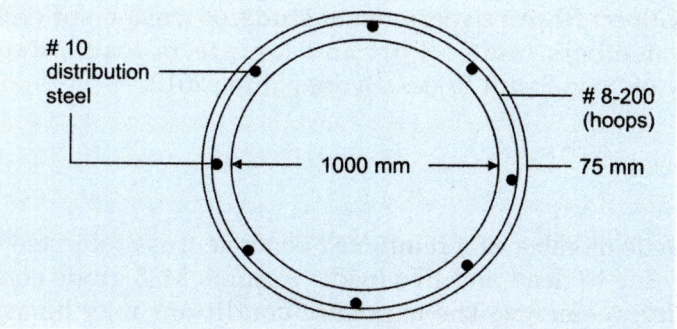

Fig. 15.4: Details of reinforcements in pipe

REFERENCES

1. IS:3370-1965 (Part-I), Indian Standard Code of Practice for Concrete Structures for Storage of Liquids, BIS, 1965, New Delhi, pp 3–14.

2. IS:3201-1965, Indian Standard Code of Practice for Design and Construction of Precast Concrete Trusses, BIS, New Delhi, 1965.

3. Krishna Raju N, *Advanced Reinforced Concrete Design*, 3 edn, CBS Publishers & Distributors, New Delhi, 2016.

4. IS:4995-1974, Criteria for Design of Reinforced Concrete Bins for the Storage of Granular and Powdery Materials, BIS, New Delhi, 1974.

5. Sargis S Safarian, Ernest C Haris, *Design and Construction of Silos and Bunkers*, Von Nostrand Reinhold Co, 1985, p 468.

6. Krishna Raju N, *Prestressed Concrete*, 5 edn, McGraw Hill Education (India) Private Ltd, New Delhi, 2012, pp 489–520.

7. IS:456-2000, Indian Standard Code of Practice for Plain and Reinforced Concrete (4th Revision), BIS, New Delhi, July 2000.

8. BS EN:1992-1-1, Euro Code-2, Design of Concrete Structures, General Rules & Rules for Buildings, British Standards Institution, London, 2004.

9. ACI:318M-11, Building code requirements for structural concrete, American Concrete Institute, Farmington Hills, Michigan, USA, 2005.

10. BS:8007-1987, Code of Practice for Design of Concrete Structures Retaining Aqueous Liquids, British Standards Institution, London, 1987, p 31.

11. Morsch E and Goodrich EP, Concrete and Steel Construction, Research Reports, London, 1909.

12. Subramanian N, *Design of Reinforced Concrete Structures*, Oxford University Press, New Delhi, 2013, pp 727–742.

13. Bischoff PH, Re-evaluation of deflection prediction for concrete beams reinforced with steel and fiber reinforced polymer bars, *Journal of Structural Engineering*, American Society of Civil Engineers, Vol 131, No. 5, 2005, pp 752–767.

14. Beeby AW and Scott RH, Cracking and deformation of axially reinforced members subjected to pure tension, Magazine of Concrete Research, London, Vol. 57, No. 10, 2005, pp 611–621.

15. Wu HQ, Gilbert RI, An experimental study of tension stiffening in reinforced concrete members under short and long term loads. Report No. R-449, University of New South Wales, Sydney, June 2012.

ASSIGNMENT

1. Design the tie member of a reinforced concrete truss subjected to a tensile force of 450 kN due to dead and live loads. Assume M35 grade concrete and Fe 415 HYSD reinforcements. The exposure conditions may be assumed as mild. Sketch the details of reinforcements in the cross-section of the tie member.

2. Design the walls of a circular water tank (with a sliding joint at base) subjected to a hoop tension of 250 kN/m. Adopt M30 grade concrete and Fe 415 grade HYSD reinforcements. The exposure conditions may be assumed as mild.

3. A reinforced concrete water pipe line of 400 mm radius is to be designed to carry water at a working pressure intensity of 70 kN/m². Design a suitable thickness and hoop steel in the pipe section. Adopt M30 grade concrete and Fe 415 HYSD bars. The design should conform to the specifications of the Indian standard codes IS:3370 and IS:456. Sketch the cross-section of the pipe showing the details of reinforcements.

4. The tension tie member of a fink type reinforced concrete truss is subjected to a tensile force of 320 kN due to the various types of loads acting on the truss. The size of the tie is fixed as 250 mm by 300 mm. Adopting M25 grade concrete and Fe 500 grade HYSD bars, design suitable reinforcements in the tie member.

5. A circular overhead water tank of 12 m diameter is covered by a spherical dome supported on a ring girder. The meridional thrust transferred from the dome to the ring girder has been estimated as 22 kN/m at an angle of 36° to the horizontal. Adopting M30 grade concrete and Fe 415 HYSD bars, design a suitable concrete section and reinforcements for the ring girder.

6. Design a suitable reinforced concrete pipe to carry water under pressure of 10 m head of water for an urban water supply system. The radius of the pipe is restricted to 600 mm. Adopt concrete of grade M30 with Fe 415 HYSD bars. The permissible stress in steel should not exceed 130 N/mm² and the direct tensile stress in concrete is restricted to 1.5 N/mm² and no cracks are permitted. Design suitable reinforcements in the pipe and sketch the details of reinforcements in the cross-section of the pipe.

REVIEW QUESTIONS

1. List with practical examples where reinforced concrete members are subjected to direct tensile stresses.

2. Explain with sketches the load deformation characteristics of reinforced concrete members subjected to direct tension.

3. Explain the terms: (a) Tensile strength (b) Tension stiffening effect (c) Hoop reinforcements with reference to reinforced concrete members subjected to direct tension.

4. Explain the reasons for adopting the criteria of limit state of serviceability in the design of reinforced concrete members subjected to direct tension.

5. What special precautions would you take in detailing of reinforcements in reinforced members subjected to direct tension?

6. Explain the salient aspects of the Indian standard code (IS:456-2000) provisions regarding anchoring of reinforcing bars in tension.

7. Explain with sketches the method used to anchor reinforcements in tension tie members of bow string girder truss.

8. What are the criteria considered while designing reinforced concrete members subjected to direct tension?

9. Discuss the classification of exposure conditions used in Indian and British standard code specifications. What are the crack widths allowed in each of the exposure conditions from durability considerations?

10. Explain the advantages of using reinforced concrete members for the construction of trusses and water supply pipes in coastal regions.

OBJECTIVE QUESTIONS

1. The tensile strength of concrete expressed as a percentage of its compressive strength is normally in the range of
   a. 20 to 25%
   b. 5 to 8 %
   c. 7 to 10%

2. In the design of reinforced concrete members subjected to direct tension, tensile strength of concrete is
   a. partially considered
   b. neglected
   c. fully included

3. According to the specifications of the Indian standard code IS:3370, part-2, the direct tensile strength of immature concrete for M30 grade concrete is
   a. 1.15
   b. 1.30
   c. 1.70

4. According to the Indian standard code IS:456-2000, the permissible stress in Fe 250 grade plain steel bars of diameter 25 mm expressed in N/mm^2, under direct tension is limited to
   a. 230
   b. 100
   c. 130

5. In designing reinforced concrete members subjected to direct tension, the limit state that is normally considered is
   a. strength
   b. serviceability
   c. both

6. In a cracked reinforced concrete tension member, the concrete between the cracks carries the tensile stresses that are transferred by the mechanism of
   a. slip
   b. bond
   c. dilatancy

7. The permissible stress in direct tension in Fe 415 grade HYSD reinforcements as per the specifications of IS:456-2000 is
   a. $110 N/mm^2$
   b. $360 N/mm^2$
   c. $30 N/mm^2$

8. According to the Indian standard code IS:456-2000, the permissible stress in concrete expressed in N/mm^2, under direct tension for M25 grade concrete when cracking is permitted is
   a. 1.5
   b. 4.0
   c. 3.2
9. The minimum nominal cover specified in Indian standard code IS:456-2000 for severe exposure conditions is
   a. 45 mm
   b. 30 mm
   c. 75 mm
10. The minimum thickness of reinforced concrete pipes used for conveying water under pressure based on practical considerations should be not less than
    a. 30 mm
    b. 75 mm
    c. 150 mm

Working Stress Method of Design

16.1 INTRODUCTION

During the early part of the 20th century, the design of structural concrete elements started with the introduction of working stress method based on the elastic theory proposed by the pioneers like Turneaure[1], Taylor[2], Morsch[3], Whitney[4] and Hool[5]. Many investigators during this period worked incessantly to establish the design principles for reinforced concrete by introducing the concept of working or permissible stresses derived by applying appropriate safety factors[6] to the ultimate strength of concrete and steel. The resulting permissible stresses[7] are well within the linear elastic range of the materials and ensured crack free structural elements.

The designs based on the working stress method although ensures safety of the structure at service or working loads, it does not provide a realistic estimate of the ultimate or collapse load of the structure in contrast to the limit state method of design[8,9]. The working stress method of design results in comparatively larger and conservative sections of the structural elements with higher quantities of steel reinforcements which will boost up the overall cost of the structure. Structural engineers have used this method of design extensively during the 20th century and presently the method is incorporated as an alternative to the limit state method of design in Annexure-B of the Indian standard code, IS:456-2000[10] for optional use.

16.2 WORKING OR PERMISSIBLE STRESSES

The permissible or working stress in concrete under compression, flexure, and bond is shown in Table 2.1 (IS:456-2000, Table 21) and the working stresses in steel reinforcement are given in Table 2.2 (IS:456-2000, Table 22). The permissible shear stresses in concrete for various grades is shown in Table 16.1 (IS:456-2000, Table 23). The maximum shear stress permissible in concrete for different grades of concrete in beams is shown in Table 16.2 (IS:456-2000, Table 24). In the case of reinforced concrete solid slabs, the permissible shear stress in concrete is obtained by multiplying the values given in Table 16.2 by a factor k whose values depend upon the thickness of the slab as shown in Table 16.3 (IS:456-2000, Clause B-5.2.1.1).

Table 16.1: Permissible shear stresses in concrete τ_c (N/mm^2) (IS:456-2000, Table 23)

$100\dfrac{A_s}{bd}$	Permissible shear stresses in concrete, τ_c (N/mm^2) Grade of concrete					
	M15	M20	M25	M30	M35	M40 and above
$\leq$ 0.15	0.18	0.18	0.19	0.20	0.20	0.20
0.25	0.22	0.22	0.23	0.23	0.23	0.23
0.50	0.29	0.30	0.31	0.31	0.31	0.32
0.75	0.34	0.35	0.36	0.37	0.37	0.38
1.00	0.37	0.39	0.40	0.41	0.42	0.42
1.25	0.40	0.42	0.44	0.45	0.45	0.46
1.50	0.42	0.45	0.46	0.48	0.49	0.49
1.75	0.44	0.47	0.49	0.50	0.52	0.52
2.00	0.44	0.49	0.51	0.53	0.54	0.55
2.25	0.44	0.51	0.53	0.55	0.56	0.57
2.50	0.44	0.51	0.55	0.57	0.58	0.60
2.75	0.44	0.51	0.56	0.58	0.60	0.62
3.00 and above	0.44	0.51	0.57	0.60	0.62	0.63

Note: A_s is that area of longitudinal tension reinforcement which continues at least one effective depth beyond the section being considered except at supports where the full area of tension reinforcement may be used provided the detailing conforms to Clause 26.2.2 and 26.2.3, IS:456-2000.

Table 16.2: Maximum shear stresses $\tau_{c,max}$ (N/mm^2) (IS:456-2000, Table 24)

Concrete grade	M15	M25	M30	M35	M40 and above
($\tau_{c,\,max}$, N/mm^2)	1.6	1.8	1.9	2.3	2.5

Table 16.3: Multiplying factor for slabs (IS:456-2000, Clause B-5.2.1.1)

Overall depth of slab (mm)	300 or more	275	250	225	200	175	150 or less
Multiplying factor k	1.00	1.05	1.10	1.15	1.20	1.25	1.30

16.3 GENERAL DESIGN PROCEDURE

In working stress method of design of structural concrete members, the cross-sectional dimensions are generally assumed based on the basic span/depth ratios given in Table 8.1. The service load moments and shear forces are computed at critical sections and the adequacy of the depth of the section is checked by using the relation,

$$d = \sqrt{\frac{M}{Qb}}$$

where,

d = effective depth of the section

M = working load moment

b = width or breadth of section

Q = a constant depending upon the permissible stresses, neutral axis depth factor (k) and lever arm coefficient (j).

Values of design constant Q for different grades of concrete and types of steel are given in Table 2.3. After satisfying the depth criterion, the area of reinforcement required in the section is evaluated by using the relation

$$A_{st} = \left(\frac{M}{\sigma_{st} jd} \right)$$

The section is reinforced with suitable number of steel bars with due regard to spacing of bars and cover requirements. The section is generally checked for resistance against shear by computing the nominal shear stress τ_v using the relation

$$\tau_v = \left(\frac{V}{bd} \right)$$

where, V = service load shear force at the section.

The permissible shear stress in concrete (τ_c) is influenced by the percentage reinforcement in the section and the grade of concrete as compiled in Table 16.1 (Table 23, IS:456-2000).

If the nominal shear stress, exceeds the permissible shear stress, suitable shear reinforcements are designed in members using the relation

$$s_v = \left(\frac{\sigma_{sv} A_{sv} d}{V_s} \right)$$

where,

s_v = spacing of stirrups
a_{sv} = cross-sectional area of stirrup legs
σ_{sv} = permissible stress in steel reinforcement
d = effective depth
V_s = working load shear force at the section
  = $(V - t_c bd)$

If $\tau_v < \tau_c$, nominal shear reinforcements are provided in beams at a spacing given by

$$s_v = \left[\frac{0.87 f_y A_{sv}}{0.4b} \right]$$

The permissible average bond stress for plain bars in tension as specified in the Indian standard code, IS:456-2000 for different grades of concrete are given in Table 16.4.

Table 16.4: Permissible bond stress in tension (IS:456-2000, Table 21)

Concrete grade	M15	M20	M25	M30	M35	M40	M45	M50
Bond stress τ_{bd} (N/mm^2)	0.6	0.8	0.9	1.0	1.1	1.2	1.3	1.4

In case of compression members, the axial load permissible on a short column reinforced with longitudinal bars and lateral ties is given by

$$P = [\sigma_{cc} A_{c} + \sigma_{sc}]$$

where,

σ_{cc} = permissible stress in concrete in direct compression (refer to Table 2.1)
A_c = cross-sectional area of concrete excluding the area of reinforcements
σ_{sc} = permissible compressive stress in reinforcement
A_{sc} = cross-sectional area of longitudinal steel bars

16.4 DESIGN OF SLABS

1. *Design example of a one-way slab*: Design a one-way slab for a residential floor using the following data:

   i. *Data*:
      Clear span = 2.5 m
      Slab supported on load bearing brick walls is 230 mm thick
      Loading: Residential floor, 2 kN/m^2
      Materials: M20 grade concrete
                       Fe 415 HYSD bars

   ii. *Allowable stresses*:
      $\sigma_{cbc} = 7\,\text{N/mm}^2$ $Q = 0.91$
      $\sigma_{st} = 230\,\text{N/mm}^2$ $j = 0.90$

   iii. *Depth of slab*:
      Assuming 0.4% of reinforcement in the slab, the value of K_t (Fig. 5.1) using Fe 415 HYSD bars, is around 1.25.
      Hence $(L/d) = (L/d)_{basic} \times K_t \times K_c$
                      $= (20 \times 1.25 \times 1) = 25$
      or $d = (2500/25) = 100$ mm
      Adopt $d = 100$ mm and overall depth = 130 mm

   iv. *Effective span*:
      Effective span is the least of
      (a) centre to centre of support = (2.5 + 0.23) = 2.73 m
      (b) clear span + effective depth = (2.5 + 0.10) = 2.60 m
      $\therefore$ Effective span $L = 2.60$ m

   v. *Loads*:
      Self-weight of slab = (0.13 × 25) = 3.25 kN/m^2
      Live load on floor   = 2.00 kN/m^2
      Floor finishes   = 0.75 kN/m^2
      Total load w = 6.00 kN/m^2
      Considering 1 m width of slab, the uniformly distributed load is 6 kN/m^2 on an effective span of 2.60 m.

   vi. *Bending moments and shear forces*:
      $M = (0.125\,wL^2) = (0.125 \times 6 \times 2.6^2) = 5.07$ kN·m
      $V = (0.5\,wL) = (0.5 \times 6 \times 2.6) = 7.80$ kN

vii. *Effective depth*:

$$d = \sqrt{\frac{M}{Qb}} = \sqrt{\frac{5.07 \times 10^6}{0.91 \times 10^3}} = 75 \text{ mm}$$

Effective depth, adopted $d = 100$ mm, hence safe.

viii. *Main reinforcements*:

$$A_{st} = \left(\frac{M}{\sigma_{st} jd}\right) = \left(\frac{5.07 \times 10^6}{230 \times 0.9 \times 100}\right) = 245 \text{ mm}^2$$

Minimum reinforcement $= (0.0012 \times 130 \times 1000) = 156 \text{ mm}^2 < 245 \text{ mm}^2$

Spacing of 10 mm diameter bars is given by

$$s = \left(\frac{1000 \, a_{st}}{A_{st}}\right) = \left(\frac{1000 \times 79}{245}\right) = 322 \text{ mm}$$

Provide 10 mm diameter bars at 300 mm centres ($A_{st} = 262 \text{ mm}^2$)

ix. *Distribution reinforcement*:

$$A_{st} = (0.0012 \times 1000 \times 130) = 156 \text{ mm}^2$$

Provide 8 mm diameter bars at 300 mm centres ($A_{st} = 167 \text{ mm}^2$).

x. *Check for shear stress*:

$$\tau_v = \left(\frac{V}{bd}\right) = \left(\frac{7.80 \times 10^3}{10^3 \times 100}\right) = 0.078 \text{ N/mm}^2$$

Assuming 50% of reinforcement to be bent up near supports, we have

$$\left(\frac{100 \, A_{st}}{bd}\right) = \left(\frac{100 \times 0.5 \times 262}{1000 \times 100}\right) = 0.131$$

From Table 23 (IS:456-2000), interpolating permissible shear stress for solid slab

$$(k\tau_c) = (1.30 \times 0.18) = 0.234 \text{ N/mm}^2 > \tau_v$$

Hence, shear stresses are within safe permissible limits.

xi. *Check for deflection control*:

Percentage reinforcement $p_t = \left(\frac{100 \times 262}{1000 \times 100}\right) = 0.262\%$

For $p_t = 0.262\%$, $K_t = 1.6$ (Fig. 4, IS:456-2000)

$\therefore \quad (L/d)_{max} = (20 \times 1.6) = 32$

$(L/d)_{provided} = (2600/100) = 26 < 32$, hence safe.

2. *Design example of a two-way slab*: Design a two-way slab for a residential floor using the following data:

i. *Data*:

Size of floor 4 m × 5 m, simply supported on all the sides on load bearing walls 230 mm thick without any provision for torsion at corners. Adopt M20 grade concrete and Fe 415 HYSD bars.

ii. *Permissible stresses*:

$\sigma_{cbc} = 7 \, \text{N/mm}^2$ $Q = 0.91$

$\sigma_{st} = 230 \, \text{N/mm}^2$ $j = 0.90$

iii. *Type of slab*:

Simply supported on all sides without any provision for torsion at corners.

$L_x = 4 \, \text{m}$ Ratio $(L_y/L_x) = 1.25$

$L_y = 5 \, \text{m}$

iv. *Depth of slab*:

From span/depth considerations:

Overall depth $D = (\text{short span}/28) = (4000/28) = 143 \, \text{mm}$

Adopt overall depth $D = 150 \, \text{mm}$

Effective depth $d = (150 - 30) = 120 \, \text{mm}$

v. *Effective span*:

Effective span is the least of the following:

(a) centre to centre of support $= (4 + 0.23) = 4.23 \, \text{m}$

(b) clear span + effective depth $= (4 + 0.12) = 4.12 \, \text{m}$

∴ Effective span $L_{xe} = 4.12 \, \text{m}$

vi. *Loads*:

Self-weight of slab $= (0.15 \times 25) = 3.75 \, \text{kN/m}^2$

Live load on floor $= 2.00$

Floor finishes $= 0.60$

Total service load w $= 6.35 \, \text{kN/m}^2$

vii. *Bending moments*:

Refer to Table 10.1 and readout the moment coefficients for the ratio

$(L_y/L_x) = 1.25$

$\alpha_x = 0.089, \ \alpha_y = 0.057$

∴ $M_x = (\alpha_x \, w \, L_{xe}^2) = (0.089 \times 6.35 \times 4.12^2) = 9.60 \, \text{kN} \cdot \text{m}$

$M_y = (\alpha_y \, w \, L_{xe}^2) = (0.057 \times 6.35 \times 4.12^2) = 6.14 \, \text{kN} \cdot \text{m}$

viii. *Check for depth*:

Effective depth

$$d = \sqrt{\frac{M}{Qb}} = \sqrt{\frac{9.60 \times 10^6}{0.91 \times 10^3}} = 102.7 \, \text{mm}$$

Effective depth for short span $= 120 \, \text{mm}$

Effective depth for long span $= (120 - 10) = 110 \, \text{mm}$

(Using 10 mm diameter bars).

ix. *Reinforcements*:

Steel for short span

$$A_{st} = \left(\frac{M}{\sigma_{st} jd}\right) = \left(\frac{9.6 \times 10^6}{230 \times 0.9 \times 120}\right) = 387 \, \text{mm}^2$$

Adopt 10 mm diameter bars at 200 mm centres (A_{st} = 393 mm²)

$$\text{Steel for long span} = \left(\frac{6.14 \times 10^6}{230 \times 0.9 \times 110}\right) = 270 \text{ mm}^2$$

Provide 10 mm diameter bars at 250 mm centres (A_{st} = 315 mm²).

x. *Shear and bond stresses*:

Shear and bond stresses in two-way slabs are negligibly small and generally within safe permissible limits. The reinforcement details are similar to that of two-way slabs designed in Chapter 10.

16.5 DESIGN OF BEAMS

1. *Design of singly reinforced concrete beams*: Design a rectangular reinforced concrete beam simply supported on masonry walls of 300 mm thick with an effective span of 5 m to support a service load of 8 kN/m and a dead load of 4 kN/m in addition to its own weight. Adopt M20 grade concrete and Fe 415 HYSD bars. Width of support of beams = 300 mm.

i. *Data*:

Effective span L = 5 m
Width of support = 300 mm
Live load = 8 kN/m
Dead load = 4 kN/m
Materials: M20 grade concrete
Fe 415 HYSD bars

ii. *Allowable stresses*:

σ_{cb} = 7 N/mm² $\qquad$ Q = 0.91
σ_{st} = 230 N/mm² $\qquad$ j = 0.90

iii. *Cross-sectional dimensions*:

Adopt width of beam b = 300 mm
Since the loading is heavy, adopt
effective depth d = (span/10) = (5000/10) = 500 mm and
overall depth D = (500 + 50) = 550 mm

iv. *Loads*:

Self-weight of slab = (0.3 × 0.55 × 25) = 4.125 kN/m
Dead load $\qquad$ = 4.000 kN/m
Live load $\qquad$ = 8.000 kN/m
Finishes $\qquad$ = 0.975 kN/m
Total load w $\qquad$ = 17.000 kN/m²

v. *Bending moments and shear forces*:

M = (0.125 wL^2) = (0.125 × 17 × 5²) = 53 kN·m
V = (0.5 wL) = (0.5 × 17 × 5) = 43 kN

vi. *Check for depth*:

$$d = \sqrt{\frac{M}{Qb}} = \sqrt{\frac{53 \times 10^6}{0.91 \times 300}} = 440 \text{ mm}$$

Effective depth d provided = 500 mm, hence adequate.

vii. *Main tension reinforcement*:

$$A_{st} = \left(\frac{M}{\sigma_{st} jd}\right) = \left(\frac{53 \times 10^6}{230 \times 0.9 \times 500}\right) = 512 \text{ mm}^2$$

Provide 2 bars of 20 mm diameter (A_{st} = 628 mm²).

viii. *Shear stress and reinforcement*:

$$\text{Nominal shear stress } \tau_v = \left(\frac{V_u}{bd}\right) = \left(\frac{43 \times 10^3}{300 \times 500}\right) = 0.28 \text{ N/mm}^2$$

$$= \left(\frac{100 A_{st}}{bd}\right) = \left(\frac{100 \times 628}{300 \times 500}\right) = 0.418$$

Refer to Table 23 (IS:456) and readout the permissible shear stress in concrete

$$\tau_c = 0.25 \text{ N/mm}^2 < \tau_v$$

Hence, shear reinforcements in the form of stirrups are required. Since τ_c is nearly equal to τ_v, provide nominal shear reinforcements given by

$$s_v = \left(\frac{A_{sv} \, 0.87 f_y}{0.4 b}\right)$$

Using 6 mm diameter two-legged stirrups

$$s_v = \left(\frac{2 \times 28 \times 0.87 \times 415}{0.4 \times 300}\right) = 168 \text{ mm}$$

Provide 6 mm diameter stirrups at 150 mm centres up to a quarter span length from supports and gradually increased to 300 mm centres towards the centre of span.

2. *Design of doubly reinforced beam*: Design a doubly reinforced concrete beam for a residential floor of a building to suit the following data.

  i. *Data*:

    Effective span = 5 m
    Dead load   = 8 kN/m
    Live load   = 12 kN/m
    Width of beam = 250 mm
    Overall depth = 500 mm
    Materials: M20 grade concrete
            Fe 415 HYSD bars
    Effective depth = 450 mm
    Cover to compression steel = 50 mm

ii. *Permissible stresses:*

$\sigma_{cb} = 7\,\text{N/mm}^2$ $\qquad$ $Q = 0.91$

$\sigma_{st} = 230\,\text{N/mm}^2$ $\qquad$ $j = 0.90$

$m = 13$ $\qquad$ $n_c = 0.284\,d$

iii. *Loads:*

Self-weight $= (0.25 \times 0.5 \times 25) = 3.125\,\text{kN/m}$

Dead load $\qquad\qquad\qquad = 8.000\,\text{kN/m}$

Live load $\qquad\qquad\qquad = 12.000\,\text{kN/m}$

Finishes etc. $\qquad\qquad = 0.875\,\text{kN/m}$

Total service load w $\qquad = 24.000\,\text{kN/m}$

iv. *Bending moments and shear forces:*

$M = 0.125\,wL^2 = (0.125 \times 24 \times 5^2) = 75\,\text{kN·m}$

$V = 0.5\,wL = (0.5 \times 24 \times 5) = 60\,\text{kN}$

v. *Resisting moment:*

Resisting moment capacity of balanced singly reinforced section is computed as

$$M_1 = (Qbd^2) = (0.91 \times 250 \times 450^2)\,10^{-6} = 46\,\text{kN·m}$$

Balance moment $\quad M_2 = (M - M_1) = (75 - 46) = 29\,\text{kN·m}$

vi. *Tension reinforcement:*

Tensile steel required for balanced singly reinforced section is given by

$$A_{st1} = \left(\frac{M_1}{\sigma_{st}\,j\,d}\right) = \left(\frac{46 \times 10^6}{230 \times 0.90 \times 450}\right) = 493\,\text{mm}^2$$

Additional steel in tension for balanced moment M_2 is given by

$$A_{st2} = \left(\frac{M_2}{\sigma_{st}(d - d_c)}\right) = \left(\frac{29 \times 10^6}{230 \times (450 - 50)}\right) = 315\,\text{mm}^2$$

Total tension steel $A_{st} = (A_{st1} + A_{st2}) = (493 + 315) = 808\,\text{mm}^2$

Provide 3 bars of 20 mm diameter ($A_{st} = 942\,\text{mm}^2$)

vii. *Compression reinforcement:*

$$A_{sc} = \left[\frac{m\,A_{st2}(d - n_c)}{(1.5m - 1)(n_c - d_c)}\right]$$

where, $\quad n_c = 0.284\,d = (0.284 \times 450) = 127.8$

$\therefore \qquad A_{sc} = \left[\frac{13 \times 315(450 - 127.8)}{(1.5 \times 13 - 1)(127.8 - 50)}\right] = 916\,\text{mm}^2$

Provide 3 bars of 20 mm diameter ($A_{sc} = 942\,\text{mm}^2$).

viii. *Shear stresses and reinforcements:*

$$\tau_v = \left(\frac{V}{bd}\right) = \left(\frac{60 \times 10^3}{250 \times 450}\right) = 0.53\,\text{N/mm}^2$$

$$= \left(\frac{100\,A_{st}}{bd}\right) = \left(\frac{100 \times 942}{250 \times 450}\right) = 0.83$$

Refer to Table 23 (IS:456) and readout the permissible shear

$$\tau_c = 0.36 \text{ N/mm}^2 < \tau_v$$

Hence, shear reinforcements are to be designed to resist the balance shear computed as

$$V_s = [V - \tau_c\,b\,d] = [60 - (0.36 \times 250 \times 450)\,10^{-3}] = 19.5 \text{ kN}$$

Using 6 mm diameter two-legged stirrups, spacing

$$s_v = \left(\frac{A_{st}\,\sigma_{sv}\,d}{V_s}\right) = \left[\frac{(2 \times 28 \times 230 \times 450)}{(19.5 \times 10^3)}\right] = 297 \text{ mm}$$

Provide 6 mm diameter two-legged stirrups at 250 mm centres at supports, gradually increasing to 300 mm centres towards the centre of span.

3. *Design of flanged beams*: Design a T-beam for an office floor using the following data.

   i. *Data*:

   Effective span = 8 m

   Spacing of T-beams = 3 m

   Loading (office floor) = 4 kN/m^2

   Slab thickness = 150 mm

   Materials: M20 grade concrete

   Fe 415 HYSD bars

   ii. *Permissible stresses*:

   $\sigma_{cb} = 7 \text{ N/mm}^2$ $\qquad Q = 0.91$

   $\sigma_{st} = 230 \text{ N/mm}^2$ $\qquad j = 0.90$

   $m = 13$

   iii. *Sectional dimensions*:

   Effective depth = d (span/15) = (8000/15) = 534 mm

   Adopt d = 550 mm and overall depth D = 600 mm and b = 300 mm

   iv. *Loads*:

   Self-weight = (0.15 × 25 × 3) $\qquad$ = 11.25 kN/m

   Live load = (4 × 3) $\qquad$ = 12.00 kN/m

   Floor finish = (0.6 × 3) $\qquad$ = 1.80 kN/m

   Self-weight of rib = (0.45 × 0.3 × 25) = 3.37 kN/m

   Plaster finishes $\qquad$ = 1.58 kN/m

   Total load w $\qquad$ = 30.00 kN/m

   v. *Bending moments and shear forces*:

   $M = 0.125\,wL^2 = (0.125 \times 30 \times 8^2) = 240$ kN·m

   $V = 0.5\,wL = (0.5 \times 30 \times 8) = 120$ kN

vi. *Check for depth*:

$$A_{st1} = \left(\frac{M_1}{\sigma_{st}\, jd}\right) = \left(\frac{240 \times 10^6}{230 \times 0.9 \times 550}\right) = 2108 \text{ mm}^2$$

Provide 4 bars of 28 mm diameter (A_{st} = 2464 mm²)

vii. *Effective flange width*:

Least of the following:

a. $b_f = [(L_0/6) + b_w + 6D_f]$
$\quad = [(8000/6) + 300 + (6 \times 150)] = 2533 \text{ mm}$

b. b_f = centre to centre of ribs = 3000 mm

Hence, b_f = 2533 mm

viii. *Check for stresses*:

Let $\quad n$ = depth of neutral axis

$(b_f n^2/2) = mA_{st}\,(d-n)$

$(2533 \times n^2)/2 = (13 \times 2464)\,(550-n)$

Solving, $\quad n = 106 \text{ mm}$

Lever arm $a = [d - (n/3)] = [550 - (106/3)] = 514.67$

Effective depth:

$$\sigma_{st} = \left(\frac{240 \times 10^6}{2464 \times 514.67}\right) = 189 \text{ N/mm}^2 < 230 \text{ N/mm}^2$$

$$\sigma_{cb} = \left(\frac{189 \times 106}{13 \times 444}\right) = 3.47 \text{ N/mm}^2 < 7 \text{ N/mm}^2$$

Hence, the stresses are within safe permissible limits.

ix. *Shear stresses and shear reinforcements*:

Maximum shear force V = 120 kN

$$\tau_c = (120 \times 10^3)/(300 \times 550) = 0.72 \text{ N/mm}^2$$

$$\left(\frac{100\,A_{st}}{b_w d}\right) = \left(\frac{100 \times 2464}{300 \times 550}\right) = 1.49$$

Refer to Table 23 (IS:456) and readout the permissible shear stress

$$\tau_c = 0.45 \text{ N/mm}^2 < \tau_v$$

Hence, shear reinforcements are to be designed to resist the balance shear given by

$$V_s = [V - (\tau_c\, b_w\, d)]$$
$$\quad = [120 - (0.45 \times 300 \times 550)\, 10^{-3}$$
$$\quad = 46 \text{ kN}$$

Using 6 mm diameter two-legged stirrups at a spacing given by

$$S_v = \left(\frac{A_{sv}\, \sigma_{sv}\, d}{V_s}\right) = \left(\frac{2 \times 28 \times 230 \times 550}{46 \times 10^3}\right) = 154 \text{ mm}$$

Provide 6 mm diameter two-legged stirrups at 150 mm centres near supports and gradually increasing to 300 mm towards the centre of span.

16.6 DESIGN OF COLUMNS AND FOOTINGS

Design a suitable RCC column of square section and a suitable footing to support an axial load of 800 kN. Size of the column is fixed as 400 mm × 400 mm. Safe bearing capacity of the soil is 200 kN/m². Adopt M20 grade concrete and Fe 415 HYSD bars.

i. *Data*:

Axial load P = 800 kN

Size of column = 400 mm × 400 mm

Safe bearing capacity of soil = 200 kN/m²

Materials: M20 grade concrete

Fe 415 HYSD bars

ii. *Permissible stresses*:

$\sigma_{cc} = 5 \, \text{N/mm}^2$ $\qquad$ $Q = 0.91$

$\sigma_{cb} = 7 \, \text{N/mm}^2$ $\qquad$ $j = 0.90$

$\sigma_{sc} = 190 \, \text{N/mm}^2$ $\qquad$ $m = 13$

$\sigma_{st} = 230 \, \text{N/mm}^2$

iii. *Main column reinforcement*:

$$P = \sigma_{sc} A_{sc} + \sigma_{cc}(A_c - A_{sc})$$
$$A_c = (400 \times 400) = 16 \times 10^4 \, \text{mm}^2$$
$$(800 \times 10^3) = 190 A_{sc} + 5[(16 \times 10^4) - A_{sc}]$$

Solving, $A_{sc} = 1081 \, \text{mm}^2$

Minimum steel = $(0.008 \times 400 \times 400) = 1280 \, \text{mm}^2$

Provide 4 bars of 22 mm diameter ($A_{sc} = 1520 \, \text{mm}^2$)

iv. *Ties*:

Greater of the diameters of:

a. $(22/4) = 5.5$ mm

b. 5 mm

Adopt 6 mm diameter ties

Pitch of the ties is the least of:

a. least lateral dimensions = 400 mm

b. 16 times the diameter of longitudinal bar = $(16 \times 22) = 352$ mm

c. 300 mm

Adopt 6 mm diameter ties at 300 mm centres.

v. *Size of footing*:

Load of column $\qquad$ = 800 kN

Self-weight of footing (10%) = 80 kN

Total load w $\qquad$ = 880 kN/m

Footing area = (880/200) = 4.4 m²
Adopt a square footing of size 2.5 m × 2.5 m
Area of footing = 6.25 m²
Upward pressure p = (800/6.25) = 128 kN/m²

vi. *Bending moments*:

0.5(2.5 × 0.4) = 1.05 m

$$M = 0.5\,wL^2 = (0.5 \times 128 \times 1.05^2) = 71 \text{ kN·m}$$

vii. *Depth of footing*:

Tensile steel required for balanced singly reinforced section is given by

$$d = \sqrt{\frac{M}{Qb}} = \sqrt{\frac{71 \times 10^6}{0.91 \times 10^3}} = 279 \text{ mm}$$

The depth required from shear considerations will be nearly two times that obtained from moment considerations.

Hence, adopt overall depth = 650 mm

Effective depth = 600 mm

viii. *Reinforcements in footing*:

$$A_{st1} = \left(\frac{M_1}{\sigma_{st}\,jd}\right) = \left(\frac{71 \times 10^6}{230 \times 0.9 \times 600}\right) = 572 \text{ mm}^2$$

But A_{st} (minimum) = (0.0012 × 650 × 10³) = 780 mm²

Provide 16 mm diameter bars at 250 mm centres (A_{st} = 802 mm²)

ix. *Check for shear stresses*:

Shear force at a distance d from the face of the column is computed as

$$V = 128 (1.05 - 0.6) = 57.6 \text{ kN}$$

$$\tau_v = \left(\frac{V}{bd}\right) = \left(\frac{57.6 \times 10^3}{10^3 \times 600}\right) = 0.096 \text{ N/mm}^2$$

$$= \left(\frac{100\,A_{st}}{bd}\right) = \left(\frac{100 \times 802}{1000 \times 600}\right) = 0.83$$

Refer to Table 23 (IS:456) and readout the permissible shear as

$$(k\,\tau_c) = (1 \times 0.18) = 0.18 \text{ N/mm}^2$$

Since $\tau_c > \tau_v$, shear stresses are within safe permissible limits.

The reinforcement details are similar to that shown in Chapter 11, comprising limit state design of columns and footings.

16.7 DESIGN OF RETAINING WALLS

1. *Cantilever type retaining wall*: A cantilever type retaining wall is to be designed to retain an earthen embankment with a horizontal top 4 m above ground level. Density of soil 18 kN/m³. Angle of repose is 30°, and the safe bearing capacity of soil is 200 kN/m². Coefficient of friction between soil and concrete is 0.5. Adopt M20 grade concrete and Fe 415 HYSD bars.

i. *Data*:

Height of embankment = 4 m
SBC of soil $p = 200$ kN/m^2
Density of soil $w = 18$ kN/m^3
Coefficient of friction = 0.5
Angle of internal friction = 30°
Materials: M20 grade concrete
Fe 415 HYSD bars

ii. *Permissible stresses*:

$\sigma_{cb} = 7$ N/mm^2 $Q = 1.91$
$\sigma_{st} = 230$ N/mm^2 $j = 0.90$

iii. *Dimensions of retaining wall*:

$$\text{Minimum depth of foundation} = \left(\frac{p}{w}\right)\left(\frac{1-\sin\phi}{1+\sin\phi}\right)^2 = \frac{200}{18}\left(\frac{1}{3}\right)^2 = 1.2 \text{ m}$$

Overall depth of wall $H = (4 + 1.2) = 5.2$ m
Thickness of base slab = $(H/12) = (5200/12) = 433$ mm
Adopt thickness of base slab = 450 mm
∴ Height of stem $h = (5.2 - 0.45) = 4.75$ m
Width of base slab $b = 0.5 H$ to $0.6 H$
$0.5H = 2.6$ m
$0.6H = 3.12$ m   ∴ Adopt $b = 3$
Toe projection = $(b/3) = 1$ m

iv. *Design of stem*:

Height of stem $h = 4.75$ m

$$\text{Maximum BM in stem} = C_p\left(\frac{wh^3}{6}\right) \text{ where, } C_p = \left(\frac{1-\sin\phi}{1+\sin\phi}\right) = \left(\frac{1}{3}\right)$$

$$= \left(\frac{1}{3}\right)\left(\frac{18 \times 4.75^3}{6}\right) = 107.17 \text{ kN} \cdot \text{m}$$

$$\therefore \text{ Effective depth} \quad d = \left(\frac{107.17 \times 10^6}{0.91 \times 1000}\right) = 343 \text{ mm}$$

Adopt effective depth $d = 400$ mm
Overall depth $D = 450$ mm
Top width of stem = 200 mm

$$A_{st} = \left(\frac{107.17 \times 10^6}{230 \times 0.90 \times 400}\right) = 1294 \text{ mm}^2$$

Adopt 20 mm diameter bars at 200 mm centres ($A_{st} = 1571$ mm^2)
Distribution steel = $(0.0012 \times 450 \times 1000) = 540$ mm^2
Provide 12 mm diameter bars at 200 mm centres ($A_{st} = 565$ mm^2)

v. *Stability calculations (pressure distribution at base):*

Heel projection = $(2 - 0.45) = 1.55$ m

The overall dimensions of wall is shown in Fig. 13.8 (refer to Chapter 13). The stability calculations for one metre run of wall is shown in Table 16.5.

Table 16.5: Stability calculation for one metre run of wall

Loads	Magnitude of load (kN)	Distance from 'a' (m)	Moment (kN·m)
$W_1 = (0.2 \times 4.75 \times 25)$	23.80	1.65	39.27
$\quad = (0.5 \times 0.25 \times 4.75 \times 25)$	14.84	1.83	27.15
$W_2 = (3 \times 0.45 \times 25)$	33.75	1.50	50.62
$W_3 = (1.55 \times 4.75 \times 18)$	132.51	0.78	103.35
Moment due to earth pressure $= C_p(wh^3/6)$			
$\qquad = (1/3)(18 \times 4.75^3)/6$			107.06
Total	$\Sigma W = 204.90$		$M = 327.45$

Distance of point of application of resultant from end a:

$$Z = \left(\frac{327.45}{204.90}\right) = 1.6 \text{ m}$$

Eccentricity $e = [Z - (b/2)] = [1.6 - (3/2)] = 0.1$ m

$(b/6) = (3/6) = 0.5 \quad \therefore e = (b/6)$

Maximum and minimum pressures at base are computed as

$$p_{\substack{max \\ min}} = \left(\frac{204.90}{3}\right)\left[1 + \left(\frac{6 \times 0.1}{3}\right)\right]$$

$p_{max} = 82.00$ kN/m^2 and $p_{min} = 54.64$ kN/m^2

vi. *Design of heel slab:*

Moment calculations are given in Table 16.6.

Table 16.6: Moment computation

Loads	Magnitude of load (kN)	Distance from 'b' (m)	Moment (kN·m)
$W_3 = (1.55 \times 4.75 \times 18)$	132.50	0.775	102.68
Self weight of heel slab $(1.55 \times 0.45 \times 25)$	17.40	0.775	13.51
Total			116.19
Deduct for upward pressure 'abih' (53.84×1.55)	83.45	0.775	64.67
Upward pressure 'ghi' $(0.55 \times 1.55 \times 13.9)$	10.77	0.516	5.55
Total deduction			70.22
Maximum working bending moment in heel slab			46.00

Hence, $A_{st} = \left(\dfrac{46 \times 10^6}{230 \times 0.9 \times 400}\right) = 555$ mm^2

Provide 12 mm diameter bars at 200 mm centres ($A_{st} = 565$ mm^2)

Distribution bars at 0.12 per cent of gross area are the same as in stem. Adopt 12 mm diameter bars at 200 mm centres.

vii. *Design of toe slab*:

Moment calculations for 1 m length of the slab are given in Table 16.7.

Loads	Magnitude of load (kN)	Distance from 'c' (m)	Moment about 'c' (kN·m)
Upward pressure 'cdif' (71.78 × 1)	71.78	0.5	35.89
Upward pressure 'jfe' (0.5 × 1 × 8.98)	4.49	0.67	3.00
Total			38.89
Deduct self-weight of toe slab (1 × 1 × 0.45 × 25)	11.2	0.5	5.60
Dead-weight of soil over toe slab (0.75 × 1 × 18)	13.5	0.5	6.75
Total deduction			12.35
Maximum service load BM in toe slab			26.54

Table 16.7: Moment calculation

Hence, $A_{st} = \left(\dfrac{26.54 \times 10^6}{230 \times 0.9 \times 400} \right) = 555 \text{ mm}^2$

Provide 12 mm diameter bars at 200 mm centres as minimum reinforcement of 0.12% is greater than the calculated quantity of reinforcement.

Provide 12 mm diameter bars at 200 mm centres as distribution reinforcement on both faces.

viii. *Check for sliding*:

Total horizontal earth pressure

$$= C_p \left(\frac{wH^2}{2} \right) = \left[\frac{1}{3} \times 18 \times \frac{5.2^2}{2} \right] = 81.12 \text{ kN}$$

Assuming $\mu = 0.5$

Maximum possible frictional force $W = (0.5 \times 204.84) = 102.4$ kN

∴ Factor of safety against sliding = $(102.4/81.12) = 1.26 < 1.5$

The wall is unsafe against sliding

Hence a shear key has to be designed.

ix. *Design of shear key*:

If p_p = intensity of passive pressure developed just in front of the shear key, the value of p_p is computed as

$$p_p = K_p p$$

where, $K_p = \left(\dfrac{1 + \sin \phi}{1 - \sin \phi} \right) = (1/K_a) = 3$

and $p = 72.8$ kN/m^2 (refer to Fig. 10.3)

∴ $p_p = (3 \times 72.8) = 218.6$ kN/m^2

If 'a' is the depth of shear key = 450 mm

Total passive force $p_p = (p_p a) = (218.6 \times 0.45) = 98.3$ kN

∴ Factor of safety against sliding is

$$= \left[\frac{\mu W + p_p}{\Sigma P}\right] = \left[\frac{102.4 + 98.3}{81.12}\right] = 2.45 > 1.5$$

Hence the retaining wall is safe against failure due to sliding. The reinforcements in the stem is extended up to the shear key.

x. *Check for shear stress at junction of stem and base slab*:

Net shear force $= (1.5\ \Sigma P - \mu W)$

$$= (1.5 \times 81.12) - 102.4 = 19.2\ \text{kN}$$

$$\tau_v = (19.2 \times 10^3)/(1000 \times 400) = 0.048\ \text{N/mm}^2$$

$$(100A_{st})/(b/d) = (100 \times 1571)/(100 \times 400) = 0.39$$

From Table 23 (IS:456), $\tau_c = 0.25\ \text{N/mm}^2 > \tau_v$

Hence shear stresses are within safe permissible limits.

The details of reinforcements in the cantilever retaining wall is similar to that shown in limit state method of design.

16.8 DESIGN OF STAIRCASES

1. Design one of the flights of a dog-legged staircase spanning between landing beams using the following data:

   i. *Data*:

   Type of staircase: Dog-legged with waist slab, treads and risers
   Number of steps in the flight = 10
   Tread $T = 300$ mm
   Riser $R = 150$ mm
   Width of landing beams = 300 mm
   Materials: M20 grade concrete
   　　　　　　Fe 415 HYSD bars

   ii. *Permissible stresses*:

   $$\sigma_{cb} = 7\ \text{N/mm}^2 \qquad Q = 1.91$$
   $$\sigma_{st} = 230\ \text{N/mm}^2 \qquad j = 0.90$$
   $$m = 13$$

   iii. *Effective span*:

   Effective span $= (10 \times 300) + 300 = 3300\ \text{mm} = 3.3\ \text{m}$
   Thickness of waist slab $= (\text{span}/20) = (3300/20) = 165\ \text{mm}$
   Adopt overall depth $D = 165\ \text{mm}$
   Effective depth $d = 140$

   iv. *Loads*:

   Dead load of slab on slope $w_s = (0.165 \times 1 \times 25) = 4.125\ \text{kN/m}$
   Dead load on horizontal span

   $$w = \left[\frac{w_s \sqrt{R^2 + T^2}}{T}\right] = \left[\frac{4.125\sqrt{150^2 + 300^2}}{300}\right] = 4.61\ \text{kN/m}$$

Dead load on one step $= (0.5 \times 0.15 \times 0.3 \times 25)$ $= 0.56$ kN/m

Load of steps per metre length $= \left(\dfrac{0.56 \times 1000}{300} \right) = 1.86$ kN/m

Finishes etc. $= 0.53$ kN/m
Total dead load $= (4.61 + 1.86 + 0.53)$ $= 7$ kN/m
Service load (liable for overcrowding) $= 5$ kN/m^2
$\therefore$ Total service load $w = (7 + 5)$ $= 12$ kN·m

iv. *Bending moments*:
   Maximum bending moment at centre of span
$$M_u = 0.125\, wL^2 = (0.125 \times 12 \times 3.3^2) = 16.34 \text{ kN·m}$$

v. *Effective depth*:

$$d = \sqrt{\frac{M}{Qb}} = \sqrt{\frac{16.34 \times 10^6}{0.91 \times 10^3}} = 279 \text{ mm}$$

   Adopt effective depth $d = 140$ mm. Hence safe.

vi. *Main reinforcement*:

$$A_{st1} = \left(\frac{M}{\sigma_{st}\, j\, d} \right) = \left(\frac{16.34 \times 10^6}{230 \times 0.90 \times 600} \right) = 572 \text{ mm}^2$$

   Provide 12 mm diameter bars at 200 mm centres ($A_{st} = 565$ mm^2/m)

vii. *Distribution reinforcement*:
   Distribution steel $= 0.12\% = (0.0012 \times 1000 \times 165) = 198$ mm^2/m
   Provide 8 mm diameter bars at 200 mm centres ($A_{st} = 251$ mm^2/m)

viii. *Details of reinforcements*:
   The details of reinforcements in the staircase are shown in Fig. 12.7 (same as that designed by limit state method).

16.9 DESIGN OF DOMES

1. General Features

Concrete domes are generally preferred to cover circular tanks and for roofs of large span structures which are circular in shape such as sports areas and churches where uninterrupted floor space is desirable. The spherical domes are supported by a ring beam at the base.

The thickness of the reinforced concrete spherical dome is generally not less than 1/500 of the diameter with values of 50 mm to 100 mm for domes in the range of 25 m to 50 m respectively. The reinforcement in the dome is made up of wire mesh and the concrete is placed in concentric rings over performed framework or the dome can be formed by gunniting using microconcrete.

2. Design Equations

Domes are designed for meridional thrust and hoop stress. Referring to Fig. 16.1, the spherical dome of radius R is subjected to an uniformly distributed load of w per unit area of surface.

Let t = thickness of dome

h = rise of dome

R = radius of dome

D = diameter of dome at base

T_1 = maximum meridional thrust per metre run

T_2 = circumferential force

$$T_1 = \left(\frac{wR}{1 + \cos\theta} \right)$$

$$T_2 = wR \left(\cos\theta - \frac{1}{1 + \cos\theta} \right)$$

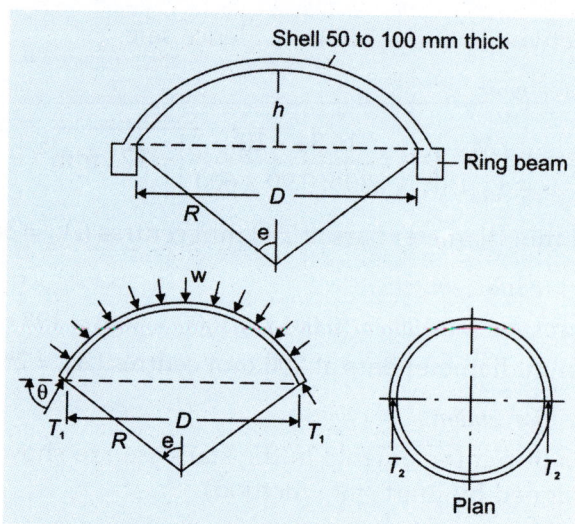

Fig. 16.1: Forces acting on spherical dome

If the meridional and circumferential stresses are low, nominal reinforcement of 0.3% cross-sectional area is provided. Ring beam is designed for hoop tension.

$$\text{Hoop tension} = \left(\frac{T_1 \cos\theta \cdot D}{2} \right)$$

The cross-sectional area of the ring beam is determined by limiting the tensile stress in the ring beam. The permissible tensile stress which depends upon the grade of concrete is given below:

Grade of concrete	M10	M15	M20	M25	M30	M35	M40
Tensile stress (N/mm²)	1.2	2.0	2.8	3.2	3.6	4.0	4.4

The tensile stress is calculated by the following equation

$$\left(\frac{F_t}{A_c + mA_{st}} \right)$$

where,

F_t = direct tension
A_c = cross-sectional area of concrete
m = modular ratio
A_{st} = cross-sectional area of steel

3. Design Example

A reinforced concrete dome of 6 m base diameter with a rise of 1.25 m is to be designed for a water tank. The uniformly distributed live load including finishes on dome may be taken as 2 kN/m². Adopting M20 concrete and grade-I steel, design the dome and ring beam. Permissible tensile stress in steel is 100 N/mm².

    i. *Data*:
        Base diameter $D = 6$ m
        Rise of dome = 1.25 m
        Live load on dome = 2 kN/m²

    ii. *Permissible stresses*:
        $\sigma_{st} = 5\,\text{N/mm}^2$
        Tensile stress in concrete = 2.8 N/mm² for M20 concrete
        $\sigma_{cc} = 5\,\text{N/mm}^2, m = 13$

    iii. *Dimensions of dome*:
        If R = radius of the dome (refer to Fig. 16.2)
        $(R - 1.25)^2 = R^2 - 3^2$ ∴ $R = 4.23$ m

        ∴ $\sin\theta = \left(\dfrac{3}{4.23}\right) = 0.7092, \cos\theta = 0.7049$

        Assume thickness of dome $t = 100$ mm.

    iv. *Loads*:
        Self-weight of dome = $(0.1 \times 24) = 2.4$ kN/m²
        Live load and finishes   = 2.0 kN/m²
        Total load w = 4.4 kN/m²

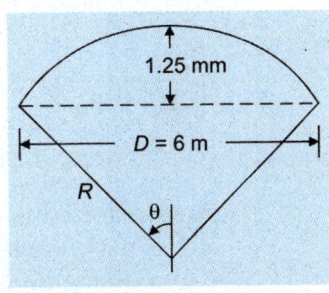

Fig. 16.2: Dimensions of dome

    v. *Stresses in dome*:

        Meridional thrust $T_1 = \left(\dfrac{wR}{1 + \cos\theta}\right) = \left(\dfrac{4.4 \times 4.23}{1 + 0.7049}\right) = 10.91$ kN/m

        ∴ Meridional compressive stress = $\left(\dfrac{10.91 \times 10^3}{1000 \times 10}\right) = 1.091$ N/mm²

        Hoop stress = $\dfrac{wR}{t}\left[\cos\theta - \dfrac{1}{1 + \cos\theta}\right] = \left(\dfrac{4.4 \times 4.23}{0.1}\right)\left[0.7049 - \dfrac{1}{1.7049}\right]$

        = 22 kN/m² = 0.022 N/mm²

vi. *Reinforcement*:

Since the stresses are very low, nominal reinforcement of 0.3% of cross-sectional area is provided.

$$\therefore \qquad A_{st} = \left(\frac{0.3}{100} \times 1000 \times 100\right) = 300 \text{ mm}^2$$

Spacing of 8 mm diameter bars $= \left(\frac{1000 \times 50}{300}\right) = 166$ mm

Adopt 8 mm diameter bars at 160 mm centres both meridionally and circumferentially.

vii. *Ring beam*:

Horizontal component of $T_1 = T_1 \cos\theta = (10.91 \times 0.7049) = 7.69$ kN/m

Hoop tension in ring beam $= \left(\frac{7.69 \times 6}{2}\right) = 23.07$ kN

$$A_{st} = \left(\frac{23.07 \times 10^3}{100}\right) = 230.7 \text{ mm}^2$$

Provide 4 bars of 10 mm diameter ($A_{st} = 314$ mm²)

If A_c = cross-sectional area of ring beam, equivalent area of concrete $= (A_c + m \cdot A_{st})$. Allowing a tensile stress of 2.8 N/mm²

$$\left[\frac{F_t}{A_c + mA_{st}}\right] = 2.8 \quad \therefore \quad \left[\frac{23.07 \times 10^3}{A_c + (13 \times 314)}\right] = 2.8$$

$$\therefore \qquad A_c = 4158 \text{ mm}^2$$

Area required is very small. However, provide a ring beam of size 150 mm × 150 mm. With 4 bars of 10 mm diameter as hoop steel and stirrups of 6 mm diameter at 150 mm centres as shown in Fig. 16.3.

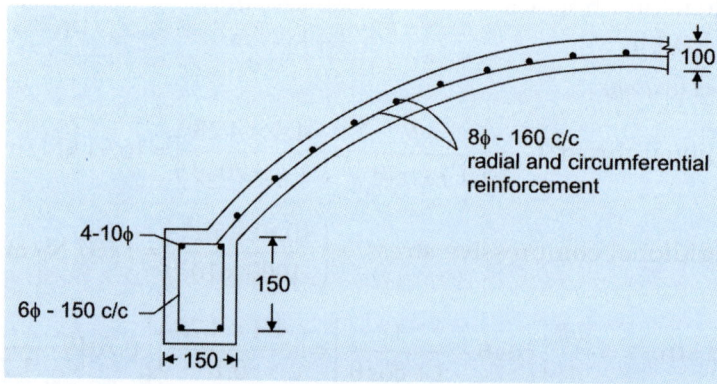

Fig. 16.3: Reinforcements in ring beam and dome

16.10 DESIGN OF WATER TANKS

1. General Features

There are three types of reinforced concrete water tanks:

    a. Tanks resting on ground

    b. Underground tanks

    c. Elevated water tanks on staging.

Water tightness is an important criterion in water tanks. Usually richer mixes with M20, M30 concrete are used. The tensile stresses permitted in concrete are restricted to control cracking in concrete as per IS:3370, part II, 1965 and are shown in Table 16.8.

Table 16.8: Permissible concrete stresses in calculations relating to resistance to cracking

Grade of concrete N/mm²	Permissible stresses (N/mm²)		Shear stress $\left(\dfrac{V}{bd}\right)$
	Direct tension	Tension due to bending	
M15	1.1	1.5	1.5
M20	1.2	1.7	1.7
M25	1.3	1.8	1.9
M30	1.5	2.0	2.2
M35	1.6	2.2	2.5
M40	1.7	2.4	2.7

For strength calculations, stresses in concrete are same as that recommended in IS:456.

Permissible stresses in steel:

Tensile stress in direct tension = 100 N/mm^2

Tensile stresses due to bending:

Steel at water face   = 100 N/mm^2

Steel away from water face   = 125 N/mm^2

2. Reinforcement details

Minimum area of steel is 0.3% gross area in sections up to 100 mm thick reduced to 0.2% in sections up to 450 mm thick. For 225 mm thickness provide 2 layers of reinforcement. The percentage of reinforcement in base or floor slabs resting directly on ground must not be less than 0.15% of the concrete section. The minimum cover to all reinforcement should be 25 mm or the diameter of the main bar whichever is greater.

3. Joints between Tank Walls and Floor

There are three types of joints between the tank walls and the floor:

    a. Flexible or free base

    b. Fixed base

    c. Hinged base

In case of free or flexible base between tank wall and base slab, the walls are free to slide and expand and the hoop tension developed in the circular walls can be calculated easily due to the hydrostatic pressure. However, for hinged and fixed bases, the coefficients for moments ring tension are given in Tables 16.9 to 16.12 as recommended in IS:3370 (part IV), 1967. These coefficients are expressed as a function of the nondimensional parameter (H^2/Dt)
where,

H = height of water tank
D = diameter of the tank
t = thickness of the tank wall

Design Examples

1. Design a circular tank with a flexible base for a capacity of 500000 litres. The depth of water is to be 4 m. Free board 200 mm. Use M20 grade concrete and grade I mild steel. Permissible direct tensile stress in concrete = 1.2 N/mm². Permissible stress in steel in direct tension = 100 N/mm². Sketch the details of reinforcement in tank walls.

   i. *Data*:

   Capacity of tank = 500000 liters
   Depth of water = 4 m
   Free board = 200 mm
   Density of water w = 10 kN/m³
   M20 concrete and grade I steel

   ii. *Permissible stresses*:

   Direct tensile strength in concrete (σ_{ct}) = 1.2 N/mm²
   Tensile stress in steel = 100 N/mm²
   Modular ratio m = 13

   iii. *Dimensions of tank*:

   Referring to Fig. 16.4a, if D = diameter of tank

   $$\text{or} \left(\frac{\pi \cdot D^2}{4} \times 4\right) = \left(\frac{500000 \times 10^3}{10^6}\right)$$

   $\therefore$ $D = 12.6\,\text{m}$

   iv. *Hoop tension and steel reinforcement*:

   $$\text{Maximum hoop tension} = \left(\frac{wH \cdot D}{2}\right) = \left(\frac{10 \times 4.2 \times 12.6}{2}\right) = 264.4\,\text{kN}$$

   $$\therefore \quad A_{st} = \left(\frac{264.6 \times 10^3}{100}\right) = 2646\,\text{mm}^2/\text{m height}$$

   Using 20 mm diameter bars

   $$\text{Spacing} = \left(\frac{1000 \times 314}{2646}\right) = 118\,\text{mm}$$

   Provide 20 mm diameter bars at 118 mm c/c.

Table 16.9: Moments in cylindrical walls: Fixed base free top (IS:3370, part IV)

Moment M_w = Coefficient × wH^3 (kN·m/m)

Positive sign indicates tension at the outside face

Coefficients at point

$\dfrac{H^2}{Dt}$	0.1H	0.2H	0.3H	0.4H	0.5H	0.6H	0.7H	0.8H	0.9H	0.10H	0.80H	0.85H	0.90H	0.95H	1.00H
0.4	+.0005	+.0014	+.0021	+.0007	-.0042	-.0150	-.0302	-.0529	-.0816	-.1205	—	—	—	—	—
0.8	+.0011	+.0037	+.0063	+.0080	+.0070	+.0023	-.0068	-.0224	-.0465	-.0795	—	—	—	—	—
1.2	+.0012	+.0042	+.0077	+.0103	+.0112	+.0090	+.0022	-.0108	-.0311	-.0602	—	—	—	—	—
1.6	+.0011	+.0041	+.0075	+.0107	+.0121	+.0111	+.0058	-.0051	-.0232	-.0505	—	—	—	—	—
2.0	+.0010	+.0035	+.0068	+.0099	+.0120	+.0115	+.0075	-.0021	-.0185	-.0436	—	—	—	—	—
3.0	+.0006	+.0024	+.0047	+.0071	+.0090	+.0097	+.0077	+.0012	-.0119	-.0333	—	—	—	—	—
4.0	+.0003	+.0015	+.0028	+.0047	+.0066	+.0077	+.0069	+.0023	-.0080	-.0268	—	—	—	—	—
5.0	+.0002	+.0008	+.0016	+.0029	+.0046	+.0059	+.0059	+.0028	-.0058	-.0222	—	—	—	—	—
6.0	+.0001	+.0003	+.0008	+.0019	+.0032	+.0046	+.0051	+.0029	-.0041	-.0187	—	—	—	—	—
8.0	.0000	+.0001	+.0002	+.0008	+.0016	+.0028	+.0038	+.0029	-.0022	-.0146	—	—	—	—	—
10.0	.0000	.0000	+.0001	+.0004	+.0007	-.0019	+.0029	+.0028	-.0012	-.0122	—	—	—	—	—
12.0	.0000	-.0001	+.0001	+.0002	+.0003	+.0013	+.0023	+.0026	-.0005	-.0104	—	—	—	—	—
14.0	.0000	+.0000	.0000	+.0000	.0001	+.0008	+.0019	+.0023	+.0001	-.0090	—	—	—	—	—
16.0	.0000	+.0000	-.0001	+.0002	-.0001	+.0004	+.0013	+.0019	+.0001	-.0079	—	—	—	—	—
20.0	—	—	—	—	—	—	—	—	—	—	+.0015	+.0014	+.0005	-.0018	-.0063
24.0	—	—	—	—	—	—	—	—	—	—	+.0012	+.0012	+.0007	-.0013	-.0053
32.0	—	—	—	—	—	—	—	—	—	—	+.0007	+.0003	+.0007	-.0008	-.0040
40.0	—	—	—	—	—	—	—	—	—	—	+.0002	+.0005	+.0005	-.0005	-.0032
48.0	—	—	—	—	—	—	—	—	—	—	.0000	+.0001	+.0005	-.0003	-.0026
56.0	—	—	—	—	—	—	—	—	—	—	.0000	+.0000	+.0004	-.0001	-.0023

Table 16.10: Ring tension in cylindrical walls: Fixed base free top (IS:3370, part IV)

Ring tension N_d = Coefficient × wHR (kN/m)

Coefficients at point

$\dfrac{H^2}{Dt}$	0.0H	0.1H	0.2H	0.3H	0.4H	0.5H	0.6H	0.7H	0.8H	0.9H	0.75H	0.80H	0.85H	0.90H	0.95H
0.4	+0.149	+0.134	+0.120	+0.101	+0.082	+0.066	+0.049	+0.029	+0.014	+0.004	—	—	—	—	—
0.8	+0.263	+0.239	+0.215	+0.190	+0.160	+0.130	+0.096	+0.063	+0.034	+0.010	—	—	—	—	—
1.2	+0.283	+0.271	+0.254	+0.234	+0.209	+0.180	+0.142	+0.099	+0.054	+0.016	—	—	—	—	—
1.6	+0.265	+0.268	+0.268	+0.266	+0.250	+0.226	+0.185	+0.134	+0.075	+0.023	—	—	—	—	—
2.0	+0.234	+0.251	+0.273	+0.285	+0.285	+0.274	+0.232	+0.172	+0.104	+0.031	—	—	—	—	—
3.0	+0.134	+0.203	+0.267	+0.322	+0.357	+0.362	+0.330	+0.262	+0.157	+0.052	—	—	—	—	—
4.0	+0.067	+0.164	+0.256	+0.339	+0.403	+0.429	+0.409	+0.334	+0.210	+0.073	—	—	—	—	—
5.0	+0.025	+0.137	+0.245	+0.346	+0.428	+0.477	+0.469	+0.398	+0.259	+0.092	—	—	—	—	—
6.0	+0.018	+0.119	+0.234	+0.344	+0.441	+0.504	+0.514	+0.447	+0.301	+0.112	—	—	—	—	—
8.0	-0.011	+0.104	+0.218	+0.335	+0.443	+0.534	+0.575	+0.530	+0.381	+0.151	—	—	—	—	—
10.0	-0.011	+0.098	+0.208	+0.323	+0.443	+0.542	+0.608	+0.589	+0.440	+0.179	—	—	—	—	—
12.0	-0.005	+0.097	+0.202	+0.312	+0.429	+0.543	+0.628	+0.633	+0.494	+0.211	—	—	—	—	—
14.0	-0.002	+0.098	+0.200	+0.306	+0.420	+0.539	+0.639	+0.666	+0.541	+0.241	—	—	—	—	—
16.0	-0.0000	+0.099	+0.199	+0.304	+0.412	+0.531	+0.641	+0.687	+0.582	+0.265	—	—	—	—	—
20.0	—	—	—	—	—	—	—	—	—	—	+0.716	+0.654	+0.520	+0.325	+0.115
24.0	—	—	—	—	—	—	—	—	—	—	+0.746	+0.702	+0.577	+0.372	+0.137
32.0	—	—	—	—	—	—	—	—	—	—	+0.782	+0.768	+0.663	+0.459	+0.182
40.0	—	—	—	—	—	—	—	—	—	—	+0.800	+0.805	+0.731	+0.530	+0.217
48.0	—	—	—	—	—	—	—	—	—	—	+0.791	+0.828	+0.785	+0.593	+0.254
56.0	—	—	—	—	—	—	—	—	—	—	+0.763	+0.838	+0.824	+0.536	+0.285

Table 16.11: Moments in cylindrical walls: Hinged base free top (IS:3370, part IV)

Moment M_w = Coefficient × wH^3 (kN·m/m)

Positive sign indicates tension at the outside face

$\dfrac{H^2}{Dt}$	0.1H	0.2H	0.3H	0.4H	0.5H	0.6H	0.7H	0.8H	0.9H	1.0H	0.75H	0.80H	0.85H	0.90H	0.95H
							Coefficients at point								
0.4	+.0020	+.0072	+.0151	+.0230	+.0301	+.0348	+.0357	+.0312	+.0197	0	—	—	—	—	—
0.8	+.0019	+.0064	+.0133	+.0207	+.0271	+.0319	+.0329	+.0292	+.0187	0	—	—	—	—	—
1.2	+.0016	+.0058	+.0111	+.0177	+.0237	+.0280	+.0296	+.0263	+.0171	0	—	—	—	—	—
1.6	+.0012	+.0044	+.0091	+.0145	+.0195	+.0236	+.0255	+.0232	+.0155	0	—	—	—	—	—
2.0	+.0006	+.0033	+.0073	+.0114	+.0158	+.0199	+.0219	+.0205	+.0145	0	—	—	—	—	—
3.0	+.0004	+.0018	+.0040	+.0063	+.0092	+.0127	+.0152	+.0153	+.0111	0	—	—	—	—	—
4.0	+.0001	+.0007	+.0016	+.0033	+.0057	+.0083	+.0109	+.0118	+.0092	0	—	—	—	—	—
5.0	.0000	+.0001	+.0006	+.0016	+.0034	+.0057	+.0080	+.0094	+.0078	0	—	—	—	—	—
6.0	.0000	.0000	+.0002	+.0000	+.0019	+.0039	+.0062	+.0078	+.0068	0	—	—	—	—	—
8.0	.0000	.0000	-.0002	.0000	+.0007	+.0020	+.0038	+.0057	+.0054	0	—	—	—	—	—
10.0	.0000	.0000	-.0002	-.0001	+.0002	+.0011	+.0025	+.0043	+.0045	0	—	—	—	—	—
12.0	.0000	.0000	-.0001	.0002	.0000	+.0005	+.0017	+.0032	+.0039	0	—	—	—	—	—
14.0	.0000	.0000	.0000	-.0001	.0001	.0000	+.0012	+.0026	+.0033	0	—	—	—	—	—
16.0	.0000	.0000	.0000	-.0001	-.0002	-.0004	+.0008	+.0022	+.0029	0	—	—	—	—	—
20.0	—	—	—	—	—	—	—	—	—	—	+.0008	+.0014	+.0020	+.0024	+.0020
24.0	—	—	—	—	—	—	—	—	—	—	+.0005	+.0010	+.0015	+.0020	+.0017
32.0	—	—	—	—	—	—	—	—	—	—	.0000	+.0005	+.0009	+.0014	+.0013
40.0	—	—	—	—	—	—	—	—	—	—	.0000	+.0003	+.0006	+.0011	+.0011
48.0	—	—	—	—	—	—	—	—	—	—	.0000	+.0001	+.0004	+.0008	+.0010
56.0	—	—	—	—	—	—	—	—	—	—	.0000	.0000	+.0003	+.0007	+.0008

Table 16.12: Ring tension in cylindrical walls: Hinged base free top (IS:3370, part IV)

Ring tension N_a = Coefficient × wHR (kN/m)

Positive sign indicates tension

$\dfrac{H^2}{Dt}$	0.0H	0.1H	0.2H	0.3H	0.4H	0.5H	0.6H	0.7H	0.8H	0.9H	0.75H	0.80H	0.85H	0.90H	0.95H
							Coefficients at point								
0.4	+0.474	+0.440	+0.395	+0.352	+0.305	+0.264	+0.215	+0.165	+0.111	+0.057	—	—	—	—	—
0.8	+0.423	+0.402	+0.381	+0.358	+0.330	+0.297	+0.249	+0.202	+0.145	+0.076	—	—	—	—	—
1.2	+0.350	+0.355	+0.361	+0.362	+0.358	+0.343	+0.309	+0.346	+0.186	+0.098	—	—	—	—	—
1.6	+0.271	+0.303	+0.341	+0.369	+0.385	+0.385	+0.362	+0.314	+0.233	+0.124	—	—	—	—	—
2.0	+0.205	+0.260	+0.321	+0.373	+0.411	+0.434	+0.419	+0.369	+0.280	+0.151	—	—	—	—	—
3.0	+0.074	+0.179	+0.281	+0.375	+0.449	+0.506	+0.519	+0.479	+0.375	+0.210	—	—	—	—	—
4.0	+0.017	+0.137	+0.253	+0.367	+0.469	+0.545	+0.579	+0.553	+0.447	+0.256	—	—	—	—	—
5.0	+0.008	+0.144	+0.235	+0.356	+0.469	+0.562	+0.617	+0.606	+0.503	+0.294	—	—	—	—	—
6.0	-0.011	+0.103	+0.223	+0.343	+0.463	+0.566	+0.639	+0.643	+0.547	+0.327	—	—	—	—	—
8.0	-0.015	+0.096	+0.208	+0.324	+0.443	+0.564	+0.661	+0.697	+0.621	+0.386	—	—	—	—	—
10.0	-0.008	+0.095	+0.200	+0.311	+0.428	+0.552	+0.666	+0.730	+0.678	+0.433	—	—	—	—	—
12.0	-0.002	+0.097	+0.197	+0.302	+0.417	+0.541	+0.664	+0.750	+0.720	+0.477	—	—	—	—	—
14.0	0.000	+0.098	+0.197	+0.299	+0.408	+0.531	+0.659	+0.761	+0.752	+0.513	—	—	—	—	—
16.0	+0.002	+0.100	+0.198	+0.299	+0.403	+0.521	+0.650	+0.764	+0.776	+0.543	—	—	—	—	—
20.0	—	—	—	—	—	—	—	—	—	—	+0.812	+0.817	+0.755	+0.603	+0.344
24.0	—	—	—	—	—	—	—	—	—	—	+0.816	+0.839	+0.793	+0.647	+0.377
32.0	—	—	—	—	—	—	—	—	—	—	+0.814	+0.861	+0.847	+0.721	+0.436
40.0	—	—	—	—	—	—	—	—	—	—	+0.802	+0.866	+0.880	+0.778	+0.483
48.0	—	—	—	—	—	—	—	—	—	—	+0.791	+0.854	+0.900	+0.820	+0.527
56.0	—	—	—	—	—	—	—	—	—	—	+0.781	+0.859	+0.911	+0.852	+0.563

v. *Thickness of tank wall*:

If t = thickness of tank wall from cracking considerations

$$\left[\dfrac{\left(\dfrac{wH \cdot d}{2}\right)}{1000t + (m-1)A_{st}}\right] = \sigma_{ct} \quad \therefore \quad \left[\dfrac{264.6 \times 10^3}{1000t + (13-1)2646}\right] = 1.2$$

$\therefore \qquad t = 188.7$ mm

Adopt 190 mm thick tank walls.

vi. *Reinforcement in tank wall*:

Spacing of hoops increased towards top. At top, minimum reinforcement is 0.3%

$$A_{st} = \left(\dfrac{0.3}{100} \times 1000 \times 190\right) = 570 \text{ mm}^2$$

$\therefore$ Spacing of hoops

(using 20 mm diameter bars) $= \left(\dfrac{1000 \times 314}{570}\right) = 550$ mm

Minimum spacing > 3 times the thickness of wall

$\qquad\qquad\qquad > 3 \times 190$

$\qquad\qquad\qquad > 570$ mm

Adopt 20 mm diameter hoop at 550 mm centres at top. Spacing at a depth of 2 m below the top.

$$A_{st} = \left[\dfrac{wHD/2}{100}\right] = \left[\dfrac{(0.5 \times 10 \times 2 \times 12.6)10^3}{100}\right] = 1260 \text{ mm}^3$$

Spacing of 20 mm bars $= \left(\dfrac{1000 \times 314}{1260}\right) = 250$ mm c/c

Distribution and temperature reinforcement is provided in the vertical direction.

Area of vertical steel $= 0.3\% = \left(\dfrac{0.3}{100} \times 1000 \times 190\right) = 570 \text{ mm}^2$

Spacing of 10 mm bars $= \left(\dfrac{1000 \times 78.5}{570}\right) = 137$ mm

Use 10 mm diameter at 135 mm c/c.

vii. *Tank floor slab*:

Provide nominal thickness of 150 mm

Minimum area of steel $A_{st} = \left(\dfrac{0.3}{100} \times 150 \times 1000\right) = 450 \text{ mm}^2$ in each direction

Provide half the reinforcement near each face

$$\therefore \qquad A_{st} = 225 \text{ mm}^2$$

Spacing of 8 mm diameter bars $= \left(\dfrac{1000 \times 50}{225} \right) = 220$

Use 8 mm diameter bars at 200 mm c/c in both directions and on each face. The reinforcement details are shown in Fig. 16.4b.

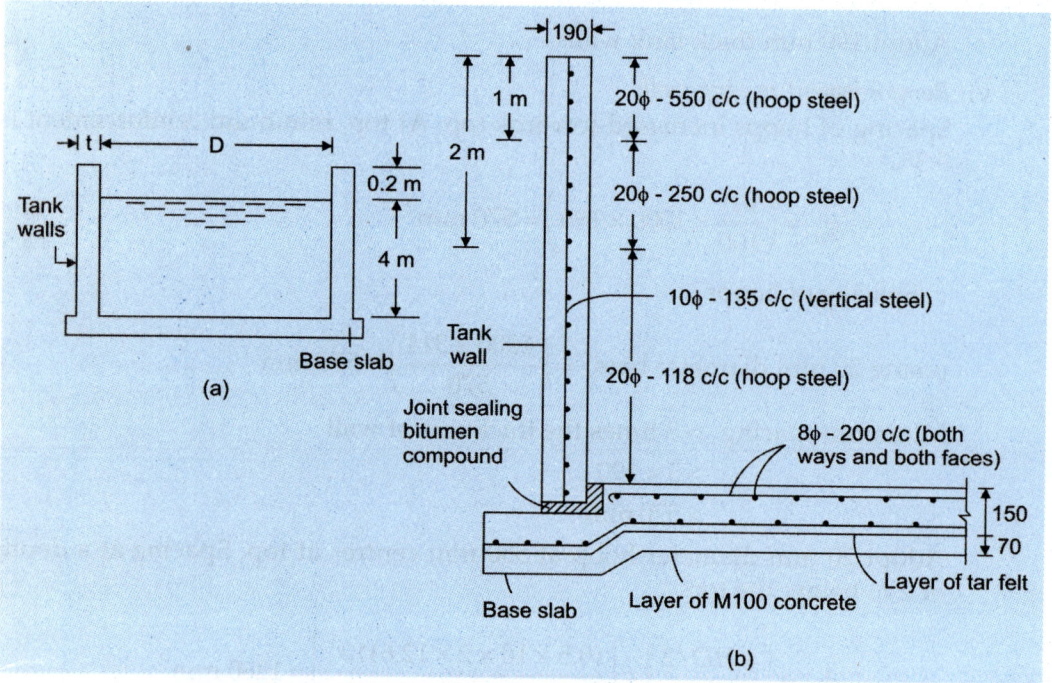

Fig. 16.4: (a) Dimensions of circular tank, and (b) Reinforcements in tank walls and base slab

2. Design a circular tank with fixed base for capacity of 400000 litres. The depth of water is to be 4 m. Free board is 200 mm. Use M20 grade concrete and grade-I mild steel. Permissible direct tensile stress in concrete is 1.2 N/mm². Permissible stress in steel in direct tension is 100 N/mm². Sketch the details of reinforcement in tank walls. Adopt IS code tables for coefficients.

   i. *Data*:

   Capacity of tank = 400000 litres

   Depth of water = 4 m

   Free board = 200 mm

   Density of water w = 10 kN/m³

   M20 concrete and grade-I steel

   ii. *Permissible stresses*:

   Direct tensile stress in concrete (σ_{ct}) = 1.2 N/mm²

Tensile stress in steel = 100 N/mm²

$\sigma_{cb} = 7\,\text{N/mm}^2$ $\qquad\qquad j = 0.841$

$Q = 14.01$ $\qquad\qquad m = 13$

iii. *Dimensions of tank*:

Referring to Fig. 16.5a, if D = diameter

$$\left(\frac{\pi D^2}{4} \times 4\right) = \left(\frac{400000 \times 10^3}{10^6}\right)$$

$\therefore \qquad\qquad D = 11.2\,\text{m}$

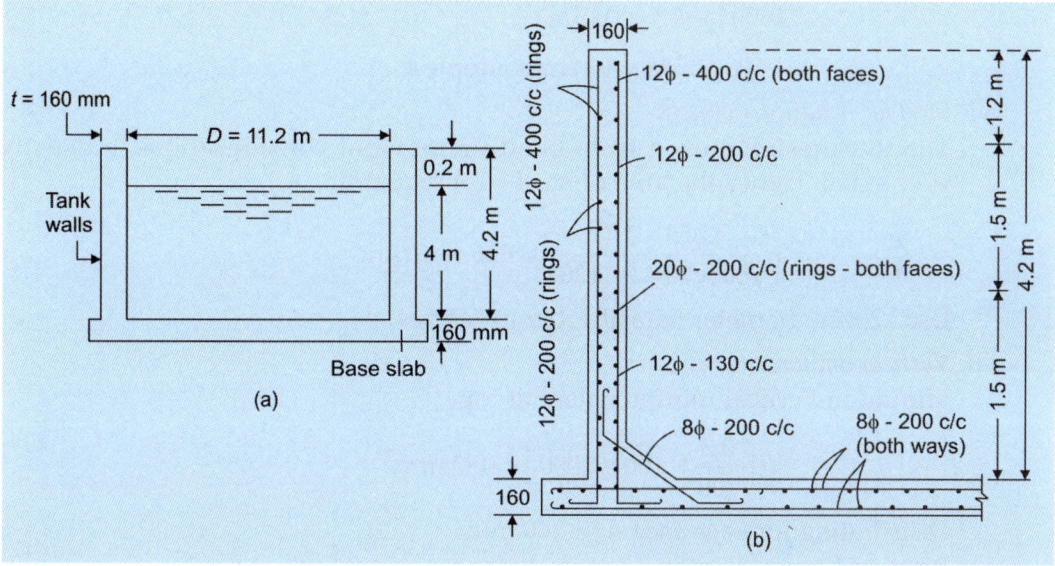

Fig. 16.5: (a) Dimensions of circular tank, and (b) Reinforcement details in circular tank fixed at base

Overall height of tank = (4 + 0.2) = 4.2 m

The thickness of walls and base slab are assumed to be 160 mm thick.

iv. *Bending moment, ring tension and shear*:

$$\left(\frac{H^2}{Dt}\right) = \left(\frac{4.2^2}{11.2 \times 0.16}\right) = 9.84 \simeq 10$$

Referring to IS:3370 (part IV), 1967, Table 14.2.

Maximum BM $\quad = -0.0122\,wH^3$ at base

$\qquad\qquad\quad = (-0.0122 \times 10 \times 4.2^3) = 9.038\,\text{kN·m/m}$

Maximum shear $= 0.158wH^2 = (0.158 \times 10 \times 4.2^2) = 27.87\,\text{kN at base}$

Maximum ring tension $= 0.608\left(\dfrac{wHd}{2}\right)$ at 0.6H from top

$$= \left(0.608 \times 10 \times 4.2 \times \frac{11.2}{2}\right) = 146.83\,\text{kN},$$

acting at 2.52 metres from top.

v. *Steel for hoop tension*:

$$\therefore \qquad A_{st} = \left(\frac{146.83 \times 10^3}{100} \right) = 1468 \text{ mm}^2/\text{m}$$

Using 20 mm bars spacing $= \left(\dfrac{1000 \times 314}{1468} \right) = 213 \text{ mm}$

Use 20 mm bars at 200 mm c/c ($A_{st} = 1571 \text{ mm}^2$)

Wall thickness required from hoop stress consideration is given by

$$\left\{ \frac{146.83 \times 10^3}{1000t + (13-1)1571} \right\} = 1.2$$

$$\therefore \qquad t = 106.6 \text{ mm} < 160 \text{ mm adopted}$$

vi. *Steel for bending moment*:

The thickness required from bending moment considerations is usually very small. Hence the area of steel (A_{st}) required

$$= \left(\frac{9.038 \times 10^6}{100 \times 0.841 \times 130} \right) = 826 \text{ mm}^2/\text{m}$$

Use 12 mm diameter bars at 130 mm centres.

vii. *Vertical reinforcement*:

Minimum vertical reinforcement at top:

$$= \left(\frac{0.3}{100} \times 160 \times 1000 \right) = 480 \text{ mm}^2$$

Distributing for each face $A_s = 240 \text{ mm}^2$

Provide 12 mm diameter bars at 400 mm c/c, on both faces.

viii. *Base slab reinforcement*:

At junction of wall and slab, provide 12 mm diameter at 130 mm centres.

At centre, top and bottom $A_s = \left(\dfrac{0.3}{100} \times 160 \times 1000 \right) = 480 \text{ mm}^2$

Providing at top and bottom $A_s = 280 \text{ mm}^2$

Use 8 mm bars at 200 c/c both at top and bottom and both ways.

ix. *Check for shear stress*:

Maximum shear at base $V = 27.87 \text{ kN/m}$

Maximum shear stress

$$\tau_v = \left(\frac{V}{bjd} \right) \text{ (refer to IS:3370)}$$

$$= \left(\frac{27.87 \times 10^3}{1000 \times 0.841 \times 130} \right)$$

$$= 0.25 \text{ N/mm}^2 < 1.7 \text{ N/mm}^2 \text{ (permissible)}$$

The details of reinforcements are shown in Fig. 16.5b.

6.11 DESIGN OF RECTANGULAR TANKS

Approximate Design Method

a. *For tanks of ratio* $(L/B) < 2$

Referring to Fig. 16.6. For the bottom height of $h = H/4$ or 1 m (whichever is more), the bending is in the vertical plane and this portion is designed as cantilever.

The corners are designed for the maximum moment obtained after moment distribution with the intensity of pressure $p = w(H - h)$.

In the absence of moment distribution, the bending moments may be computed by the following approximate relations:

Bending moment at centre of span

$= \left(\dfrac{pB^2}{16}\right)$ (producing tension on outer face)

Bending moment at ends of span

$= \left(\dfrac{pB^2}{12}\right)$ (producing tension on water face)

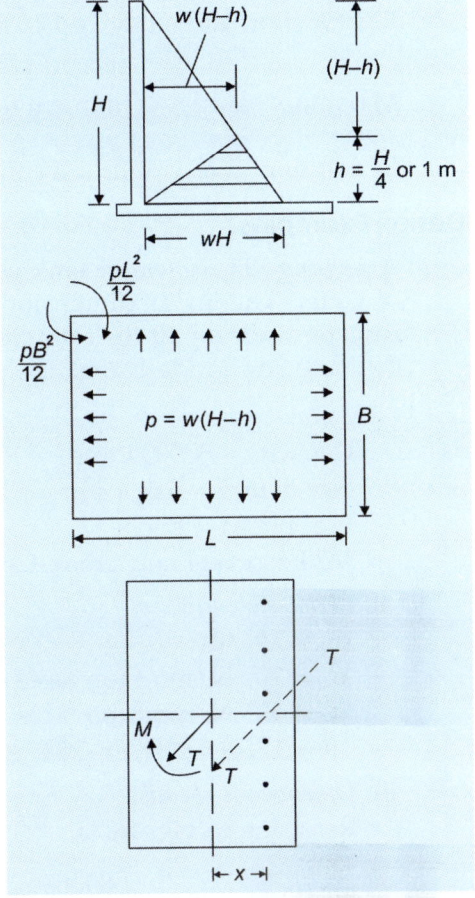

Fig. 16.6: Forces acting on rectangular tank walls

In addition to the bending moments, the walls are subjected to direct tension given by:

Direct tension on long walls $T_L = w(H - h)\, B/2$

Direct tension on short walls $T_b = w(H - h)\, L/2$

Design moment $= (M - Tx)$

For BM, $A_{st} = \left[\dfrac{M - Tx}{\sigma_{st}\, jd}\right]$

$\therefore$ $A_{st} = (A_{st1} + A_{st2})$

For direct tension $A_{st2} = (T/\sigma_{st})$

b. *Ratio of* $(L/B) > 2$

In this case, long walls are assumed to bend vertically and hence designed as cantilevers. Short walls are assumed to bend horizontally supported on long walls above $H/4$ or 1 m from bottom.

Bending moment for long walls $= \left(\dfrac{wH^3}{6}\right)$

BM for short walls (above 1 m from base) $= \dfrac{w(H - h)B^2}{16}$

Maximum cantilever moment for short wall $= \left(\dfrac{wHh \cdot h}{2 \times 3}\right) = \left(\dfrac{wH \cdot h^2}{6}\right)$

In addition, direct pulls are considered for long and short walls.

Design Examples

1. A rectangular RC water tank with an open top is required to store 80000 litres of water. The inside dimensions of tank may be taken as 6 m × 4 m. The tank rests on walls on all the four sides. Design the side walls of the tank using M20 concrete and grade-I steel.

   i. *Data*:

      Capacity = 80000 litres
      Size of tank = 6 m × 4 m
      Free board = 15 cm
      M20 concrete and grade-I steel.

   ii. *Permissible stresses*:

      $\sigma_{cb} = 7\,\text{N/mm}^2$
      $\sigma_{st} = 100\,\text{N/mm}^2$ (on faces near water face)
      $\sigma_{st} = 125\,\text{N/mm}^2$ (on faces away from water face)
      $m = 13, Q = 1.41, j = 0.84$

   iii. *Dimensions of tank*:

      Referring to Fig. 16.7a,

      height of water $= \left(\dfrac{80000 \times 10^3}{600 \times 400}\right) = 335\,\text{cm}$

      Free board = 15 cm
      Height of side walls = (335 + 15) = 350 cm = 3.5 m
      $(L/B) = 6/4 = 1.5 < 2$
      ∴ Walls designed at continuous slab subjected to water pressure above $(H/4)$ or 1 m from bottom.
      ∴ $p = w(H - h)$ at $xx = (10 \times 2.5) = 25\,\text{kN/m}^2$

   iv. *Moments in side walls*:

      The moments in side walls is determined by moment distribution.
      Fixed end moment:
      Long walls

      $$\left(\dfrac{p \cdot L^2}{12}\right) = \left(\dfrac{25 \times 6^2}{12}\right) = 75\,\text{kN} \cdot \text{m}, \left(\dfrac{pB^2}{12}\right) = \left(\dfrac{25 \times 4^2}{12}\right) = 34\,\text{kN} \cdot \text{m}$$

      $$\left(\dfrac{p \cdot L^2}{8}\right) = \left(\dfrac{25 \times 6^2}{8}\right) = 112.5\,\text{kN} \cdot \text{m}, \left(\dfrac{pB^2}{8}\right) = \left(\dfrac{25 \times 4^2}{8}\right) = 50\,\text{kN} \cdot \text{m}$$

      ∴ Moment distribution is shown in Fig. 16.7b.

Moment at support = 59 kN·m

At centre (long walls) = (112 − 59) = 53 kN·m

At centre (short walls) = (50 − 59) = −9 kN·m

The bending moment diagram is shown in Fig. 16.7(c)

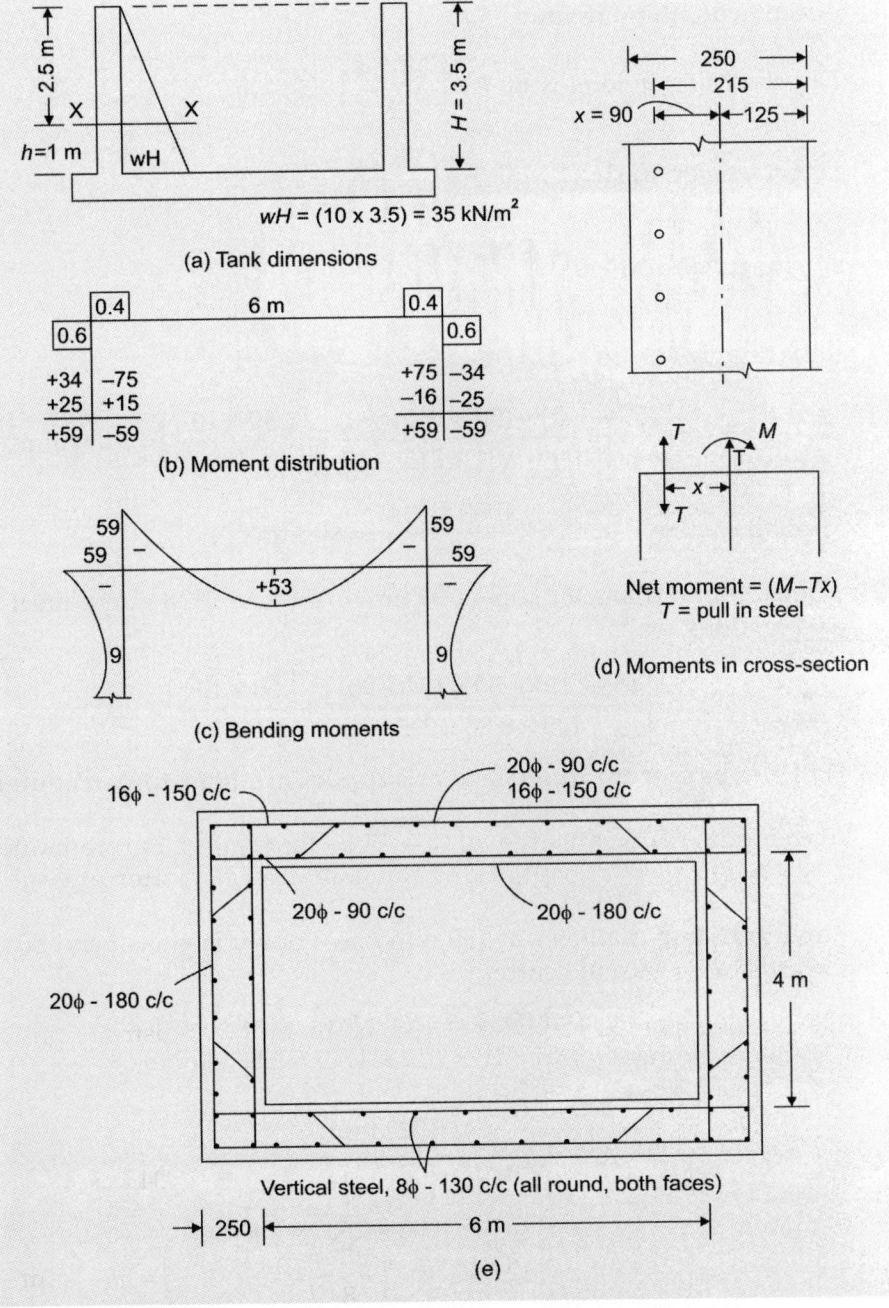

Fig. 16.7: Reinforcement details in water tank ($L/B < 2$)

v. *Steel for hoop tension*:
Maximum moment = 59 kN·m

$$\therefore \qquad d = \sqrt{\frac{59 \times 10^6}{1.41 \times 1000}} = 204 \text{ mm} \quad (\text{at } XX)$$

Adopt overall depth = 250 mm and
effective depth = 215 mm

Direct tension in long wall $T = \left(\dfrac{25 \times 4}{2}\right) = 50$ kN

Direct tension in short wall $T = \left(\dfrac{25 \times 6}{2}\right) = 75$ kN

$$A_{st} \text{ (long wall corners)} = \left[\frac{M - Tx}{\sigma_{st}dj}\right] + \frac{T}{\sigma_{st}}$$

Referring to Fig. 16.7(d),

$$\therefore \qquad A_{st} = \left[\frac{(59 \times 10^6) - (50 \times 10^3 \times 90)}{100 \times 0.84 \times 215}\right] + \left(\frac{50 \times 10^3}{100}\right) = 3480 \text{ mm}^2$$

Spacing of 20 mm bars $= \left(\dfrac{1000 \times 314}{3480}\right) = 90$ mm c/c

Adopt 20 mm diameter bars at 90 mm c/c (A_{st} = 3928 mm²). Steel at centre of span (long walls)

$$= \left[\frac{(53 \times 10^6 - 50 \times 10^3 \times 90)}{125 \times 0.86 \times 215}\right] + \left[\frac{50 \times 10^3}{125}\right] = 2500 \text{ mm}^2$$

Half the bars from inner face at support are bent toward outer face at centre providing an area of $\left(\dfrac{3928}{2}\right) = 1964$ mm². For remaining area (2500 − 1964) = 536 mm².

Provide 16 mm diameter at 150 mm c/c. For short walls bend 50% of bars towards outer face at centre.

vi. *Steel for cantilever moment (for 1 m height from bottom)*:
Cantilever moment = (3.5 × 10 × 1/2 × 1/3) = 5.833 kN·m

$$\therefore \qquad A_{st} = \left(\frac{5.833 \times 10^6}{100 \times 0.84 \times 215}\right) = 323 \text{ mm}^2$$

Minimum steel = 0.3% $= \left(\dfrac{0.3}{100} \times 1000 \times 250\right) = 750$ mm²

Steel on each face $= \left(\dfrac{750}{2}\right) = 375$ mm²

$$\text{Spacing of 8 mm bars} = \left(\frac{1000 \times 50}{375}\right) = 130 \text{ mm c/c}$$

Adopt 8 mm diameter bars at 130 mm c/c on both faces as shown in Fig. 16.7e.

2. A reinforced concrete water tank resting on ground is 6 m × 2 m, with a maximum depth of 2.5 m. Using M20 concrete and grade-I steel design the tank walls.

   i. *Data*:

   Size of tank = 6 m × 2 m

   Depth of tank = 2.5 m

   M20 concrete and grade-I steel.

   ii. *Permissible stresses*:

   $\sigma_{cb} = 7 \text{ N/mm}^2$

   $\sigma_{st} = 100 \text{ N/mm}^2$ (on faces near water face)

   $\sigma_{st} = 125 \text{ N/mm}^2$ (on faces away from water face)

   $m = 13, Q = 1.41, j = 0.84$

   iii. *Dimensions of tank*:

   $L = 6 \text{ m}, B = 2 \text{ m}$

   $$\text{Ratio}\left(\frac{L}{B}\right) = \frac{6}{2} = 3 > 2$$

   Long walls are designed as vertical cantilevers and short walls as spanning horizontally between long walls.

   Maximum bending moment at base,

   $$\text{long wall} = \left(\frac{wH^3}{6}\right) = \left(\frac{10 \times 2.5^3}{6}\right) = 26.04 \text{ kN} \cdot \text{m}$$

   $$d = \sqrt{\frac{M}{Qb}} = \sqrt{\frac{26.04 \times 10^6}{1000 \times 1.41}} = 136 \text{ mm}$$

   Using 16 mm diameter bars and 25 mm clear cover.

   Overall depth = 170 mm

   Effective depth = 137 mm

   $$\therefore \quad A_{st} = \left(\frac{26.04 \times 10^6}{100 \times 0.84 \times 137}\right) = 2262 \text{ mm}^2$$

   $$\text{Spacing of 16 mm bars} = \left(\frac{100 \times 201}{2262}\right) = 88.8 \text{ mm}$$

   Adopt 16 mm diameter bars at 85 mm c/c. Spacing towards the top is increased to 170 mm c/c for the top 1 m.

   Intensity of pressure 1 m above base

   $$p = w(H - h) = 10(2.5 - 1) = 15 \text{ kN/m}^2$$

Direct tension in long walls $T = \left(\dfrac{15 \times 2}{2} \right) = 15 \text{ kN}$

$\therefore \qquad A_{st} = \left(\dfrac{15 \times 10^3}{100} \right) = 150 \text{ mm}^2$

But minimum area $= 0.3\% = \left(\dfrac{0.3}{100} \times 170 \times 1000 \right) = 510 \text{ mm}^2$

Spacing of 10 mm diameter bars $= \left(\dfrac{1000 \times 79}{510} \right) = 154 \text{ mm}$

Since steel is provided on both faces, provide 10 mm diameter bars at 300 mm c/c on both faces.

iv. *Design of short walls*:

$$p = 15 \text{ kN/m}^3$$

Effective span of horizontally spanning slab $= (2 + 0.17) = 2.17 \text{ m}$
Bending moment (corner section)

$$= \left(\dfrac{pL^2}{12} \right) = \left(\dfrac{15 \times 2.17^2}{12} \right) = 5.886 \text{ kN} \cdot \text{m}$$

Tension transferred per metre height of short wall $= (15 \times 1) = 15 \text{ kN}$

$\therefore \qquad A_{st} = \left[\dfrac{M - Tx}{\sigma_{st} jd} + \dfrac{T}{\sigma_{st}} \right]$

$\qquad = \left[\dfrac{5.886 \times 10^6 - 15 \times 10^3 \times (137 - 85)}{100 \times 0.84 \times 137} \right] + \left[\dfrac{15 \times 10^3}{100} \right] = 594 \text{ mm}^2$

Spacing of 10 mm bars $= \left(\dfrac{1000 \times 79}{594} \right) = 132 \text{ mm}$

Adopt 16 mm bars at 130 mm centres. At mid span section, the bending moment is $(pL^2/24)$ and hence provide 10 mm diameter bars at 260 mm centres away from water face.

v. *Design for cantilevering effect of short wall*:

Maximum BM $= (1/2 \times 10 \times 2.5 \times 1 \times 1/3) = 4.166 \text{ kN·m}$
Effective depth (using 10 mm bars) $= (170 - 25 - 10 - 5) = 130 \text{ mm}$

$$A_{st} = \left(\dfrac{4.166 \times 10^3}{100 \times 0.84 \times 130} \right) = 380 \text{ mm}^2$$

But 0.3% of gross area $= \left(\dfrac{0.3}{100} \times 170 \times 1000 \right) = 510 \text{ mm}^2$

$\therefore$ Spacing of 10 mm diameter bars

$$= \left(\dfrac{1000 \times 79}{510} \right) = 154 \text{ mm}$$

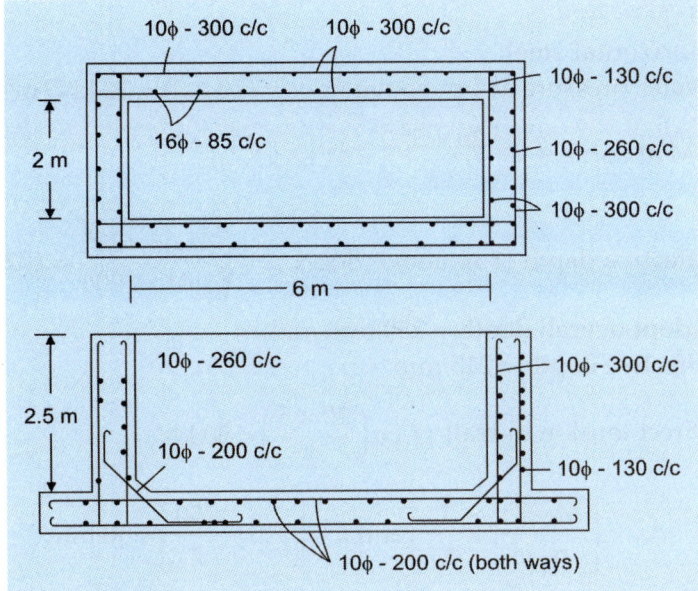

10φ - 300 c/c   10φ - 300 c/c

10φ - 130 c/c

16φ - 85 c/c

10φ - 260 c/c

2 m

10φ - 300 c/c

6 m

10φ - 260 c/c   10φ - 300 c/c

2.5 m

10φ - 200 c/c   10φ - 130 c/c

10φ - 200 c/c (both ways)

Fig. 16.8: Reinforcement details in rectangular tank ($L/B < 2$)

Steel is provided on both faces, so adopt 10 mm diameter bars at 300 mm c/c on both faces in the vertical direction. The details of reinforcements are shown in Fig. 16.8.

3. Design the side walls of a square of RCC tank of capacity 70000 litres of water. Depth of water in the tank is 2.8 m. Free board is 0.2 m. Adopt M20 concrete and grade-I steel. Tensile stresses in steel is limited to 100 N/mm² at water face and 125 N/mm² away from face. Sketch the details of reinforcements in the walls of the tank.

   i. *Data*:
     Capacity = 70000 litres
     Depth of water in tank = 2.8 m
     Free board = 0.2 m
     M20 concrete and grade-I steel

  ii. *Permissible stresses*:
     $\sigma_{cb} = 7\,\text{N/mm}^2$
     $\sigma_{st} = 100\,\text{N/mm}^2$ (on faces near water face)
     $\sigma_{st} = 125\,\text{N/mm}^2$ (on faces away from water face)
     $m = 13, Q = 1.41, j = 0.84$

 iii. *Dimensions of tank*:
     Referring to Fig. 16.9a,
     let L = side of the square tank

$$(L^2 \times 2.8) = \left(\frac{70000 \times 10^3}{10^6} \right)$$

∴ $L = 5\,\text{m}$
Adopt a square tank of size 5 m × 5 m × 3 m.

iv. *Design of side wall*:

a. Horizontal steel:

Water pressure at 1 m above floor = $(10 \times 2) = 20$ kN/m^2

BM at corners $= \left(\dfrac{20 \times 5^2}{12} \right) = 41.66$ kN $\cdot$ m

Effective depth (1 m above floor) $d = \sqrt{\dfrac{41.66 \times 10^6}{1.41 \times 1000}} = 172$ mm

Adopt overall depth = 250 mm and
Effective depth = 215 mm

Direct tension in wall $(T) = \left(\dfrac{20 \times 5}{2} \right) = 50$ kN

$$A_{st} = \left[\dfrac{M - Tx}{\sigma_{st} jd} + \dfrac{T}{\sigma_{st}} \right] \text{ and } x = \left[215 - \dfrac{250}{2} \right] = 90 \text{ mm}$$

$$\therefore \ A_{st} = \left[\dfrac{41.66 \times 10^6 - 50 \times 10^3 \times 90}{100 \times 0.84 \times 215} \right] + \left[\dfrac{56 \times 10^3}{100} \right] = 2558 \text{ mm}^2$$

Spacing of 20 mm bars $= \left(\dfrac{1000 \times 314}{2558} \right) = 120$ mm c/c

b. Vertical steel:

Cantilever moment up to 1 m height from bottom
$= (1/2 \times 30 \times 1 \times 1/3) = 5$ kN·m

$$\therefore \ A_{st} = \left(\dfrac{5 \times 10^6}{100 \times 0.84 \times 215} \right) = 276 \text{ mm}^2$$

Minimum steel $= \left(\dfrac{0.3}{100} \times 250 \times 1000 \right) = 750$ mm^2

Spacing of 10 mm bars $= \left(\dfrac{1000 \times 78.5}{750} \right) = 100$ mm c/c

Adopt 10 mm bars at 200 mm c/c at both faces.

c. Horizontal steel at mid-span section (away from water face):

Maximum moment $= \left(\dfrac{pL^2}{16} \right) = \left(\dfrac{20 \times 5^2}{16} \right) = 31.25$ kN·m

$$\therefore \ A_{st} = \left[\dfrac{31.25 \times 10^6 - 50 \times 10^3 \times 90}{125 \times 0.86 \times 215} \right] + \left[\dfrac{50 \times 10^3}{125} \right] = 1657 \text{ mm}^2$$

Half the bars from inner face at support are bent toward outer face at centre providing an area of $(0.5 \times 2558) = 1279$ mm^2.

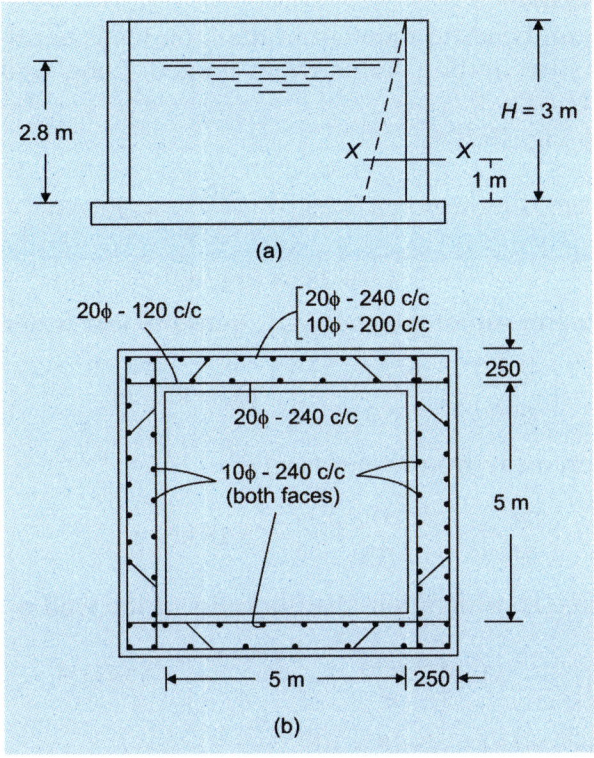

Fig. 16.9: (a) Dimensions of tank, and (b) Reinforcement details in square tank

For remaining area = (1657 − 1279) = 378 mm^2

Provide 10 mm bars at a spacing $= \left(\dfrac{100 \times 78.5}{378} \right) = 200$ mm c/c

The reinforcement details in the tank walls are shown in Fig. 16.9b.

4. Design a RC tank of internal dimensions 10 m × 3 m × 3 m. The tank is to be provided underground. The soil surrounding the tank is likely to get wet. Angle of repose of soil in dry state is 30° and in wet state is 6°. Adopt suitable working stresses. Soil weights 20 kN·m^3. Adopt M20 concrete and grade-I steel.

   i. *Data*:

     Size of tank = 10 m × 3 m × 3 m

     Tank surrounded by soil on all sides

     Angle of repose of soil: dry state = 30°, wet state = 6°

     Density of soil = 20 kN/m^3

  ii. *Permissible stresses*:

     $\sigma_{st} = 100$ N/mm^2

     Permissible stress in tension due to bending in concrete

     $\sigma_{ct} = 1.7$ N/mm^2 (cracking consideration)

     $Q = 1.41, j = 0.84, m = 13$

iii. *Design of long walls:*

The maximum bending moment in the long walls occur for the case, tank empty and surrounding soil is water logged. Long walls are designed as cantilevers ($L/B > 2$).

Referring to Fig. 16.10a.

$$\text{Pressure exerted by wet soil} = wh\left[\frac{1-\sin\phi}{1+\sin\phi}\right] = (20 \times 3)\left[\frac{1-\sin 6°}{1+\sin 6°}\right]$$

$$p = 48.64 \text{ kN/m}^2$$

Considering 1 m run of the wall M_{max} (tension near water face)

$$= \left(\frac{ph^2}{33.5}\right) = \left(\frac{48.64 \times 3^2}{33.5}\right) = 13.06 \text{ kN} \cdot \text{m}$$

M_{max} (tension away from water face)

$$= \left(\frac{ph^2}{15}\right) = \left(\frac{48.64 \times 3^2}{15}\right) = 29.18 \text{ kN} \cdot \text{m}$$

From cracking consideration, the thickness of the wall is determined.

$$M = \left[\frac{\sigma_{ct}bD^2}{6}\right] = \left[\frac{1.7bD^2}{6}\right] = 0.28bD^2$$

$$\therefore (0.28 \times 1000 \times D^2) = (29.18 \times 10^6)$$

$$\therefore \qquad D = \left(\frac{29.18 \times 10^6}{0.28 \times 1000}\right) = 323 \text{ mm}$$

Provide overall depth $D = 350$ mm

Effective depth $= (350 - 40) = 310$ mm

$$A_{st} = \left(\frac{29.18 \times 10^6}{100 \times 0.84 \times 310}\right) = 1120 \text{ mm}^2$$

$$\text{Spacing of 18 mm diameter bars} = \left(\frac{254 \times 1000}{1120}\right) = 226 \text{ mm c/c}$$

Provide 18 mm diameter vertical bars at 220 mm centres.

Steel for BM = 13.06 kN·m

$$A_{st} = \left(\frac{13.06 \times 10^6}{100 \times 0.84 \times 310}\right) = 502 \text{ mm}^2$$

$$\text{Spacing of 12 mm bars} = \left(\frac{113 \times 1000}{502}\right) = 225 \text{ mm c/c}$$

Horizontal reinforcement in long walls

$$= \left(\frac{0.3}{100} \times 350 \times 1000\right) = 1050 \text{ mm}^2/\text{m}$$

Providing on both faces, area of steel for each = 525 mm²/m
Use 10 mm diameter bars at 150 mm c/c horizontally on both faces.

iv. *Design of short walls:*

Short walls are designed as spanning between the long walls.

Intensity of earth pressure at bottom $p = 48.64$ kN/m²

Maximum moment at corners = $(p \cdot L^2/12) = (48.64 \times 3.35^2)/12$
$$= 45.48 \text{ kN·m}$$

$$\therefore \quad d = \sqrt{\frac{45.84 \times 10^6}{1.41 \times 1000}} = 180 \text{ mm}$$

Actual thickness of walls provided = 350 mm

$$\therefore \quad A_{st} = \left(\frac{45.48 \times 10^6}{100 \times 0.84 \times 310}\right) = 1746 \text{ mm}^2$$

Spacing of 18 mm diameter bars $= \left(\frac{254 \times 1000}{1746}\right) = 145$ mm c/c

Adopt 18 mm diameter bars at 145 mm centres on both faces. The spacing may be increased to 300 mm centres toward the top.

Vertical reinforcement = 10 mm diameter bars at 150 mm c/c on both faces.

v. *Design of roof slab:*

Load: Dead load (100 mm thick slab) = 2.4 kN/m²
Live load   = 1.5 kN/m²
Finishes   = 0.6 kN/m²
Total load   = 4.5 kN/m²

Maximum BM = $(0.125 \times 4.5 \times 3.35^2)$
$$= 6.31 \text{ kN·m}$$

$$d = \sqrt{\frac{6.31 \times 10^6}{1.41 \times 1000}} = 66.89 \text{ mm}$$

Adopt overall depth = 100 mm
Effective depth = 75 mm

$$A_{st} = \left(\frac{6.31 \times 10^6}{100 \times 0.84 \times 75}\right) = 952 \text{ mm}^2$$

Spacing of 16 mm bars $= \left(\frac{201 \times 1000}{952}\right) = 211$ mm

Provide 16 mm diameter bars at 200 mm c/c

Distribution steel $= \left(\frac{0.3 \times 100 \times 1000}{100}\right) = 300$ mm²

Spacing of 8 mm bars $= \left(\frac{50 \times 1000}{300}\right) = 166$ mm

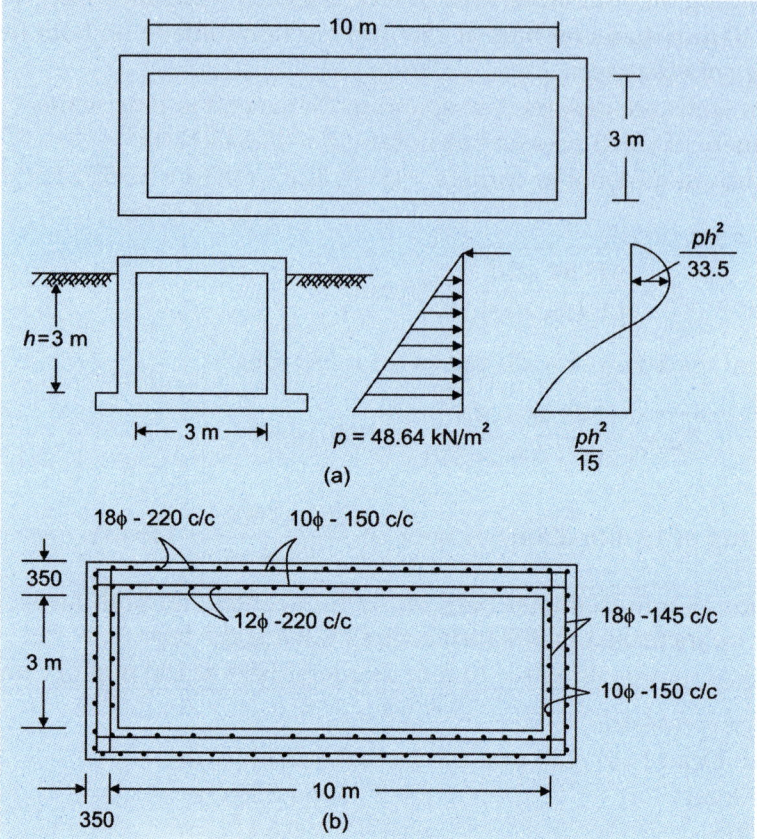

Fig. 16.10: (a) Moments in tank walls, and (b) Reinforcements in tank walls

Adopt 8 mm bars at 160 mm c/c.

The details of reinforcements are shown in Fig. 16.10b.

5. Design an overhead flat bottomed RCC cylindrical water tank to store 100 kl of water. The top of the tank is covered with a dome. Height of staging is 12 m above ground level. Provide 2 m depth of foundation. Intensity of wind pressure may be taken as 1.5 kN/m². Safe bearing capacity of soil at site is 100 kN/m² Adopting M20 grade concrete and Fe415 grade tor steel design the following:

   a. Size of tank

   b. Ring beam at junction of dome and side walls

   c. Side walls of tank

   d. Bottom ring girder

   e. Tank floor slab

   f. Bracing at 4 m intervals

   g. RC columns assuming six column supports

   h. Foundation for the tank.

i. *Data*:

Capacity of tank = 100 kl
Height of supporting tower = 12 m
Number of supporting columns = 6
Depth of foundations = 2 m below ground level
Safe bearing capacity of soil = 250 kN/m²

ii. *Permissible stresses*:

M20 grade concrete and Fe 415 grade tor steel for calculations relating to resistance to cracking (IS:3310).

$\sigma_{ct} = 1.2\,\text{N/mm}^2$ $\qquad\qquad$ $\sigma_{st} = 150\,\text{N/mm}^2$

For strength calculations, the stresses in concrete and steel are the same as that recommended in IS:456 code.

$\sigma_{cc} = 5\,\text{N/mm}^2$ $\qquad\qquad$ $m = 13$
$\sigma_{cb} = 7\,\text{N/mm}^2$ $\qquad\qquad$ $Q = 0.897$
$\qquad\qquad\qquad\qquad\qquad$ $j = 0.90$

iii. *Dimensions of tank*:

Let D = inside diameter of the tank
Assuming an average depth = 0.6 D, we have:

$$\left(\frac{\pi D^2}{4} \times 0.6\,D\right) = 100 \times 10^3 \text{ (litres)} = 100\,\text{m}^3$$

∴ $\qquad\qquad D = 6\,\text{m}$

Height of storage = (0.6 × 6)   = 3.6 m
Free board $\qquad\qquad\qquad$ = 0.2 m
Height of cylindrical portion = 3.8 m

$$\text{Depth of dome } = \left[\frac{1}{6} \times 6\right] = 1\,\text{m}$$

Diameter of dome = 6 m
Spacings of bracings = 4 m
The salient dimensions of the tank and the staging is shown in Fig. 16.11.

iv. *Design of top dome*:

Thickness of dome slab t $\qquad$ = 100 mm
Live load on dome $\qquad\qquad$ = 1.5 kN/m²
Self-weight of dome = (0.1 × 24) = 2.4 kN/m²
Finishes $\qquad\qquad\qquad\qquad$ = 0.1 kN/m²
Total load w $\qquad\qquad\qquad$ = 4.0 kN/m²

If R = radius of dome
$\quad D$ = diameter at base = 6 m
$\quad r$ = central rise = 1 m

$$R = \left[\frac{(D/2)^2 + r^2}{2r}\right] = \left[\frac{3^2 + 1^2}{2 \times 1}\right] = 5\,\text{m}$$

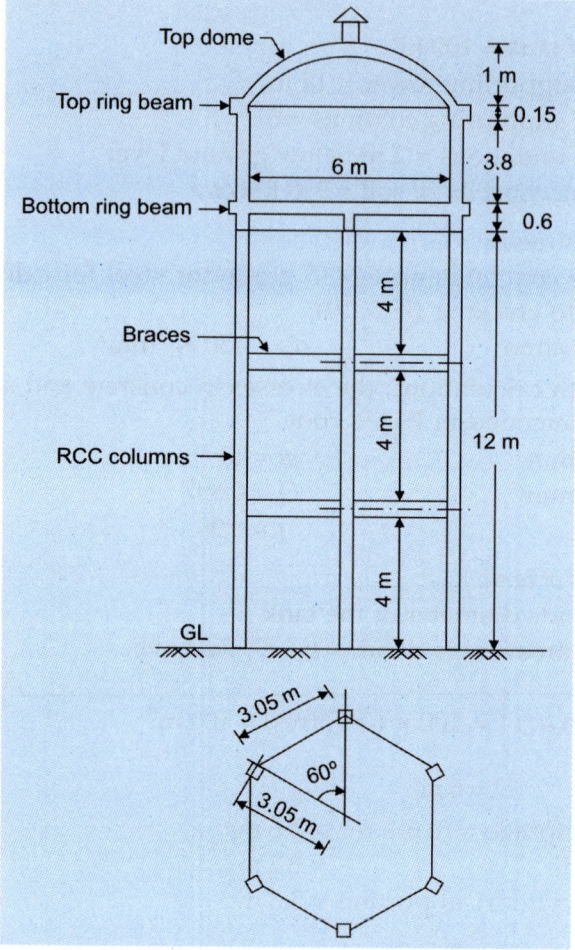

Fig. 16.11: Salient dimensions of tank

$\cos \theta = (4/5) = 0.8$

$\therefore \theta = 36°50'$

Meridional thrust $T_1 = \left(\dfrac{wR}{1 + \cos \theta} \right)$

$$= \left(\dfrac{4 \times 5}{1 + 0.8} \right) = 11.1 \text{ kN/m}$$

Circumferential force $= wR \left[\cos \theta - \left(\dfrac{1}{1 + \cos \theta} \right) \right]$

$$= (4 \times 5) \left[0.8 - \left(\dfrac{1}{1.8} \right) \right] = 5 \text{ kN/m}$$

Meridional stress $= \left(\dfrac{11.1 \times 10^3}{1000 \times 100} \right) = 0.11 \text{ N/mm}^2 < 5 \text{ N/mm}^2$

$$\text{Hoop stress} = \left(\frac{5 \times 10^3}{1000 \times 100}\right) = 0.05 \text{ N/mm}^2 < 5 \text{ N/mm}^2$$

The stresses in dome are within safe permissible limits. Providing nominal reinforcements of 0.3%

$$A_{st} = \left(\frac{0.3 \times 100 \times 1000}{100}\right) = 300 \text{ mm}^2$$

Provide 8 mm diameter bars at 160 mm centres both circumferentially and meridionally.

v. *Design of top ring beam*:

$$\text{Hoop tension } F_t = \left(\frac{T_1 \cos\theta D}{2}\right) = \left(\frac{11.1 \times 0.8 \times 6}{2}\right) = 27 \text{ kN}$$

$$A_{st} = \left(\frac{27 \times 10^3}{150}\right) = 180 \text{ mm}^2$$

Provide 4 bars of 8 mm diameter with 6 mm diameter stirrups at 150 mm centres (A_{st} = 200 mm²).

If A_c = cross-sectional area of ring beam, we have

$$\left(\frac{27 \times 10^3}{A_c + (m-1) A_{st}}\right) = 1.2$$

or

$$\left(\frac{27 \times 10^3}{A_c + (m-1) 200}\right) = 1.2$$

Solving, A_c = 20340 mm²

Adopt a ring beam of size 150 mm × 150 mm.

vi. *Design of cylindrical tank wall*:

Maximum hoop tension at base of wall is given by

$$F_t = \left(\frac{w \cdot H \cdot D}{2}\right) = \left(\frac{10 \times 3.6 \times 6}{2}\right) = 108 \text{ kN/m}$$

$$A_{st} = \left(\frac{T}{\sigma_{st}}\right) = \left(\frac{108 \times 10^3}{150}\right) = 720 \text{ mm}^2/\text{m}$$

Provide 12 mm diameter bars at 150 mm centres at the bottom of the tank (A_{st} = 754 mm²)

If t = thickness of tank wall at bottom of tank, then

$$\left[\frac{T}{A_c + (m-1) A_{st}}\right] = \sigma_{ct}$$

or

$$\left[\frac{108 \times 10^3}{100t + (12 \times 754)}\right] = 1.2$$

Solving, the thickness $t = 81$ mm

Adopt 100 mm thick walls uniform up to the top of the tank.

Minimum reinforcement = 0.24%

$$= \left(\frac{0.24 \times 100 \times 1000}{100} \right) = 240 \text{ mm}^2/\text{m}$$

Provide 12 mm diameter hoops at 300 mm centres for the top one metre. For the middle one metre, adopt a spacing of 225 mm c/c.

Distribution steel $A_{st} = \left(\frac{0.2 \times 100 \times 1000}{100} \right) = 200 \text{ mm}^2$

Provide 12 mm diameter bars at 300 mm centres in the vertical direction.

vii. *Tank floor slab*:

The tank floor slab is circular and fixed at the periphery to the circular ring beam.

Load on the circular slab w = (weight of water) + (self-weight of slab assumed at 300 mm thick)

$$= (10 \times 3.6) + (0.3 \times 24)$$
$$= 43.2 \text{ kN/m}^2$$

a. Maximum radial and circumferential moments:

Positive moment at centre of span

$$M_{rp} = \left(\frac{3}{16} wr^2 \right) = \left(\frac{3}{16} \times 43.2 \times 3^2 \right) = 73 \text{ kN} \cdot \text{m}$$

Negative moment at support

$$M_{rn} = \left(\frac{wr^2}{8} \right) = \left(\frac{43.2 \times 3^2}{8} \right) = 50 \text{ kN} \cdot \text{m}$$

Circumferential moment

$$M_c = \left(\frac{w \cdot r^2}{16} \right) = \left(\frac{43.2 \times 3^2}{16} \right) = 26 \text{ kN} \cdot \text{m}$$

Effective depth of the slab

$$d = \sqrt{\frac{M}{Qb}} = \sqrt{\frac{73 \times 10^6}{1.009 \times 1000}} = 268 \text{ mm}$$

Adopt $d = 270$ mm and overall depth = 300 mm

b. Reinforcements in circular slab:

$$A_{st} \text{ (centre of span)} = \left[\frac{73 \times 10^6}{190 \times 0.89 \times 270} \right] = 1599 \text{ mm}^2$$

$$A_{st} \text{ (supports)} = \left[\frac{50 \times 10^6}{150 \times 0.88 \times 270} \right] = 1402 \text{ mm}^2$$

$$A_{st} \text{ (circumferential moment)} = \left[\frac{26 \times 10^6}{150 \times 0.88 \times 270}\right] = 730 \text{ mm}^2$$

Provide 16 mm diameter bars at 120 mm centres both ways at bottom and for a length of 1.2 m from supports at top, radially and circumferentially.

viii. *Design of cylindrical tank wall*:

    a. Total load on ring beam:

Weight of water	= 1000 kN
Load from dome = $(2\pi Rrw) = (2 \times \pi \times 5 \times 1 \times 4)$	= 126 kN
Weight of top ring beam = $(0.15 \times 0.15 \times 24 \times \pi \times 6.15)$	= 10.4 kN
Weight of cylindrical wall = $(\pi \times 6.1 \times 0.1 \times 3.8 \times 24)$	= 175 kN
Weight of floor slab = $(\pi \times 3^2 \times 0.3 \times 24)$	= 203.5 kN
Weight of bottom ring beam	
Rib section of 300×400 mm = $(0.3 \times 0.4 \times \pi \times 6.1 \times 24)$	= 55.1 kN
Total vertical load on beam W	= 1570 kN

Uniformly distributed load per metre on girder

$$= \left(\frac{W}{\pi D}\right) = \left(\frac{1570}{\pi \times 6.1}\right) = 82 \text{ kN/m}$$

    b. Moments and shear forces in ring beam:

Assuming six columns supporting the ring beam, the moments are as follows:

Negative BM at supports = 0.0148 WR
$$= (0.0148 \times 1570 \times 3.05) = 71 \text{ kN·m}$$

Positive BM at centre of supports = 0.0075 WR
$$= (0.0075 \times 1570 \times 3.05 = 36 \text{ kN·m}$$

Torsional moment = 0.0015 WR
$$= (0.0015 \times 1570 \times 3.05) = 7.5 \text{ kN·m}$$

Shear force at support $V = \left[\dfrac{\text{Total load}}{2 \times \text{number of columns}}\right]$

$$= \left(\frac{1570}{2 \times 6}\right) = 131 \text{ kN}$$

Shear force at section of maximum torsion

$$V = \left[131 - \left(\frac{82 \times 3.05 \times \pi \times 12.73}{180}\right)\right] = 76 \text{ kN}$$

    c. *Design of support section*:

Bending moment $M = 71$ kN·m

Shear force $V = 131$ kN

Effective depth $= \sqrt{\dfrac{M}{Qb}} = \sqrt{\dfrac{71 \times 10^6}{1.14 \times 300}} = 455$ mm

Adopt $d = 550$ mm and overall depth $= 600$ mm

$$\therefore A_{st} = \left(\frac{71 \times 10^6}{150 \times 0.88 \times 550} \right) = 978 \text{ mm}^2$$

Provide 4 bars of 20 mm diameter ($A_{st} = 1256$ mm^2)

$$\tau_v = \left(\frac{V}{bd} \right) = \left(\frac{131 \times 10^3}{400 \times 550} \right) = 0.595 \text{ N/mm}^2$$

$$\left(\frac{100 A_{st}}{bd} \right) = \left(\frac{100 \times 1256}{400 \times 500} \right) = 0.57$$

From Table 23, IS:456 code, $\tau_c = 0.31$ N/mm$^2 < \tau_v$
Hence shear reinforcements are required.

Shear resisted by concrete $= \left(\dfrac{0.31 \times 400 \times 550}{1000} \right) = 68$ kN

Balance shear $= (131 - 68) = 63$ kN
Using 10 mm diameter two-legged stirrups, the spacing

$$s_v = \left(\frac{150 \times 2 \times 79 \times 550}{63 \times 10^3} \right) = 206 \text{ mm}$$

Adopt 10 mm diameter, two-legged stirrups at 200 mm centres.

d. *Design of centre of span section:*
Bending moment $M = 36$ kN·m

$$A_{st} = \left(\frac{36 \times 10^6}{190 \times 0.89 \times 550} \right) = 345 \text{ mm}^2$$

Minimum quantity of steel

$$A_s = \left(\frac{0.85 \, bd}{f_y} \right) = \left(\frac{0.85 \times 400 \times 550}{415} \right) = 450 \text{ mm}^2$$

Provide 2 bars of 20 mm diameter ($A_{st} = 628$ mm^2)

e. *Design of section subjected to maximum torsion and shear:*
Torsional moment $T = 7.5$ kN·m
Shear force $V = 76$ kN
Bending moment $M = 0$
Overall depth $D = 600$ mm
Width of section $b = 400$ mm

$$M_s = T \left[\frac{1 + (D/b)}{1.7} \right] = 7.5 \left[\frac{1 + (600/400)}{1.7} \right] = 11 \text{ kN} \cdot \text{m}$$

$$M_e = (M + M_t) = (0 + 11) = 11 \text{ kN·m}$$

$$A_{st} = \left(\frac{11 \times 10^6}{190 \times 0.89 \times 550}\right) = 118 \text{ mm}^2$$

But minimum reinforcement = 450 mm²

Provide 2 bars of 20 mm diameter.

$$V_e = V + 1.6\,(T/b)$$
$$= [76 + 1.6(7.5/0.4)] = 106 \text{ kN}$$

$$\tau_{ve} = \left(\frac{106 \times 10^3}{400 \times 550}\right) = 0.48 \text{ N/mm}^2$$

$$= \left(\frac{100\,A_{st}}{bd}\right) = \left(\frac{100 \times 450}{400 \times 550}\right) = 0.20$$

$$\therefore \quad \tau_c = 0.20 \text{ N/mm}^2 < \tau_{ve}$$

Hence shear reinforcements are required. Using 10 mm diameter two-legged stirrups with side covers of 25 mm and top and bottom covers of 50 mm, we have

$$b_t = 350 \text{ mm}, d_1 = 500 \text{ mm and } A_{sv} = (2 \times 79) = 158 \text{ mm}^2$$

$$s_v = \left[\frac{A_{sv}\sigma_{sv}}{(\tau_{ve} - \tau_c)b}\right] = \left[\frac{158 \times 150}{(0.48 - 0.20)400}\right] = 211 \text{ mm}$$

Adopt 10 mm diameter two-legged stirrups at 200 mm centres.

ix. *Supporting tower*:

   a. Load on columns:

     Total load from ring beam = 1570 kN

     Load on each column = (1570/6) = 262 kN

     Self-weight of column (400 mm × 40 mm) = (0.4 × 0.4 × 12 × 25)
           = 47 kN

     Self-weight of braces (350 mm × 450 mm) = (2 × 0.35 × 0.45 × 3.05 × 24)
           = 23 kN

     Total load on each column (tank full) = (262 + 47 + 23) = 332 kN

     Total axial load on each column (tank empty) = (1570 − 1000)/6 + 47 + 23 = 165 kN

   b. Wind forces:

     Intensity of wind pressure = 1.5 kN/m²

     Reduction coefficient for circular shape = 0.7

       – Wind force on top of dome and cylindrical wall (including bottom ring beam) = (0.7 × 1.5 × 5.35 × 6.2) = 35 kN

       – Wind force on columns = (6 × 0.4 × 12 × 1.5) = 43 kN

       – Wind force on braces = (2 × 6.1 × 4.5 × 1.5) = 8.2 kN

       ∴ Total horizontal wind force = (35 + 43 + 8.2) = 86.2 kN

Assuming contra flexure points at mid height of columns and fixity at the base due to raft foundations, the moment at the base of columns

$$M = (0.5 \times 86.2 \times 4) = 173 \text{ kN·m}$$

If M_1 = moment at the base of columns due to wind loads
$$= (35 \times 14.675) + (43 \times 6) + (8.2 \times 6)$$
$$= 821 \text{ kN·m}$$

If V = reaction developed at base of exterior columns

$$M_1 = \Sigma M + \frac{V}{r_1} \Sigma r^2$$

where, $r_1 = 3.05 \cos 30° = 2.64$ m
$$r_2 = 4(2.64)^2 = 27.87$$

$$\therefore \quad 821 = 173 + \left(\frac{V}{2.64} \times 27.87 \right)$$

Solving, $V = 62$ kN

$\therefore$ Total load on leeward column at base
$$P = (332 + 62) = 394 \text{ kN}$$

Moment in each column at base
$$M = (173/6) = 29 \text{ kN·m}$$

Eccentricity $e = (M/P) = \left(\dfrac{29 \times 10^6}{394 \times 10^3} \right) = 74$ mm

Since, eccentricity is small, direct stresses are predominant. Using 8 bars of 16 mm diameter equally spaced on all faces:
$$A_{sc} = (8 \times 201) = 1608 \text{ mm}^2$$
$$A_c = [(400 \times 400) - 1608] + [1.5 \times 13 \times 1608] = 189748 \text{ mm}^2$$
$$I_e = \left(\frac{400 \times 400^3}{12} \right) + (2 \times 1.5 \times 13 \times 3 \times 201 \times 150^2)$$
$$= 265 \times 10^7 \text{ mm}^4$$

Direct compressive stress $\sigma'_{cc} = \left(\dfrac{394 \times 10^3}{189748} \right) = 2.07 \text{ N/mm}^2$

Bending stress $\sigma'_{cb} = \left(\dfrac{29 \times 10^6 \times 200}{265 \times 10^7} \right) = 2.18 \text{ N/mm}^2$

Permissible stress in concrete is increased by 33.33% while considering wind effects.

Hence $\qquad \left(\dfrac{\sigma'_{cc}}{\sigma_{cc}} + \dfrac{\sigma'_{cb}}{\sigma_{cb}} \right) < 1$

$$\left(\frac{2.07}{5 \times 1.33} + \frac{2.18}{7 \times 1.33} \right) = 0.54 < 1$$

The stresses are within safe permissible limits. Adopt 6 mm diameter ties at 250 mm centres.

x. *Design of bracings*:

Moment in brace = (2 × moment in column × sec 30°)
$$= (2 \times 29 \times 1.15) = 67 \text{ kN·m}$$

Section of brace = (350 × 450) mm
$$b = 350 \text{ mm}$$
$$d = 400 \text{ mm}$$

Moment of resistance of section
$$M_1 = (0.897 \times 350 \times 400^2)/10^6$$
$$= 50 \text{ kN·m}$$

Balance moment $M_2 = (M - M_1)$
$$= (67 - 50)$$
$$= 17 \text{ kN·m}$$

$$A_{st1} = \left(\frac{50 \times 10^6}{230 \times 0.9 \times 400} \right) = 604 \text{ mm}^2$$

$$A_{st2} = \left(\frac{17 \times 10^6}{230 \times 0.9 \times 350} \right) = 235 \text{ mm}^2$$

∴ $$A_{st} = (604 + 235) = 839 \text{ mm}^2$$

Provide 3 bars of 20 mm diameter at top and bottom since wind direction is reversible
$$(A_{st} = 942 \text{ mm}^2)$$

Length of brace = (2 × 3.05 × sin 30) = 3.05

Maximum shear force in brace:

$$= \left(\frac{\text{Moment in brace}}{1/2 \times \text{length of brace}} \right)$$

$$= \left(\frac{67}{0.5 \times 3.05} \right) = 44 \text{ kN}$$

$$\tau_v = \left(\frac{44 \times 10^3}{350 \times 400} \right) = 0.31 \text{ N/mm}^2$$

$$= \left(\frac{100 A_{st}}{bd} \right) = \left(\frac{100 \times 942}{350 \times 400} \right) = 0.67$$

From Table 23, IS:456, $\tau_c = 0.33 \text{ N/mm}^2$.

Since $\tau_c > \tau_v$, provide nominal shear reinforcements.

Using 6 mm diameter two-legged stirrup

$$\text{Spacing } s_v = \left(\frac{A_{sy} f_y}{0.4b} \right) = \left(\frac{2 \times 28 \times 415}{0.4 \times 350} \right) = 166 \text{ mm}$$

Adopt 6 mm diameter, two-legged stirrups at 160 mm c/c.

xi. *Design of foundations*:

    a. Circular girder:

        A circular girder with a raft slab is provided for the tower foundations.

        Total load on foundation = $(332 \times 6) = 1992$ kN

        Self-weight of foundation at 10%   =   198

        Total load   = 2190 kN

        Safe bearing capacity of soil at site = 100 kN/m^2

$$\therefore \text{ Area of foundation } = \left(\frac{2190}{100}\right) = 21.90 \text{ m}^2$$

        If b = width of footing required

        $(\pi \times 6.1 \times b) = 21.90$

        or $b = 1.14$ m

        Adopt width of footing = 1.5 m and a circular girder with a width of 500 mm

        Total load on ring girder $W = 2190$ kN

$$\text{Load per metre run of girder } = \left(\frac{2190}{\pi \times 6.1}\right) = 115 \text{ kN/m}$$

        Since $\tau_v > \tau_c$, shear reinforcements are required.

        Balance shear = $[182.5 - (0.3 \times 500 \times 500)/1000] = 107.5$ kN

        Using 10 mm diameter, two-legged stirrups spacing

$$s_v = \left(\frac{230 \times 2 \times 78.5 \times 500}{107.5 \times 10^3}\right) = 167 \text{ mm}$$

        Adopt a spacing of 160 mm centres.

        Steel required for mid-span section

$$A_{st} = \left(\frac{50 \times 10^6}{230 \times 0.9 \times 500}\right) = 484 \text{ mm}^2$$

        But minimum steel $= \left(\dfrac{0.85\,bd}{f_y}\right)$

$$= \left(\frac{0.85 \times 500 \times 500}{415}\right) = 512 \text{ mm}^2$$

        Provide 2 bars of 20 mm diameter ($A_{st} = 628$ mm^2)

        Equivalent shear

$$V_e = (V + 1.6\, T/b)$$
$$= [105 + 1.6(10/0.5)] = 137 \text{ kN}$$

$$\tau_v = \left(\frac{137 \times 10^3}{500 \times 500}\right) = 0.548 \text{ N/mm}^2$$

$$\left(\frac{100\,A_{st}}{bd}\right) = \left(\frac{100 \times 628}{500 \times 500}\right) = 0.25$$

From Table 23, IS:456, $\tau_c = 0.22 \text{ N/mm}^2$

Since $\tau_v > \tau_c$, shear reinforcements are required

Balance shear $= [137 - (0.22 \times 500 \times 500)/1000] = 82 \text{ kN}$

Using 10 mm diameter two-legged stirrups, the spacing

$$s_v = \left(\frac{230 \times 2 \times 78.5 \times 500}{12 \times 10^3} \right) = 220 \text{ mm}$$

b. **Raft slab:**

Maximum projection of raft slab from the face of column = 500 mm

$$\text{Soil pressure} = \left(\frac{2190}{\pi \times 6.1 \times 1.5} \right) = 76 \text{ kN/m}^2$$

Consider 1 m width of slab along the circular arc:

$$\text{Maximum BM} = \left(\frac{76 \times 0.5^2}{2} \right) = 9.5 \text{ kN} \cdot \text{m}$$

$$\text{Effective depth } d = \sqrt{\frac{9.5 \times 10^6}{0.897 \times 1000}} = 103 \text{ mm}$$

Adopt $d = 120$ mm and an overall depth = 150 mm

$$A_{st} = \left(\frac{9.5 \times 10^6}{230 \times 0.9 \times 120} \right) = 382 \text{ mm}^2$$

Provide 12 mm diameter bars at 200 mm centres ($A_{st} = 565 \text{ mm}^2$)

$$\text{Distribution steel} = \left(\frac{0.12 \times 1000 \times 150}{100} \right) = 180 \text{ mm}^2$$

Provide 6 mm diameter bars at 150 mm centres ($A_{st} = 217 \text{ mm}^2$)

Maximum shear force in raft slab at a distance of 120 mm from face of girder $= (76 \times 0.38 \times 1) = 28.88 \text{ kN}$

$$\text{or} \qquad \tau_v = \left(\frac{28.88 \times 10^3}{1000 \times 320} \right) = 0.09 \text{ N/mm}^2$$

$$\left(\frac{100 A_{st}}{bd} \right) = \left(\frac{100 \times 565}{1000 \times 120} \right) = 0.47$$

From Table 23, IS:456 code, $k_s \tau_c = (1.30 \times 0.28)$
$$= 0.364 \text{ N/mm}^2 > \tau_v$$

Hence shear stresses in the slab are within safe permissible limits. The details of reinforcements in various structural components of the tank are shown in Figs 16.12–16.14.

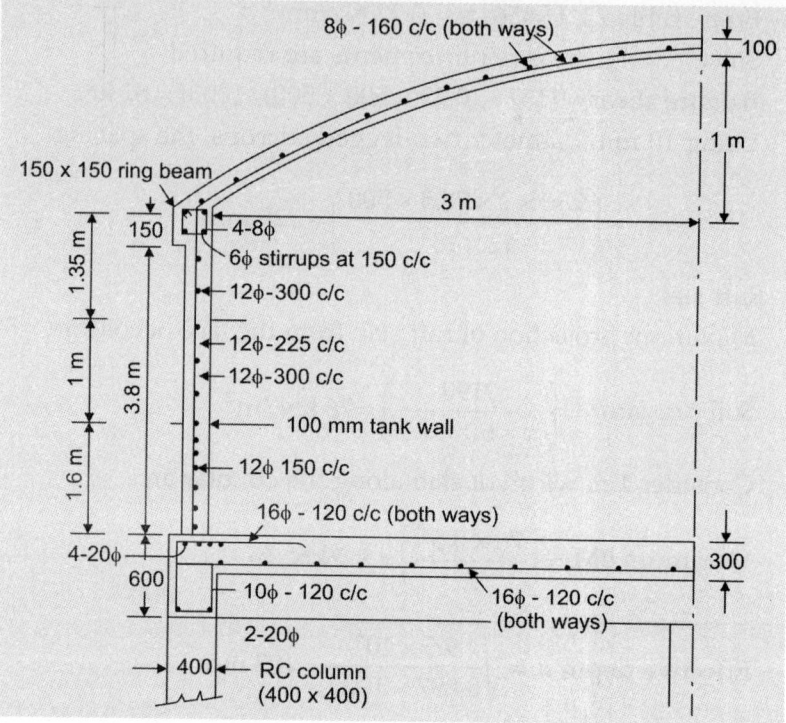

Fig. 16.12: Reinforcement details in dome, tank walls and ring girder

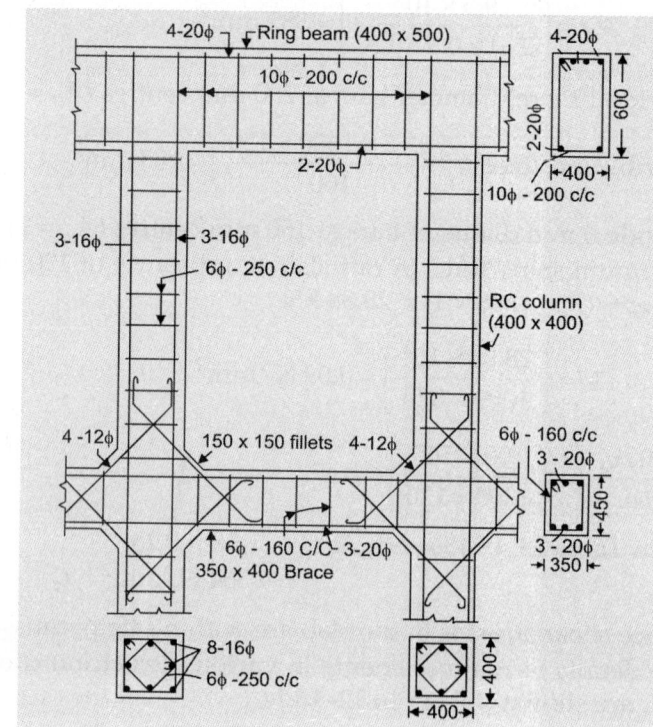

Fig. 16.13: Reinforcement details in ring girder, columns and braces

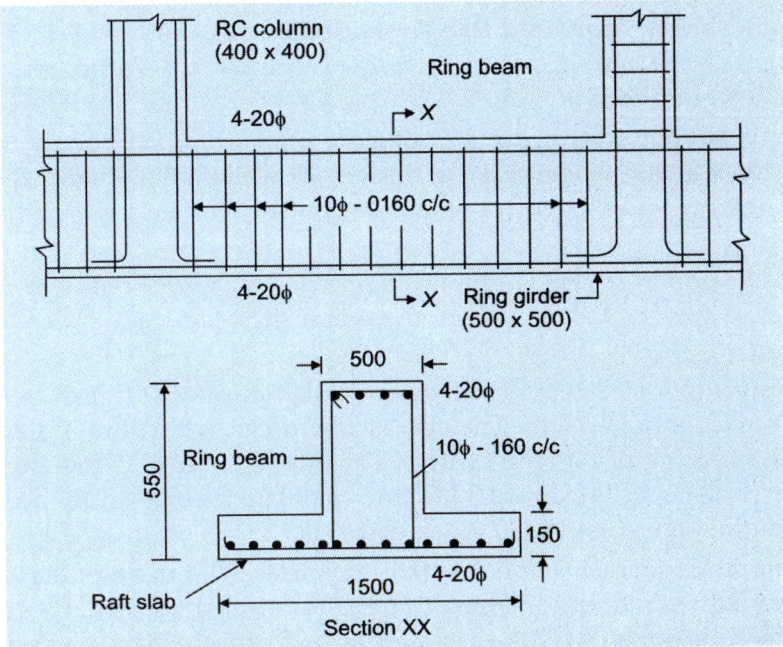

Fig. 16.14: Reinforcement details in ring beam and raft slab

REFERENCES

1. Turneaure F, Cyclopedia of Civil Engineering, American Technical School, 1908.
2. Frederick W Taylor and Sanford E Thompson, *A Treatise on Concrete, Plain & Reinforced*, 1 edn, 1905; 2 edn, 1912; 3 edn, 1916.
3. Morsch E and Goodrich EP, *Concrete and Steel Construction*, 3 edn, 1909–1910.
4. Charles A Whitney and George A Hool, Concrete designers' manual, 1921.
5. George A Hool, *Reinforced Concrete Construction–Fundamental Principles*, Vol 1, 1912.
6. Bate SCC, Why limit state design? *Concrete*, London, March 1968, pp 103–8.
7. Faber O and Bowie PG, *Reinforced Concrete, Theory & Practice*, Vols 1 and 2, 1912–1920.
8. Rowe RE, Cranston WB and Best BC, *New Concepts in the Design of Concrete, Structural Engineer*, Vol 43, 1965, pp 339–403.
9. Park R and Paulay T, *Reinforced Concrete Structures*, John Wiley & Sons Inc, New York, 1975.
10. IS:456-2000, Indian Standard Code of Practice for Plain and Reinforced Concrete, (4th Revision), BIS, New Delhi, 2000, p 37.

ASSIGNMENT

1. A simply supported slab has a clear span of 2.1 m and is supported on brick walls 40 cm thick along the edges. If the live load on the slab is 4 kN/m², and the floor finish weighs 0.6 kN/m², design the slab using M20 grade concrete and Fe 415 HYSD bars.

2. Design a simply supported slab supported on masonry walls 20 cm thick and having a clear span of 2.75 m. Live load is 4 kN/m². Adopt σ_{cb} = 7 N/mm², σ_{st} = 230 N/mm² and m = 13.

3. A simply supported corridor slab of clear span 3 m is supported on brick walls 40 cm thick to one side and 20 cm thick brick wall on the other. Adopt materials M20 concrete and ribbed tor steel. Design the slab and sketch the reinforcements.

4. A simply supported slab is supported on brick masonry walls 250 mm thick and has a clear span of 4.25 m. Design the slab using M20 grade concrete and ribbed tor steel.

5. A simply supported reinforced concrete slab of overall thickness 180 mm has an effective span of 3.5 m. The slab is reinforced with 10 mm diameter HYSD bars at a spacing of 240 mm centres. The effective depth is 160 mm. If the dead weight of slab and finishes is 5 kN/m², calculate the maximum permissible live load on the slab. Adopt M20 grade concrete.

6. A reinforced concrete slab of overall thickness 200 mm and effective span 4 m is provided with 10 mm diameter Fe 415 tor steel spaced at 100 mm centres at an effective depth of 180 mm. If the self weight of slab and finishes is 5.5 kN/m², estimate the maximum permissible live load on the slab assuming M20 grade concrete.

7. Design a two-way RCC slab for a room having clear dimensions of 4 m × 4.5 m. The slab is supported on masonry walls 30 cm thick on all the four sides and the corners are held down. The live load on the slab inclusive of finishes may be taken as 2 kN/m². Adopt M20 grade concrete and Fe 415 HYSD bars.

8. Design a two-way RCC slab for a warehouse building having clear dimensions of 5 m × 6 m. The slab is continuous on all the four edges being supported on RC beams 30 cm wide. The live load on the slab inclusive of finishes is 8.5 kN/m². Using M20 concrete and Fe 415 HYSD bars, sketch the details of reinforcement.

9. Design a singly reinforced beam to carry a classroom floor over a clear span of 7 m. The beams are supported on 400 mm thick brick walls and are spaced 2.75 m centres. The thickness of slab is 100 mm. Adopt M20 grade concrete and Fe 415 HYSD bars.

10. A singly reinforced simply supported beam 200 mm wide by 500 mm overall depth supports a uniformly distributed live load of 40 kN/m over an effective span of 5 m. The beam is reinforced with 5 bars of 20 mm diameter at an effective depth of 450 mm. Assuming that 2 bars can be bent up near supports, design suitable shear reinforcements near the support section. Adopt M20 grade concrete and Fe 415 HYSD bars.

11. A singly reinforced concrete beam of effective span 6 m has a rectangular section 300 mm wide by 800 mm overall depth. The beam is reinforced with 4 bars of 32 mm diameter at an effective depth of 750 mm. The dead load on the beam inclusive of self weight is 12 kN/m. If M20 grade concrete and Fe 415 HYSD bars are used in the beam, estimate the maximum permissible live load on the beam.

12. A singly reinforced concrete beam of effective span 5 m has a rectangular section 250 mm wide by 600 mm deep. The beam is reinforced with 4 bars of 10 mm diameter at an effective depth of 550 mm. The dead load on the beam inclusive of self-weight is 6 kN/m. Calculate the maximum permissible live load on the beam. Adopt M20 grade concrete and Fe 415 HYSD bars.

13. Design a balanced singly reinforced concrete section having an effective depth twice that of width to support a uniformly distributed live load of 10 kN/m over an effective span of 6 m cover to tensile steel is 50 mm. Adopt M20 grade concrete and Fe 415 grade tor steel. Sketch the details of reinforcements in the beam.

14. A doubly reinforced concrete beam of overall dimensions 300 mm × 600 mm is simply supported over an effective span of 5 m and has to support a uniformly distributed live load of 25.7 kN/m. The covers to the centres of compressive and tensile reinforcements from the edges is 40 mm. Design the steel reinforcements in the beam using M20 grade concrete and Fe 415 HYSD bars.

15. A doubly reinforced RC beam of overall dimensions 300 mm × 600 mm is simply supported over an effective span of 5 m and has to support a uniformly distributed load (inclusive of self-weight) of 30 kN/m. The cover to the centre of tensile and compressive reinforcements from the nearby edges is to be 40 mm. Determine the necessary tensile and compressive reinforcements. Adopt M20 grade concrete and Fe 415 HYSD bars.

16. A reinforced concrete T-beam with an effective flange width of 145 cm, slab thickness of 100 mm and rib width 250 mm has an overall depth of 50 cm. The beam is reinforced with 6 mild steel bars of 25 mm diameter with a cover of 6 cm to the centroid of the steel. If M20 concrete and Fe 415 HYSD bars are used, calculate the moment of resistance of the section.

17. A reinforced concrete T-beam having an effective flange width of 2.5 m, thickness of slab is 150 mm, width and depth of rib being 300 mm and 500 mm respectively is reinforced with 4 bars of 25 mm diameter at an effective depth of 600 mm. If the grade of concrete is M20 and Fe 415 HYSD bars are used, estimate the maximum permissible live load on the beam over an effective span of 8 m.

18. The floor of an office building is made up of T-beam and slab having the following details:

    Effective span = 7.5 m

    Effective width of flange = 2 m

    Thickness of flange = 150 m

    Width of rib = 300 mm

    Depth of rib = 300 mm

    Tension steel = 8 bars of 32 mm diameter

    Effective depth = 600 mm

    Materials: M20 grade concrete and Fe 415 grade tor steel. Estimate the moment of resistance of the cross-section and calculate the maximum permissible live load on the beam.

19. Design an isolated footing for an RCC column 450 mm × 450 mm in section to support an axial load of 1100 kN. The safe bearing capacity of the soil under the footing is 200 kN/m². Use M20 grade concrete and Fe 415 HYSD bars. Sketch the details of the reinforcements in the footing.

20. The cross-section of a short RCC column is 300 mm × 300 mm. Design suitable reinforcements if it has to support a load of 300 kN at an eccentricity of 20 mm with respect to one of the symmetrical axis. Concrete M20 grade and Fe 415 HYSD bars are used. Sketch the details of the reinforcements in the cross-section.

21. An eccentrically loaded RCC column 500 mm × 500 mm in cross-section is reinforced with 8 bars of 28 mm diameter, out of which four bars are located in the compression face and an equal number of the tension side with an effective cover of 50 mm. The column supports an eccentric load of 500 kN at an eccentricity of 300 mm. If the modular ratio is 13, determine the stresses developed in the steel and concrete.

22. A rectangular column 300 mm × 500 mm in section is required to support an axial load of 900 kN. Design suitable reinforcements in the column. Also design a suitable footing if the safe bearing capacity of the soil is 200 kN/m². Adopt M20 concrete and Fe 415 HYSD bars.

23. Design the stem of an RCC cantilever type retaining wall to retain earth level with the top of the wall to a height of 5 m. The density of soil may be taken as 14 kN/m³ and the angle of repose is 30°. Use M20 grade concrete and Fe 415 HYSD bars. Sketch the details of reinforcements in the wall showing the curtailment of bars.

24. A reinforced concrete cantilever retaining wall has to retain a granular fill to a height of 4 m above ground level and level with the top of the wall. The depth of the foundation may be assumed as 1.2 m below the ground level. The unit weight of fill is 14 kN/m³ and angle to repose is 30°. The safe bearing capacity of the foundation soil is 200 kN/m². The base slab is to be 400 mm thick. Design the stem and show the curtailment of reinforcement in the wall slab. Adopt M20 grade concrete and Fe 415 HYSD bars.

25. Design a counterfort type retaining wall to the following particulars:
    Height above ground level = 7 m
    SBC of soil = 200 kN/m²
    Angle of repose = 30°
    Weight of soil = 18 kN/m³
    Spacings of counterforts = 3 m (centres)
    Weight of concrete = 24 kN/m³
    Adopt M20 grade concrete and Fe 415 HYSD bars.

26. Design a staircase flight for a school building to suit the following data:
    Height between floors = 3.75 m
    Midlanding is cantilevered out and the width is 1.5 m
    Number of steps not to exceed 10 in any flight
    Tread T = 300 mm, Rise R = 160 mm
    Materials: M20 grade concrete and Fe 415 HYSD bars.

27. A staircase room measures 4 m × 2.1 m and the height between the floors is 3 m. Design a suitable dog-legged staircase with midlanding slab. Tread = 300 mm and rise is 150 mm. *Materials*: M20 grade concrete and Fe 415 HYSD bars. Sketch the details of reinforcements in one of the flights.

28. Design the flight of staircase for an office building to suit the following data:
    Centre to centre of landing beams = 4.4 m
    Tread = 300 mm, Rise = 160 mm
    Materials: M20 grade concrete and Fe 415 HYSD bars.
    Design the staircase flight which is liable for overcrowding and sketch the details of reinforcements in the flight.

29. The staircase room of a building measures 4.5 m × 5.5 m. The height between the floors is 4 m. The width of the staircase should be 1.5 m. *Materials*: M20 grade concrete and Fe 415 grade tor steel. Design a suitable staircase assuming the live load as 5 kN/m².

30. Design a suitable dog-legged staircase in a public building to be located in a staircase room 6 m long and 3 m wide.
    Height between floors = 3.60 m
    Live load on stairs = 4 kN/m²
    The stairs are supported on beams over walls and sides of steps are built into the wall by 120 mm. Use M20 grade concrete and Fe 415 HYSD bars.

31. A two-storeyed building has the height between floors as 3.90 m. The size of staircase room is 5.25 m × 4.25 m. The width of staircase is 1.5 m. The risers are 150 mm and treads are 250 mm. Live load on floor is 4 kN/m². The stairs are supported on beams on walls 400 mm thick and steps are built into walls by 200 mm. Using M20 grade concrete and Fe 415 HYSD bars, design the staircase and sketch the details of reinforcements.

32. Design a reinforced concrete spherical dome on a base radius of 15 m and rise of 5 m for supporting a total uniformly distributed (inclusive of self-weight) load of 3 kN/m³. Adopt M20 grade concrete and Fe 415 HYSD bars.

33. A spherical dome is to be designed to cover a circular tank 20 m diameter with a central rise of 3 m. The uniformly distributed live load including finishes on dome may be taken as 2 kN/m². Adopting M20 grade concrete and Fe 415 HYSD bars, design the dome and a suitable ring beam.

34. Design a reinforced concrete dome of base diameter 24 m. The rise of dome is one-eighth of the diameter. Thickness of dome is 100 mm. The live load on dome may be taken as 1.5 kN/m². Design a suitable ring beam for the dome. Adopt M20 grade concrete and Fe 415 HYSD bars.

35. A circular water tank resting on ground with a sliding base is required to store 350000 litres of water. The bearing capacity of the soil is 100 kN/m². M20 grade concrete and Fe 415 HYSD bars are available at site. Design the tank walls and base slab and sketch the details of reinforcements.

36. A rectangular reinforced concrete water tank with an open top is required to store 80000 litres of water. The inside dimensions of the tank may be taken as 6 m × 4 m. The tank rests on all walls on all the four sides. Design the side walls of the tank using the following data:

Permissible compression stress in concrete = 7 N/mm^2.

Permissible tensile stress in steel = 100 N/mm^2, on faces near the water side and 125 N/mm^2 on faces away from the water side. Modular ratio is 13.

37. A reinforced concrete water tank resting on ground is 5 m × 2 m with a maximum depth of 3 m. Using M20 grade concrete and Fe 415 HYSD bars, design the tank walls. Sketch the details of reinforcements in the tank walls.

38. Design a circular tank with fixed base for capacity of 500000 litres. The depth of water is to be 5 m. Free board is 300 mm. Use M20 grade concrete and Fe 415 HYSD bars. Permissible direct tensile stress in concrete is 1.2 N/mm^2. Permissible stress in steel in direct tensile is 100 N/mm^2. Sketch the details of reinforcements in tank walls. Adopt IS:3370 tables for coefficients.

39. A square RCC tank is required to store 60000 litres of water. Depth of water in the tank is 3 m. Free board is 0.2 m. Adopt M20 grade concrete and Fe 415 HYSD bars. Tensile stresses in steel is limited to 100 N/mm^2 at water face and 125 N/mm^2 away from water face. Sketch the details of reinforcements in the walls of the tank.

40. Design an RCC tank of internal dimensions 8 m × 2 m × 2 m. The tank is to be provided underground. The soil surrounding the tank is likely to get wet. Angle of repose of soil in dry state is 30° and in wet state is 6°. Adopt suitable working stresses. Soil weighs 20 kN/m^3. Adopt M20 grade concrete and Fe 415 HYSD bars.

41. A circular reinforced concrete water tank of 5 m diameter and 3 m height is supported by a tower consisting of six reinforced concrete columns on a circle of 5 m diameter. The tower height is 8 m with a bracing at a height of 4 m from the ground. The tank is designed to hold water up to a depth of 2.75 m. The self-weight of the tank with water is estimated to be 1000 kN. Intensity of wind pressure of tank is 1.5 kN/m^2. Using M20 grade concrete and Fe 415 grade tor steel, design the supporting tower of the tank.

42. Design an overhead flat bottomed RCC cylindrical water tank to store 75000 litres of water. The top of the tank is covered with a spherical dome. Height of staging is 16 m above ground level. Intensity of wind pressure is 1 kN/m^2. The safe bearing capacity of soil at site is 150 kN/m^2. Bracings are provided at intervals of 4 m. Adopting M20 grade concrete and Fe 415 grade tor steel, design the various structural components of the tank.

REVIEW QUESTIONS

1. Distinguish the differences between the working stress method, ultimate load design and the limit state method of design.

2. Specify the limit state, where the working stress method is useful? Substantiate with reasons.

3. What will be the differences in the reinforcements of a given singly reinforced beam section using the working stress and limit state methods?

4. In what way the working load method is deficient when compared with the limit sate method of design?

5. What is the role of modular ratio m in the design of reinforced concrete structural elements using the working stress method?

6. List the various deficiencies of the working stress method of design.

7. Why is it undesirable to design over-reinforced sections in the working stress method?

8. Why does the IS code specify an effectively higher modular ratio for compression reinforcement as compared to tension reinforcement?

9. In the design of structural concrete elements, which method results in larger cross-sectional dimensions?

10. Is it economical to use the working stress method in comparison with the limit state method when steel with higher strength is used?

OBJECTIVE QUESTIONS

1. Structural elements designed by working stress method of design ensures the desired factor of safety at
   a. ultimate or collapse state
   b. serviceability limit state
   c. both at service and collapse states

2. The area of steel in a concrete section designed to resist the working moment in a reinforced concrete beam is
   a. directly proportional to the effective depth of the beam
   b. inversely proportional to the permissible stress in steel
   c. inversely proportional to the working moment

3. In the working stress method of design, the permissible shear stress in concrete
   a. does not depend upon the characteristic strength of concrete
   b. depends upon the grade of reinforcement
   c. depends upon the percentage reinforcement ratio in the cross section

4. The modular ratio of concrete depends upon
   a. the ultimate compressive strength of concrete
   b. the cross-sectional area of the structural concrete member
   c. permissible compressive stress in concrete

5. In designing tension members by the working stress method, the permissible tensile stress in concrete depends upon
   a. the area of concrete only
   b. both the area of concrete and steel
   c. the grade of concrete

6. In case of reinforced concrete cantilever type retaining walls, the main reinforcements in the wall will be minimum at
   a. the top of the wall
   b. middle of the wall
   c. bottom of the wall

7. The walls of a reinforced concrete circular tank with a free base should be designed for
   a. flexure
   b. combined flexure and direct tension
   c. hoop tension

8. In the working stress method of design of footings for reinforced concrete columns, the depth of the footing should be designed on the basis of
   a. flexure only
   b. shear only
   c. flexure and shear both

9. The main reinforcements in the toe slab of a cantilever retaining wall are located at the
   a. soffit of the slab
   b. both at top and bottom of the slab
   c. top of the slab

10. The reinforcements in the spherical dome of a reinforced concrete tank is controlled by
   a. flexural stresses
   b. stresses due to circumferential thrust
   c. shrinkage and minimum reinforcement considerations

Subject Index

Author Index